Growth of New Technology Ventures in China's Emerging Market

NEW HORIZONS IN INTERNATIONAL BUSINESS

Series Editor: Peter J. Buckley
Centre for International Business,
University of Leeds (CIBUL), UK

The New Horizons in International Business series has established itself as the world's leading forum for the presentation of new ideas in international business research. It offers pre-eminent contributions in the areas of multinational enterprise – including foreign direct investment, business strategy and corporate alliances, global competitive strategies, and entrepreneurship. In short, this series constitutes essential reading for academics, business strategists and policy makers alike.

Titles in the series include:

Multinational Firms' Location and the New Economic Geography
Edited by Jean-Louis Mucchielli and Thierry Mayer

Free Trade in the Americas
Economic and Political Issues for Governments and Firms
Edited by Sidney Weintraub, Alan M. Rugman and Gavin Boyd

Economic Integration and Multinational Investment Behaviour
European and East Asian Experiences
Edited by Pierre-Bruno Ruffini

Strategic Business Alliances
An Examination of the Core Dimensions
Keith W. Glaister, Rumy Husan and Peter J. Buckley

Investment Strategies in Emerging Markets
Edited by Saul Estrin and Klaus E. Meyer

Multinationals and Industrial Competitiveness
A New Agenda
John H. Dunning and Rajneesh Narula

Foreign Direct Investment
Six Country Case Studies
Edited by Yingqi Annie Wei and V.N. Balasubramanyam

Japanese Multinationals in Europe
A Comparison of the Automobile and Pharmaceutical Industries
Ken-ichi Ando

International Joint Venture Performance in South East Asia
Craig C. Julian

Governance, Multinationals and Growth
Edited by Lorraine Eden and Wendy Dobson

European–American Trade and Financial Alliances
Edited by Gavin Boyd, Alan M. Rugman and Pier Carlo Padoan

Transformations in Global Governance
Implications for Multinationals and Other Stakeholders
Edited by Sushil Vachani

Growth of New Technology Ventures in China's Emerging Market

Edited by

Haiyang Li

Assistant Professor, Jesse H. Jones Graduate School of Management, Rice University, USA

NEW HORIZONS IN INTERNATIONAL BUSINESS

Edward Elgar

Cheltenham, UK • Northampton, MA, USA

Published by
Edward Elgar Publishing Limited
Glensanda House
Montpellier Parade
Cheltenham
Glos GL50 1UA
UK

Edward Elgar Publishing, Inc.
136 West Street
Suite 202
Northampton
Massachusetts 01060
USA

A catalogue record for this book
is available from the British Library

Library of Congress Cataloguing in Publication Data

Growth of new technology ventures in China's emerging market / edited by
Haiyang Li.
p. cm. — (New horizons in international business)
Includes bibliographical references and index.
1. High technology industries—China. 2. New business enterprises—China.
I. Li, Haiyang, 1970– II. Series.

HC430.H53G82 2006
338.4′760951—dc22 2005055952

ISBN-13: 978 1 84542 119 9
ISBN-10: 1 84542 119 1

Printed and bound in Great Britain by MPG Books Ltd, Bodmin, Cornwall

Contents

Contributors

David Ahlstrom is a professor of management at The Chinese University of Hong Kong, Faculty of Business, where he has taught for ten years. He received his PhD and MPhil degrees in management and international business from New York University's Stern School of Business, and an MBA from the University of Hawaii in Honolulu. Professor Ahlstrom's research interests include venture capital and entrepreneurship, management in emerging economies, and the practice of social influence in organizations.

Jonathan Brookfield received a PhD in management from the University of Pennsylvania (Wharton). He is currently an assistant professor at Texas A&M University, where he teaches classes on international business and strategy. His research interests include political risk, comparative political economy, industrial organization, business groups, firm boundaries, industrial districts, Chinese business organization and the Asian business environment.

Garry D. Bruton is an associate professor of entrepreneurship and strategy at the M.J. Neeley School of Business at Texas Christian University in Fort Worth, Texas. He received his PhD from Oklahoma State University. Professor Bruton has published over 60 academic articles in some of the leading academic publications for small business and entrepreneurship, including the *Academy of Management Journal*, *Strategic Management Journal*, *Journal of Business Venturing*, *Entrepreneurship Theory & Practice*, and *Journal of Small Business Management*. His research interests include various aspects of entrepreneurship in emerging markets, venture capital, high technology start-ups, and the management of technology and innovation.

Tsang-Sing Chan is Chair Professor of Marketing and Associate Vice President at Lingnan University, Hong Kong. He earned his DBA degree at Indiana University. His current research interests include comparative consumer behaviour and management localization issues.

Jin Chen is a professor at the College of Management, Zhejiang University, China. He received his PhD on Management Engineering in China and has been a visiting scholar of MIT. His research areas are managing technological innovation, R&D management and science and technology policy in developing countries. He has published 12 books on management of technology and innovation, and his papers have appeared in *IEEE Transaction on Engineering Management*, *Technovation*, *International Journal of Manpower*, *International Journal of Innovation and Technology Management*, and *Industry and Higher Education*.

Weiqing Gao is doing his post-doctoral research at the Renmin University Business School, Renmin University of China. His current research interest is in the area of corporate strategy and resource and capability development. He holds a PhD from the Renmin University of China.

Xudong Gao is an assistant professor of Strategy and International Management at the School of Economics and Management, Tsinghua University. His current research interests include technological capability development and competitive strategy. He holds a PhD from the MIT Sloan School of Management.

Quanlin Gu is a research fellow in the Institute for Business Management Research of Peking University. He received his MA in Economics from McMaster University. His research interests focus on industrial organization and firm dynamics in economic transition.

Michael A. Hitt is a distinguished professor and holds the Joe B. Foster Chair in Business Leadership and the C.W. and Dorothy Conn Chair in New Ventures at Texas A&M University. His research interests focus on international strategy, especially international strategic alliances and learning in emerging markets, managing the firm's resource portfolio, and strategic entrepreneurship.

R. Michael Holmes Jr. is a doctoral student at Texas A&M University. His major is strategic management and his minor is entrepreneurship. He received his undergraduate degree in quantitative finance and economics from the University of Alabama. Currently, his research interests include emerging markets, strategic alliances, entrepreneurship and behavioural decision making.

Susanna Khavul is an assistant professor at the London Business School and holds a doctorate in strategic management from Boston University. Her research focuses on explaining the innovative performance of high technology firms operating in a global economy. She is the winner of several international research awards and consults to industry and government on innovation, technology commercialization and start-up finance.

Dan Li is an assistant professor of international business in the Kelley School of Business at Indiana University. She received her PhD in

Management from Texas A&M University. Her research focuses on the management of multinational enterprises, predominantly the areas of international strategic alliances.

Haiyang Li is an assistant professor of strategic management and innovation in the Jesse H. Jones Graduate School of Management, Rice University. He earned his doctoral degree from City University of Hong Kong. His research interests focus on product innovation and business strategies of new technology ventures (particularly in transitional economies), strategic alliances and multinational corporations' R&D investment in emerging markets.

Hongbin Li is an assistant professor in the Department of Economics, The Chinese University of Hong Kong. He received his PhD in economics from Stanford University. His research interests include development economics, transition and the Chinese economy.

Jun Li is an assistant professor of strategic management at the Whittemore School of Business and Economics, University of New Hampshire. He received his doctoral degree from Texas A&M University. His research focuses on the management team and governance in entrepreneurial firms, technology-based new ventures, and multinational corporations in transitional economies.

Mingfang Li is a professor of management at California State University, Northridge. His teaching area focuses on strategic management, technology and innovation, and global management. His research focuses on corporate governance, technology and innovation, and emerging economies. He has published in outlets such as *Strategic Management Journal* and *Asia Pacific Journal of Management*. He received his PhD from Virginia Tech in 1990.

Yuan Li is a professor in innovation management and strategic management at the School of Management, Xi'an Jiaotong University, China. His research interest includes choice of innovation mechanisms, innovation in strategic alliance and strategic flexibility. He has been previously published in journals such as the *International Journal of Production Research*, *European Journal of Operation Research* and *Decision Support Systems*.

Toyah Miller is a doctoral student at Mays Business School at Texas A&M University. She received her undergraduate degree in Management Information Systems from Baylor University. Prior to initiating her PhD studies, Toyah was a management consultant in the telecommunications and entertainment industry. Currently, her research interests are in corporate governance, diversification strategy, corporate reputation and entrepreneurship.

Mike W. Peng is the Provost's Distinguished Professor of Global Strategy at the University of Texas at Dallas. He holds a PhD degree from the University of Washington. He is the author of over 30 refereed journal articles and three books, including, most recently, *Global Strategy* (2006).

Yajun Wu is an associate professor of strategic management at the Guanghua School of Management, Peking University, China. He received his PhD from Peking University in 1999. Yajun Wu's teaching areas include competitive strategies and corporate strategies, strategic management and environmental management. His research focuses on strategic management and environmental policies. His research has appeared in a variety of academic outlets.

Kuang S. Yeh is a professor of organization management and human resources at National Sun Yat-Sen University in Kaohsiung, Taiwan. His research interest is in organizational networks, corporate governance and business ethics. Professor Yeh has a BA in political science from National Taiwan University, an MPA from University of Delaware, and a PhD in Organization Theory and Policy Analysis from Carnegie Mellon University.

Shaker A. Zahra holds the Robert Buuck Chair in Entrepreneurship and is a professor in the department of strategic management and organization at Carlson School of Management at the University of Minnesota. His research focuses on entrepreneurial and innovation strategies in high technology and information industries. He received his PhD degree in business administration from the University of Mississippi and an honorary PhD from Jonkoping International Business School, Sweden.

Weiying Zhang is a professor and associate dean of Guanghua School of Management, Peking University. He earned his DPhil. in Economics from Oxford University. His research interests include industrial organization, firm theory and economic transition.

Yan Zhang is an assistant professor of strategic management in the Jesse H. Jones Graduate School of Management at Rice University. She received her PhD in strategic management from the Marshall School of Business at the University of Southern California. Her research focuses on CEO succession, corporate governance, global strategic alliances and multinational companies' operations in emerging markets.

Yongbin Zhao is a PhD candidate at the School of Management, Xi'an Jiaotong University, China. His research interests include new product development management, entrepreneurship and the effect of human resource on technological innovation.

Congcong Zheng is an assistant professor in San Diego State University. Her research area is on decision making and capability development in entrepreneurial firms, especially in regard to their decision to enter risky international markets. She is working on a cross-country project investigating the behaviour of entrepreneurial firms from emerging markets. She received her PhD degree from London Business School.

Li-An Zhou is an associate professor in the Guanghua School of Management, Peking University. He received his PhD in economics from Stanford University. His research interests focus on political economy, comparative economics and industrial organization.

Acknowledgments

Most of the chapters included in this book are based on the presentations at the Research Roundtable, 'New venture growth strategies in China's high technology industries: Internal growth versus strategic alliances with foreign direct investment', which I chaired, at the conference on China–US Relations: Past, Present, and Future, in College Station, Texas, in November 2003. To further the discussions on this emerging but important topic and call for more papers, I organized a Professional Development Workshop titled 'Growth strategies of China's High Technology Ventures' at the Academy of Management Annual Conference in New Orleans in August 2004. Chapters included in this book (except two reprints: Chs 5 and 7) all went through a blind review process with at least one-round revise and resubmit.

I would like to thank a number of people for their support and assistance in editing this book. Special thanks goes to Dr Richard Ewing, Wei Zhao, Emily Yaung Ashworth and Lorraine Eden for their encouragement and support for publishing this edited volume. I thank Tim Holcomb, Dan Li, Jun Li and Yan Zhang for their comments and suggestions on some of the chapters. I also would like to thank participants at the Research Roundtable at Texas A&M University and the Professional Development Workshop at the Academy of Management Conference in New Orleans.

The production of this edited book has not been possible without the assistance of my former secretary Joyce Thomas from Texas A&M University and my current secretary Elaine Morris from Rice University. They carefully made the corrections in the copy-edited chapters and prepared the entire manuscripts in camera-ready format. Their patience, goodwill and editorial assistance are greatly appreciated. I am also grateful to Alan Sturmer, Senior Acquisitions Editor, and Karen McCarthy, Deputy Managing Editor, from Edward Elgar Publishing for their valuable suggestions and support in the publication process.

Lastly, I thank all the contributors to this volume. They were patient, responsive to reviewers' comments and my rewriting requests, and tolerant of minor criticisms. As readers will recognize, some of the contributors are senior researchers while others are young enthusiasts. I hope that they are as proud of this book as I am.

September 2005
Haiyang Li
Jesse H. Jones Graduate School of Management
Rice University
Houston, Texas

PART I

Overview of new technology ventures in China

1. Growth of new technology ventures in China: an introduction

Haiyang Li and Michael A. Hitt

CHINA AS THE MOTHER OF EMERGING MARKETS

In recent years, emerging market firms have become important players in global markets. Firms based in emerging markets accounted for $862 billion of accumulated foreign-direct investment in 2003 (UNCTAD, 2004). While there are other large and important emerging markets such as India, Brazil and Russia, the mother of all emerging markets is China. A recent report published by Boston Consulting Group (BCG, 2004) states that China is emerging as the industrial power base of the future. For the US and Europe, China already represents a large and fast-growing source of industrial products. For example, imports from China represent nearly 20 per cent of all imports from low-cost countries into the US and Germany. Also, China is predicted to be the world's fifth largest source of outward foreign direct investment during 2004–7 (Buckley, 2005) and China's gross domestic product is expected to exceed that of the US by 2040 (Wilson & Purushothaman, 2003). Therefore, within approximately 35 years, China will become the largest and most powerful economic force in the world.

China's current economic power affords its firms considerable opportunities in global markets. Many of China's firms are able to extract more knowledge from foreign partners and are better able to exploit that knowledge than firms from other emerging market countries. These firms are in a stronger bargaining position in partnerships with multinational corporations. Also Chinese firms often have a competitive advantage over foreign multinational firms when operating in other emerging market countries. This is because they have an understanding of the needs of emerging market consumers and how to operate effectively in these countries. Their considerable resource power also gives them significant opportunities. Indeed, several technology firms from China are starting to build a global reputation. For example, Lenovo (the former Legend Group), China's top computer maker founded in 1984 with 11 people and US$25 000 initial investment, recently acquired IBM's personal computer business for almost

US$2 billion. Therefore China is a more mature emerging market than many other emerging markets. Perhaps the only emerging market with relatively similar economic potential and competitive strength is that of India (Hitt, Li & Worthington, 2005).

More importantly, different from other emerging markets, China has significantly improved its R&D investments over past years. From 1998 to 2003, China increased its R&D expenditure from US$6.72 billion (0.70 per cent of its GDP) to US$18.76 billion (1.31 per cent of its GDP), a 279 per cent increase within a five-year period (China's National Bureau of Statistics, 2004a). China is the most R&D intensive country among all emerging markets (ranked seventh in the world).

ENTREPRENEURSHIP AND GROWTH OF NEW TECHNOLOGY VENTURES IN CHINA

Entrepreneurial and new venture activity is important in emerging markets because it is a critical engine of economic growth and development. In general, there are four forces underlying the rise of entrepreneurship in China, particularly in the nation's technology industries. The first is the emergence of market force. Since its economic reform began in 1979, China's economy has been changing from a centrally planned system to a market-oriented system. This significant institutional transition is characterized by highly uncertain and dynamic markets (Li & Atuahene-Gima, 2001). As Nee (1989:666–7) argued, with the emergence of a more open market in China, administrative redistributors lose power and firms have increasing discretion over the terms of exchange for their goods and services. Also markets provide powerful incentives for firms because firms can retain a great share of the surplus through market transitions. Thus reforms in the form of liberalization and more open markets in China introduce fundamental and comprehensive changes to 'the formal and informal rules of the game that affect organizations as players' (Peng, 2003:275), which encourage entrepreneurial and new venture activities.

The second one is the force of redistributive institutions (for example, government agencies at different levels). China's economic reform involves an incremental process that has been led and controlled by the government. While market-oriented reforms result in the power decline of redistributive institutions, they continue to play an important role in economic activities because the development of market institutions is under way but far from complete. To make the markets thrive and prosper, the Chinese government encourages various businesses, especially private firms, to engage in competition. Government at various levels provides favourable environments in which creditable commitment to market-oriented policies is backed by relatively reliable administrative procedures. This is in contrast with the pre-reform era in which private entrepreneurship had been actively suppressed or eradicated by the state (Nee, 1989, 1992).

Especially, the Chinese government plays a crucial role in facilitating entrepreneurship in technology industries. In the mid-1980s, impressed by the successful model of Silicon Valley in the US and concerned that their country would be henceforth disadvantaged scientifically, Chinese scientists appealed to the central government, urging that China focus on the development of new and high technologies. Then Chinese leader, Deng Xiaoping, responded rapidly, and the central government announced the 863 Program (initiated in March 1986) that formalized China's intent to develop new scientific and technological capabilities through the creation of national technology development zones (Li, Schoonhoven & Zhang, 2005). The first national technology development zone in China was established in 1988 in Zhongguancun, Beijing. Over the next several years, an additional 52 national technology development zones were founded throughout China (indeed all provinces in China except for the three westernmost provinces, Qinghai Province, Tibet, and Ningxia Autonomous Regions, contain at least one technology development zone).

The establishment of high technology development zones in China has stimulated the significant growth of technology ventures (within the zones). From 1996 to 2003, the number of technology ventures in the zones increased from 13 722 to 32 857, the number of venture employees increased from 1.29 million to 3.95 million, and the venture sales revenue increased from US$28.1 billion to US$255.3 billion (China's National Bureau of Statistics, 2004b). Technology ventures in these zones have enjoyed many preferential benefits. For example, in the Beijing Technology Development Zone, technology ventures may get their first three years of tax waived, and can have a 50 per cent reduction in tax over the subsequent three years. In addition, the municipal government of Beijing provided technology ventures in the Zone with certain preferential treatment in land use. The fee for the transfer of the land use rights and the urban infrastructure construction fees are either exempted or collected at a reduced rate.

The third force is foreign direct investment. With a population of 1.3 billion and an immense supply of low-wage workers, China presents important growth opportunities for many foreign investors (Hitt *et al.*, 2005). Indeed China has overtaken the United States to become the largest recipient of foreign direct investment since 2002 (UNCTAD, 2003). The entry of foreign direct investments has not only facilitated China's market-oriented reforms, but also increased new venture activities in China's technology-based industries. The growth potential of technology-based industries has attracted increasing numbers of foreign entrants. According to a recent report published by China's National Bureau of Statistics (2004b), foreign firms making direct investments have become the major force in the growth of high technology industries in China. For example, in 2003, 33.6 per cent of technology firms in technology development zones were foreign firms (including Sino–foreign joint ventures and wholly owned subsidiaries), which accounted for 83 per cent of China's export of high technology products (62 per cent by foreign wholly owned subsidiaries and 21 per cent by Sino–foreign joint ventures). In contrast, state-owned firms only accounted for 10 per cent of China's export of high technology products.

More importantly, starting from the mid-1990s, many multinational corporations have been exploring the Chinese market by establishing huge R&D facilities there (Li, Holmes & Hitt, 2005; Li & Zhong, 2003; Walsh, 2003). In China, for example, there are approximately 300 foreign-funded R&D facilities, most of which were founded within the last three years (Buckley, 2005). These R&D facilities will help Chinese firms climb further up the technological ladder and increase technology entrepreneurship in this nation.

The last force is the change in Chinese cultural values. Though China's reforms initially were focused on the economy, the transition to the market economy has been changing Chinese socialist cultural values and ideologies. As Chang and MacMillan (1991:379) observed, 'more and more Chinese people are aware that starting their own businesses is the best way to assure their own survival and the survival of the nation'. Thus, entrepreneurs in China are represented not only by peasants and workers, but also by intellectuals. This is in contrast to the Chinese traditional values held by intellectuals in that they did not respect businesspeople and they were unlikely to become involved in businesses. Indeed, over the past two decades, intellectuals from universities, research institutes and government agencies have founded thousands of technology ventures across the country. For example, in Zhongguancun Science Park in Beijing, up to 2003, there were 784 technology firms founded by employees or former employees from universities (for example, Peking University and Tsinghua University) or research institutes (for example, institutes under the Chinese Academy of Science).

The four forces (market, institutional, foreign investment and cultural) we discussed above are not independent but are closely intertwined to affect entrepreneurial and new venture activities in China. Because the economic transition process may require a long time, both the market forces and the redistributive institutions coexist for coordinating the economy, as both of them have allocative control over key factor resources. Li and Atuahene-Gima (2001:1125) specifically argued that in China's transition economy the redistributive institutions interact with market forces to affect firm outcomes. The economic transition process is also closely associated with foreign investment and cultural changes. On the one hand, economic transition has created opportunities for foreign investments and resulted in cultural and social value changes. On the other hand, the increasing role of foreign investments in China and the changes in cultural values have significantly stimulated the economic transition process. Therefore, to better understand the rise of entrepreneurship and the growth of new technology ventures in China, a comprehensive framework is needed that integrates the market, institutional, foreign investment and cultural factors.

SIGNIFICANT ISSUES ADDRESSED IN THIS VOLUME

The contributors to this volume make an initial attempt in this direction. They link new venture concepts and theories based on developed market economies to a transition market context. More importantly, by addressing the four forces through multiple disciplinary lenses, this book offers a better understanding of how new technology ventures in China's emerging market can grow successfully. The volume is divided into three sections: new ventures and entrepreneurship in China, environment and new venture strategies, and entry of foreign ventures and internationalization of local ventures in China.

The first section is an overview of new ventures in China. It covers three issues. In Chapter 2 Li and Miller's work comprehensively reviews the development of research on new ventures in emerging markets during the period 1990–2004 and they examine the major theoretical perspectives adopted by researchers during this period. Chapter 3, by Gu, Li, Zhang and Zhou, empirically examines the determinants of the survival of technology ventures in China's Zhongguancun Science Park. This study sheds light on the changing role of government and market power in determining firm dynamics in an economic transition context. In Chapter 4, Ahlstrom, Yeh and Bruton's study focuses on venture capital in China. They argue that, while venture capitalists have directed increasing attention to technology ventures in China, the challenges encountered by private venture capitalists in China differ strikingly from those in more mature markets. This study details some of the opportunities and challenges in China and the authors offer recommendations for the way venture capitalists might address those challenges.

The second section, on environment and new venture strategies, offers a diverse range of topics including growth strategy, environment–strategy interaction, entrepreneurial strategy making and innovation, as well as organizational reward and control practices. Peng's study (Chapter 5) focuses on entrepreneurs in transition economies, describing three major entrepreneurial strategies, prospecting, networking and boundary blurring. Li, Gao and Wu's chapter examines strategies for domestic technology ventures in China's emerging market when facing a global competitive environment. They argue that Chinese culture must be a major consideration in understanding firm strategy formulation and implementation. They further identify four key strategic elements: localization, resource-expansion oriented growth, strategic divergence, and collaboration for competition. In Chapter 7, Li, Zhang and Chan examine the relationship between entrepreneurial strategy making and performance in Chinese technology ventures. They find that the entrepreneurial strategy making–performance relationship is not universal. Instead, a firm can benefit from entrepreneurial strategy making only when the environment is highly uncertain and when the firm has strong marketing competences. Gao and Gao's chapter applies the concepts of barriers to appropriability and opportunities for improvement to explore the possibilities and conditions for local Chinese telecommunication firms to compete against multinational corporations by developing innovation capabilities and proprietary technologies. Their case studies offer new insights into local firms' technology strategies. Lastly, Li, Li, Li and Zhao's study (Chapter 9) investigates the effects of organizational reward and control on Chinese technology firms' innovation and consequently on their financial performance. Results from the analysis of 194 technology firms in China indicate that non-material reward is positively associated with both radical and incremental innovations, whereas material reward seems to have no effect on incremental innovation and to restrict radical innovation in Chinese technology firms. The authors also find that process control encourages radical innovation, but discourages incremental innovation. Contrarily, output control is negatively related to radical innovation, but positively related to incremental innovation. Their study shows that radical innovation contributes to Chinese technology firms' financial performance while incremental innovation does not.

The third section of the book focuses on the international element of new ventures in China. The unique feature of this section is that it not only examines the entry of foreign ventures but also addresses the globalization of local Chinese ventures. In Chapter 10, Brookfield and Li focus on corporate political activities and the ways in which foreign companies have dealt with issues of political risk in the Chinese environment. Using a case study of six foreign ventures in China, they argue that location can be used as a political strategy. In Chapter 11, Zahra examines variations in the learning approaches used by French, Swedish and US new ventures in the

Chinese market and the profoundly important effects of home national culture on the choice of learning approaches. While much of the literature on R&D globalization has focused on multinational corporations from developed markets, Chen and Holmes's study draws upon the resource-based view and organizational learning perspective to examine R&D globalization strategies of technology ventures in China. They find that Chinese technology ventures tend to emphasize access to advanced technology from foreign markets when they pursue R&D globalization. Finally, Zheng and Khavul propose in Chapter 13 that internationalization represents a type of risky organizational change for firms. With a sample of Chinese technology ventures, they find support for their key argument that the probability of a firm engaging in internationalization behaviour is related to its previous performance and its aspiration level.

CONCLUSIONS

New technology ventures in China are continuing to grow. They not only play a crucial role in the Chinese domestic market but also play an increasing role in the global market. However, we need to increase our understanding of how these Chinese technology ventures have been growing so fast and how these technology ventures *learn in order to catch up with* foreign firms in technology-based industries. Also we still know little about how market, institutional, foreign investment and cultural forces jointly affect the growth of these ventures. In this book, we aim to contribute to the strategic management and entrepreneurship literatures by providing insights into the nature of new technology ventures in China. We hope that this book serves as a foundation for an effort to sharpen our understanding of how economic transition affects new venture behaviours and outcomes in the context of technology-based industries.

REFERENCES

Boston Consulting Group (2004), 'Capturing global advantage', BCG Report, April.

Buckley, P. (2005), 'An investigation of the strategies and motivations of outward investing Chinese enterprises', presentation at the Academy of International Business Conference, Quebec, July.

Chang, W. and I.C. MacMillan (1991), 'A review of entrepreneurial development in People's Republic of China', *Journal of Business Venturing*, **6**, 375–9.

China's National Bureau of Statistics (2004a), *China S&T Statistics Data Book*, Beijing.

China's National Bureau of Statistics (2004b), *China Statistics Yearbook on High Technology Industry*, Beijing.
Hitt, M.A., H. Li and W.J. Worthington (2005), 'Emerging markets as learning laboratories: learning behaviors of local firms and foreign entrants in different institutional contexts', *Management and Organization Review*, **1** (3), 353–80.
Li, H. and K. Atuahene-Gima (2001), 'Product innovation strategy and performance of new technology ventures in China', *Academy of Management Journal*, **44** (6), 1123–34.
Li, J. and J. Zhong (2003), 'Explaining the growth of international R&D alliances in China', *Managerial and Decision Economics*, **24**, 10–15.
Li, H., R.M. Holmes and M.A. Hitt (2005), 'Why do multinational corporations establish R&D facilities in an emerging market? Theory and empirical evidence from China', paper presented at the Academy of Management Annual Conference, Honolulu, Hawaii.
Li, H., K.B. Schoonhoven and Y. Zhang (2005), 'The growth of entrepreneurial clusters: New technology development zones in China, 1988–2000', paper presented at the Academy of Management Annual Conference, Honolulu, Hawaii.
Nee, V. (1989), 'A theory of market transition: from redistribution to markets in state socialism', *American Sociological Review*, **54**, 663–81.
Nee, V. (1992), 'Organizational dynamics of market transition: hybrid forms, property rights, and mixed economy in China', *Administrative Science Quarterly*, **37**, 1–27.
Peng, M. (2003), 'Institutional transitions and strategic choices', *Academy of Management Review*, **28**, 275–96.
UNCTAD (2003), *World investment report*, New York and Geneva: United Nations Conference on Trade and Development.
UNCTAD (2004), *World investment report: The shift toward services*, New York and Geneva: United Nations Conference on Trade and Development.
Walsh, K. (2003), *Foreign high-tech R&D in China: Risks, rewards, and implications for U.S.–China relations*, Washington, DC: The Henry L. Stimpson Center.
Wilson, D. and R. Purushothaman (2003), 'Dreaming with BRICs: The path to 2050', Goldman Sachs Global Economics Paper number 99.
World Investment Report (2003), *United Nations Conference on Trade and Development* (*UNCTAD*), United Nations: New York and Geneva, pp. 40–41.

2. New ventures in emerging markets: comprehensive review and future directions

Haiyang Li and Toyah Miller

INTRODUCTION

Both researchers and practitioners have paid considerable attention to the significant role of new ventures (that is, firms younger than eight years) in the global economy (Oviatt & McDougall, 1994; Zahra, Ireland & Hitt, 2000). The increasing growth of new ventures has not only added more products and services to the market and stimulated market competition, but also provided more employment opportunities and improved the quality of life of people (McDougall & Robinson, 1990; Vesper, 1980). Thus developing new ventures has been viewed as both a revitalization tool for developed economies and a driving force for emerging markets (Cooper, Willard & Woo, 1986; Li & Atuahene-Gima, 2001; Zhao & Aram, 1995).

Recent literature has suggested that emerging markets are a rich landscape of opportunity allowing new ventures to capitalize on growth (Li, 2001). An emerging market is generally defined as one that often has low per capita income but with a rapid pace of economic development, and government policies favouring economic liberalization and a free market economy (see Hoskisson, Eden, Lau & Wright, 2000). According to Hoskisson *et al.* (2000), emerging markets not only include developing economies in Latin America, the Middle East, Southeast Asia and Africa, but also transition economies such as China, Central and Eastern Europe, and the former Soviet Union. These transition economies are changing from formal planned economies to market economies based on privatization and emphasis on entrepreneurial activities (Zahra, Ireland, Gutierrez & Hitt, 2000).

Compared to their counterparts in developed markets, new ventures in emerging markets face significant challenges from their fast-changing environment. For example, in developed market economies such as the US, new ventures can acquire critical resources (for example, managerial

Table 2.1 Number of contributions, by year

Journal	1990	1991	1993	1994	1995
Academy of Management Executive					
Academy of Management Journal					
Academy of Management Review					
Entrepreneurship: Theory & Practice					1
Journal of Business Venturing	1	2	2	3	2
Journal of International Business Studies					
Journal of Management Studies					
Journal of World Business					
Organization Science					
Strategic Management Journal					
Grand Total	1	2	2	3	3

Note: There was no publication in 1992.

resources and financial resources) from the markets. In contrast, new ventures in emerging markets have great difficulties in obtaining critical resources from markets because in these economies, banks and strategic factor markets are severely underdeveloped (Li & Atuahene-Gima, 2002; Peng & Heath, 1996). Also limitations with regard to the infrastructure (such as transportation and communication) in these markets create challenges for new ventures' resource acquisition. While research on new ventures in emerging markets has been flourishing over the past 15 years, this research is isolated, fragmented and scattered, and must be assimilated if it is to be useful. Periodic reviews of the new venture literature can contribute to integration of the research, yet there has not been a critical review of new ventures in emerging markets. To address the gap in the literature, the present study aims systematically to review and assess the theoretical and methodological aspects of existing studies on new ventures in emerging markets and then provide directions for future research.

We started by surveying 13 top academic and scholarly journals including: *Academy of Management Executive (AME)*, *Academy of Management Journal (AMJ)*, *Academy of Management Review (AMR)*, *Administrative Science Quarterly (ASQ)*, *Entrepreneurship Theory & Practice (ETP)*, *Journal of Business Ventures (JBV)*, *Journal of International Business Studies (JIBS)*, *Journal of Management (JoM)*, *Journal of Management Studies (JMS)*, *Journal of World Business (JWB)*, *Management Science (MS)*, *Organization Science (OS)*, *and Strategic Management Journal (SMJ)*. Articles that focused on new ventures in emerging markets were included in our survey. We focused on the period 1990–2004. The rationale

1996	1997	1998	1999	2000	2001	2002	2003	2004	Grand Total
				1	2				3
				1	1				2
				4					4
3									4
	1		3			2	1	3	20
	1				1				2
			1		1				2
				2					2
		1						1	2
					1	1			2
3	2	1	4	8	6	3	1	4	43

for choosing the year of 1990 as a cut-off date was that in 1989 former socialism countries including the Soviet Union and Central and Eastern Europe collapsed and their economies started to be privatized and marketized. The process of privatization and marketization significantly encourages entrepreneurial activities in these countries. We chose the year of 1990 to reflect the earliest work in this area since the events of 1989. We further checked the work published in the above journals in 1989 and did not find any article which had a focus on new ventures in emerging markets. Each article in the above journals was individually reviewed to select those which focused on new ventures in emerging markets. While scholars have defined new ventures in different ways, many of them have adopted the cutting-off point of less than eight years old (for example, Lee, Lee and Pennings, 2001, Li & Atuahene-Gima, 2001).

Table 2.1 shows that 43 articles were published on the subject between the years 1990–2004. *JBV* had the greatest number of publications, numbering 20. This is not surprising given this journal's focus on new ventures and entrepreneurship. *AMR* and *ETP* followed, at some distance, with four publications. Note that, in 2000, *AMR* had a special issue on privatization in emerging markets in which all four papers on this subject were published. *AME* had three publications. *AMJ*, *JIBS*, *JMS*, *JWB*, *OS* and *SMJ* had two publications apiece. Interestingly, among the journals we have surveyed, we could not locate any related work on this subject in such journals as *ASQ*, *MS*, and *JoM* during this period.

The remainder of the chapter is organized into three sections as follows: (a) current theoretical perspectives used in this context are reviewed,

(b) methodological issues are summarized, and (c) directions for future research are recommended.

THEORETICAL PERSPECTIVES ON NEW VENTURES IN EMERGING MARKETS

Hoskisson *et al.* (2000) suggested that there are three leading theories which guide strategy research on emerging markets: institutional theory, transaction cost economics/agency theory, and resource-based view. While our review has shown the prominence of institutional theory and the resource-based view in research on new ventures in emerging markets, we also found that there are other theories which play an important role in this research context, such as social capital and networking perspective and resource dependence theory. However, the application of transaction cost economics/agency theory in this area is very limited. In the next sections, we examine how these theories are applied in emerging market contexts and whether these contexts permit development of new concepts or modification of old ones.

Institutional Theory

Institutional theory suggests that institutions influence individuals, groups and organizations within it through the way it is structured and managed (Meyer & Rowan, 1977; Scott, 1995). There are two different broad approaches of institutional theory: the economic approach and the sociological approach. The economic approach suggests that, because of market imperfection, institutions can facilitate market interactions by reducing both transaction and information costs. As North (1990) has argued, institutions provide the rules of the game that regulate economic activities. The sociological approach suggests that institutions influence individuals, groups and organizations within it through the way it is structured and managed (Meyer & Rowan, 1977). Organizations are then driven by the demands of institutions to incorporate practices that may not be efficient (Meyer & Rowan, 1977). Organizations' actions and outcomes are trained by their social environments (DiMaggio & Powell, 1983; Meyer & Rowan, 1977). Organizational actions, which are consistent with these expectations, have a greater chance of increasing their legitimacy. DiMaggio and Powell (1983) posit that organizations are malleable to the institutional forces around them which urge them towards conformity, where isomorphism or the similar resemblance of organizations exists owing to the common constraints they face. Organizations that work within the practices and rules of the institution will gain legitimacy and ultimately survival.

Institutional theory is pre-eminent in research on new ventures in emerging markets because 'government and societal influences are stronger in these emerging economies than in developed economies' (Hoskisson *et al.*, 2000:252). The political, economic, legal and cultural environments of emerging economies play an important role in affecting entrepreneurial ventures. Instead of taking the institution for granted, as is often done in research in Western societies, institutional theory suggests that new ventures face many challenges from unsupportive cultural attitudes, underdeveloped legal structures, governmental corruption, high taxation and lack of resources, leading to restrictions in new venture actions. In addition, many cultures from emerging markets are hostile towards entrepreneurs, fearing subsequent exploitation. Cultures may hold the role of an entrepreneur in low esteem.

Institutional context can drive strategy and also shed light on how new ventures emerge, compete, grow and survive (Khanna & Palepu, 1997). Peng and Heath (1996) argued that firms in transition economies are more likely to use network-based strategies rather than internal growth strategies because of institutional constraints. In another paper, Peng (2001) specifically evaluated how institutions in emerging markets influence new venture strategies. He argued that, given institutional constraints such as lack of property rights, new ventures can create value through three important strategies, prospecting, networking and boundary blurring. Tan (1996) argued that entrepreneurial strategies of innovativeness, proactiveness and risk taking are spurred by regulatory hostility, dynamism and complexity in China as compared to Western economies.

In his seminal study, Stinchcombe (1965) noted that new ventures tend to have a high failure rate because their stable links to clients, supporters or customers have not yet been established when they begin their operations. Thus legitimacy building becomes a critical issue for new ventures. Legitimacy increases the probability that a new venture can appeal to such resources as human capital and financial resources. In emerging economies where legal systems are weak and the private sector is often stereotyped as exploitive, legitimacy of new ventures is threatened. The newness of emerging market firms creates special needs for legitimacy (Hitt *et al.*, 2000). Strategic behaviour can be seen as a response to the illegitimate status of entrepreneurs in China (Tsang, 1996). Hostile institutional forces may decrease the effectiveness of generic strategies, leading entrepreneurs to seek out creatively other strategies. Venture capitalists were more willing to connect Chinese firms to joint venture partners for legitimacy (Hitt *et. al.*, 2000). More specifically, Ahlstrom and Bruton (2002) noted that the laws and government policies toward private entrepreneurial firms in China are subject to a broad and varied interpretation of a range of officials who enact and enforce them. In the search for legitimacy, the private firms in China follow different types of strategies such as coopting officials by providing stock or other interest in the firm to the officials or by hiring local officials or their relatives.

Institutions also affect behaviours of venture capitalists in emerging markets. Bruton and Ahlstrom (2003) observed that venture capitalists adhere to institutional norms in selecting firms, monitoring funds and determining exit from funds. For example, the absence of strong regulatory and normative institutions makes it more likely venture capitalists will fund firms near to their offices. Also, because accounting information is less reliable, other means such as relationships with the entrepreneurs or others that know the entrepreneurs better are crucial. Moreover, for venture capitalists, exit opportunities through initial public offering (IPO) are very limited because the financial market is in a nascent stage.

Research has made strides towards understanding how unique cultural characteristics of institutions affect new ventures in emerging markets. Culture may influence the degree of entrepreneurship in an organization. Institutional factors including economic, political/legal and social factors may moderate the relationship between a country's culture and its entrepreneurial orientation. Countries differ in their support for entrepreneurship. Takyi-Asiedu (1993) argued that many countries may be similar on cultural dimensions, yet have different propensities and tolerances toward entrepreneurship. For example, as culture and ideology barriers have been lifted in China since its reform, entrepreneurship begins to flourish (Chang & MacMillan, 1991). In another study, Begley and Tan (2001) applied the theory of face adapted from Earley (1997) to predict interest in entrepreneurship in East Asian versus Anglo-Saxon countries. They found that social status will positively predict interest in entrepreneurship better in East Asia than in the West and that shame at business failure will negatively predict interest in entrepreneurship better in East Asia than the West. Since new ventures are risky, people in East Asian society that are aware that venture failure means loss of work may be especially conservative in assessing their feasibility.

The Resource-based View

The resource-based perspective (Barney, 1991; Wernerfelt, 1984) suggests that resources and capabilities create a competitive advantage when they are firm-specific, rare, durable and difficult to imitate or replace. Resources are typically defined as inputs to the production process, while capabilities are the ability to coordinate and deploy those resources to perform tasks. Resources may be tangible (for example, equipment, finance), or intangible (for example, brand name, trade secrets). Capabilities may consist of sub-routines and master routines (for example, product development, distribution) that integrate sub-routines into performance (Nelson & Winter, 1982). Strategy then becomes the effective use of resources in efforts to gain profits, or rents, in excess of a break-even point (Penrose, 1959).

The resource-based view suggests that new ventures' idiosyncratic resources and capabilities define a competitive advantage (Barney, 1991). What resources and capabilities stand out in shaping new venture performance in emerging markets? Research in this direction is very limited and existing studies tend to apply the concepts from the developed markets in the emerging market context. For example, drawing upon Eisenhardt and Schoohoven's (1990) work, Bruton and Rubanik (2002) found that founding team size of new technology ventures is positively related to venture performance in Russia because it provides greater financial resources available to the technology ventures in transitional economies. In Russia, there are virtually no funds available to technology start-ups except those that are internally generated. Thus, the more members involved in the founding team, the more likely they can gather greater financial capital. Lee, Lee and Pennings (2001) examined the influence of internal capabilities on firm performance by using data from Korean technological start-ups. They found that internal capabilities operationalized as entrepreneurial orientation, technological capabilities and financial resources invested during the development period are important predictors of new venture performance.

From a strategic point of view, innovation focus is critical for technology ventures in emerging markets. Li (2001) argued that product innovation may contribute to new venture performance in an emerging market in two ways. First, new ventures may rely on product innovation to exploit aggressively the growth opportunities in the market. Second, new ventures can use product innovation to gain competitive advantages, thus buffering environmental threats to their survival and growth. With a sample of new technology ventures in China, Li and Atuahene-Gima (2001) found that product innovation strategy is positively related to new venture performance. Similarly, with a sample of technological start-ups in Russia, Bruton and Rubanik (2002) demonstrated that technological innovativeness can contribute to new venture growth.

Resource Dependence Theory

The basic premise of resource dependence theory (Pfeffer & Salancik, 1978), drawn from work in sociology and political science, is that organizational behaviour can be explained by looking at the organization's context. Organizations are not internally self-sufficient because they depend on resources from external environments. Organizations are dependent on a resource environment that can impose constraints and create uncertainty. Resource dependence theory has two broad tenets (c.f. Li & Atuahene-Gima, 2001). The first tenet is that firms attempt to manage uncertainty and mitigate the effects of external forces in order to enhance their performance. The second tenet is that firms are constrained by and depend on other organizations that control critical resources for them. Thus managers make strategic choices concerning interorganizational and other external relationships in an attempt at 'altering the system of constraints and dependence confronting the organization' (Pfeffer & Salancik, 1978:267).

Resource dependence theory is particularly appropriate for research on new ventures in emerging markets because both new ventures and their context in emerging markets are characterized by lack of resources. As Hitt *et al.* (2000:451) argued, 'Given poorly developed financial markets, weak institutions for distribution of capital, and volatility in economic development, capital generally has low availability and high costs in emerging market countries.' These conditions produce a resource gap between firms in emerging markets and those in developed markets. New ventures in emerging markets are challenged with attaining and maintaining valuable, rare and inimitable resources to sustain their competitive advantages. Managers of these ventures are forced to find creative ways to secure and train human capital (Diomande, 1990).

Steensma *et al.* (2000) drew on resource dependence theory to examine small manufacturing enterprises' (SMEs) propensity to form technology alliances. They argued that, in contrast to their larger counterparts, SMEs experience resource dependence and must secure critical resources and know-how from larger, and potentially more powerful, trading partners. They found that technological uncertainty has a stronger and positive effect on SMEs' propensity to pursue technology alliances in countries with higher levels of uncertainty avoidance, such as Mexico.

Using resource dependency theory, Li and Atuahene-Gima (2001) examine the contingency effect of product innovation strategy on performance in Chinese new technology ventures. They argue that the effectiveness or ineffectiveness of product innovation strategy in China depends on the perceived environmental conditions and the relationship-based strategies adopted. Institutional support and environmental turbulence enhance the effectiveness of the ventures' product innovation strategy while strategic alliances for product development hinder the positive effect of product innovation strategy on new venture performance.

Ahlstrom and Bruton (2001) used resource dependence theory to explain how private Chinese firms build legitimacy. Organizational legitimacy implies that a firm needs to get support for its existence, goals and activities from external constituencies. In China, *guanxi*, or connections with key individuals, is useful in building legitimacy through the reciprocal obligatory relationships between individuals.

Social Capital and Network Perspectives

Social capital is the sum of the resources, actual or virtual, that accrue to an individual or group by virtue of possessing a durable network of more or less institutionalized relationships of mutual acquaintance and recognition (Coleman, 1988). Granovetter (1985) suggests that economic actions and outcomes of firms are embedded in and influenced by their social relations with outside actors. Social capital inherent in social ties or networks enhances actors' abilities to achieve their goals. According to Coleman (1988), members of a closely knit network can trust each other to honour obligations, which diminishes the uncertainty of their exchanges and enhances their ability to cooperate in the pursuit of their interests (c.f. Gargiulo & Benassi, 2000:184). Social capital refers to the 'goodwill that is available to individuals or groups. Its source lies in the structure and content of the actor's social relations' (Adler and Kwon, 2002:23). Social ties and networks can enable firms to gain access to resources and information, facilitate the flow of knowledge among partners, and obtain external legitimacy and status (Gulati, 1998). This is particularly beneficial to young, resource-deficient firms such as new ventures. External ties can enable them to acquire resources from external sourcing, and thus increase the effectiveness of their strategies based on specialization in innovation.

While social capital and networks exist in any economies, they are especially important in emerging markets, for two major reasons. First, given institutional constraints in emerging markets, networks can act as substitutes for governmental support or legitimacy, resource access and market reputation (Xin & Pearce, 1996). Networks (such as managerial ties) in emerging markets can compensate for the lack of market-supporting institutions such as transparent laws and regulations (Peng & Heath, 1996). Second, many cultures in emerging markets emphasize the roles of networks in doing business. For example, Chinese culture uses the concept of *guanxi* to describe social networks. The terms *blat* in Russia and *chaebol* in Korea refer to the same type of social networks.

Existing work applying social capital and network theory to new ventures in emerging markets has largely centred on obtaining access to resources. Zhao and Aram (1995) argued that Chinese businesses understand the importance of forming dependable relations. In a case study of six young Chinese firms, they found that the range and intensity of networks influenced entrepreneurial success and attainment of needed external resources. Networks may lead to legitimacy and access to capital. For example, Li and Atuahene-Gima (2002) found that Chinese new ventures' strategic alliances with foreign firms on downstream activities (that is, the adoption of agency business activities) can help these ventures acquire both financial and technological resources. Social capital also affects investment decisions of venture capitalists. Batjargal and Liu (2004) found that Chinese entrepreneurs' social capital has significant effects on investment selection decisions of venture capitalists in interaction with growth potential and technology/products of the venture. Strong ties between entrepreneurs and venture capitalists have significant direct effects on investment process decisions such as contractual covenants, investment delivery and venture valuation.

Despite the significant value of networking in new ventures in emerging markets, its limitation should also be acknowledged. Lee, Lee and Pennings (2001) examined the influence of internal capabilities (entrepreneurial orientation, technological capabilities and financial resources) and external networks (partnership-based and sponsorship-based networks) on firm performance by using data from Korean technology-based ventures. Their results show that internal capabilities are important predictors of new venture performance. Among external networks, however, only the linkages with venture capital companies were positively related to new venture performance. Indeed, Li and Atuahene-Gima (2001) found that political networking (a firm's resource allocation to cultivate relationships with government officials, banks and administrative and other regulative agencies) played no significant role in Chinese technology ventures. The reason is that transaction costs associated with building political connections attenuate benefits for new technology ventures. Political networking activities may drain the finances of the new ventures as well as hamper their efficient management. These results suggest that the value of relationship-based resources in emerging markets is declining as the market becomes more free and competitive.

Psychological Theories

Social psychology and social cognitive theories have also been applied in the studies of new ventures in emerging markets. Social cognitive theories illustrate how individuals characterize their external environment (Shaver & Scott, 1991). Many researchers have shown that personal characteristics of the entrepreneurs can affect new venture outcomes although these relationships are fairly weak (McClelland, 1965; Solomon & Winslow, 1988). Supported by social psychology research (Locke & Latham, 1990), Baum, Locke and Smith (2001) suggested that personal characteristics are powerful predictors of performance when combined with the adequate competencies, motivation and strategy. As a result, cognition processes mediate the relationship between interpretation of opportunities and action. Social cognitive theories also explain how entrepreneurial potential can be tested (Krueger & Brazeal, 1994). Krueger and Brazeal (1994) use Shapero's (1982) model of the entrepreneurial event and Ajzen's (1991) theory of planned behaviour, arguing that both perceived desirability and self-efficacy are antecedents to the propensity of entrepreneurs to act or start a new venture. Cognitive processes like self-efficacy can help cope with hostile entrepreneurial environments (Luthans, Stajkovic and Ibrayeva, 2000). In a study of 168 entrepreneurs in Singapore, Lee and Tsang (2001) found that experience, networking activities, number of partners, internal locus of control and need for achievement had a positive impact on venture growth.

Other researchers have also explored the ways in which culture influences social cognitive processes such as entrepreneurial orientation. Entrepreneurial orientation (Lumpkin and Dess, 1996) consists of autonomy, innovativeness, risk taking, proactiveness and competitive aggressiveness. These dimensions are internal processes that mediate the relationship between the environment and new venture entry. Entrepreneurial orientation may be specific to entrepreneurs through their context, experience, education and disposition. Lee and Peterson (2000) created the cultural model of entrepreneurship stating that a society's propensity to generate autonomous, risk-taking, innovative and competitively aggressive and proactive entrepreneurs depends on the cultural foundations. Wright, Hoskisson, Busenitz and Dial (2000) showed that entrepreneurial orientation (EO) is gained not only through managerial incentives, but also through entrepreneurial mindsets. Some studies find that EO is different among the population in emerging markets. Ray (1994) found that risk is contextually determined, and entrepreneurs have higher risk taking responses than non-entrepreneurs in Singapore. Holt (1997) similarly found a great difference between values of Chinese entrepreneurs and managers of state-owned/controlled enterprises where entrepreneurs are more self-determined and independent.

METHODOLOGICAL ISSUES

In reviewing the research methodologies of the studies conducted on new ventures in emerging markets, a number of aspects were examined including the fieldwork period, data collection methods, sample size and sampled countries.

Overall, research on new ventures in emerging markets is very limited and scattered. For the ten journals we surveyed (*ASQ, MS* and *JoM* were not included because they have not published any paper on this subject), there were only 43 publications during the 15-year period 1990–2004. On average, there was only 0.29 of a paper on this topic per journal per year. Except for *JBV*, the journals started to publish work on this subject in the middle of the 1990s or even in the early 2000s. For example, *ETP* had the first publication on this topic in 1995, and *OS* had the first one in 1998. *AMJ*, *AMR* and *AME* did not have their first one until 2000 and *SMJ* even started in 2001. Clearly, research on new ventures in emerging markets is at an embryonic stage and research attention to this subject is extremely limited. This is in contrast to the significant role of new ventures in these markets and their overall contributions to the global market.

Data Collection Method

Data collection is challenging in emerging markets where governments are often very protective of their data and barriers to communication with entrepreneurs exist. For example, a recent regulation issued by the Chinese government notes that organizations or individuals from foreign countries are not allowed to conduct independently any social survey activities by themselves in China. Instead, they have to entrust a certified domestic survey institute with conducting these activities. The certification for doing foreign-related surveys will be screened and approved by the National Statistics Bureau. This regulation poses significant constraints on scholars from other countries who are interested in China-related issues. Thus collaborating with local researchers may be a key means of gaining access to data sources (Li & Atuahene-Gima, 2001). A similar point was also made by Lee and Miller (1999) in the context of Korea. Also, in emerging economies, respondents have difficulties in understanding terms and concepts familiar to those in developed markets (Hoskisson *et al.*, 2000). It is important for researchers to check and make sure that language terms are understood properly by respondents.

As shown in Table 2.2, while different data collection methods have been employed, a survey with questionnaires was the most frequently used data collection method, mainly for reasons associated with convenience and economy. Next come case studies and personal interviews, which were used more extensively for small sample sizes. The heavy reliance on surveys with questionnaires as a principal data collection method, however, despite its cost advantages, limits our in-depth understanding of the subject of

*Table 2.2 Data collection methods**

Data collection type	Number of times used
Survey/questionnaire	15
Case study	8
Interview	7
Database	2

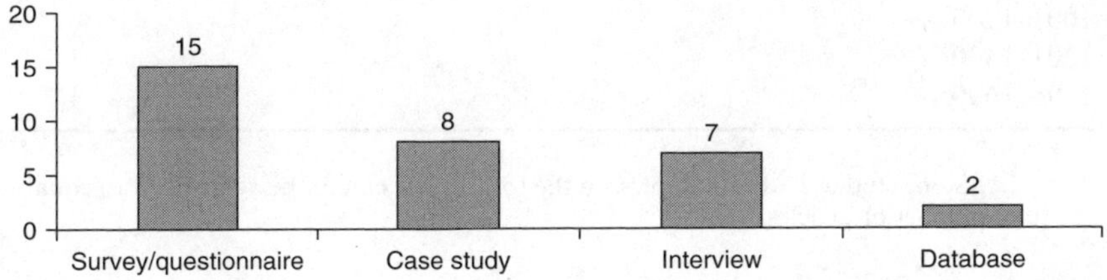

Note: * Some studies used multiple data collection methods, so this number cannot be aggregated to equal the total number of articles.

interest. Given the embryonic nature of this area, it may be important to use a personal approach (such as case studies and personal interviews) in order to gain useful insights into this interesting phenomenon. For example, the methodology proposed by Eisenhardt (1989) for theory development through case studies may be appropriate in the context of emerging markets.

Sample Size

The sample size ranged between three and 1470 firms. As shown in Table 2.3, the majority of research has sample sizes less than 100. Small sample size was largely influenced by the data collection methods available where many of the case studies and personal interviews yield small sample sizes. Small sample size may also result from a low response rate for questionnaire surveys in emerging markets. For example, in Korea, Lee, Lee and Pennings (2001) reported a response rate of 14.8 per cent. In China, Luo and Peng (1999) reported response rates of around 25 per cent from mail survey. While these response rates are not great, they are reasonable given the fact that most studies required the direct involvement of founders, CEOs or other senior executives. Indeed, in a study of new ventures conducted in the US, McDougall *et al.* (1994) had an 11 per cent response rate.

*Table 2.3 Sample sizes**

Size range	Total number
1–100	15
101–200	6
201–400	2
401–600	1
601–800	0
801–1000	1
1001–1200	0
1201–1400	1
1401–1600	1

Note: * Some studies had no samples, so the total above cannot be aggregated to equal the total number of articles.

Nonetheless, small sample size may affect the validity and generalizability of previous findings. Scholars have to find creative ways of improving response rate for survey-based research. Li and Atuahene-Gima (2001, 2002) reported a response rate of over 60 per cent in China during a series of studies where they adopted a face-to-face interview method (on-site interview) for data collection in order to improve the response rate. This method of data collection is thought to be better than traditional mail surveys in emerging markets because it offers a chance for the researchers to clarify questions, and enhance the respondents' understanding of the issues under study (Li & Atuahene-Gima, 2002). Lee and Miller (1999) also noted the importance of this method for data collection in Korea even though it is costly and time consuming.

Sampled Countries

Table 2.4 shows the countries which were covered in previous research. In total, the studies covered 16 emerging markets, which is 25 per cent of the 64 emerging markets (that is, the most comprehensive list of these markets to date) identified by Hosikisson *et al.* (2000). Obviously, research on new ventures in emerging markets has not covered different markets in a relatively broad way. The studies have largely focused on several key countries, including China (9), Korea (5), Russia (4) and Singapore (4). The majority of the studies only focused on a single country, while there were very few with a focus on several emerging markets. Several others went on to compare emerging markets to the developed markets.

*Table 2.4 Number of sampled countries**

Area sample	Number of samples
Asian countries	18
Africa/Middle East	2
Central and Eastern Europe	7
Latin America	1
Country	**Number of samples**
China	9
Korea	5
Russia	4
Singapore	4
Indonesia	2
Hungary	2
Czech Republic	1
Nigeria	1
Philippines	1
Taiwan	1
Thailand	1
Mexico	1
Belarus	1
Ukraine	1
India	1
Poland	1

Note: * Some studies sampled multiple countries.

China has been a promising area for researchers. This is not surprising given that China has more than 25 per cent of the world's population and its economic growth has been sustained at 8 per cent every year. Different from other transition economies which adopted a shock therapy, the Chinese government has carefully planned actions to move the country to a market economy while stabilizing the political base. In a recent Goldman Sachs global economics report, Wilson and Purushothaman (2003) predicted that China's economy could be larger than the US in 2041, and larger than all other countries as early as 2016. It should be noted that, with China's economic transition, entrepreneurial activities have been flourishing across the nation. In particular, stimulated by the government's preferential policies (for example, sales tax, land use and equipment importing) for high technology industries, there are 53 high technology science parks which have been established cross the country since 1988. In each science park, thousands of technology new ventures have been founded. For example, in the Zhongguancun Science Park in Beijing (founded in 1988), up to 2002, there were more than 9500 firms with 400 000 employees and industrial outputs of US$22.8 billion. Therefore, given the growth of technology new ventures in China, there are many interesting issues which should be explored regarding how Chinese technology ventures deal with institutional voids (the situations in which institutional framework and strategic factor markets have not been underdeveloped) and survive and achieve their competitive advantages (Li & Atuahene-Gima, 2002).

While future research will probably continue in China and a few other countries, research findings based on a single country may have limited generalizability to other emerging markets. This is because emerging markets are not homogeneous and they have different starting points (Hoskisson *et al.*, 2000; Khanna & Palepu, 1997). In particular, China has its unique culture and had been controlled by centrally planned communism for many years. Findings based on the Chinese context may have difficulties in generalizing into other emerging markets such as Russia, India or Brazil. Thus comparative studies with samples from different emerging markets will help us better understand new ventures in this context, though few have been done.

Industry Coverage

The studies we surveyed covered a wide range of industrial sectors, ranging from technology, manufacturing, trading and retail, to services (see Table 2.5). Some 46 per cent of the empirical studies have focused on technology industries including electronics, semiconductors and technical consulting. This is consistent with the fact that many emerging markets have emphasized technology development. Technology is an important factor in aiding the growth of emerging economies. Governments from emerging markets actively search for extensive technologies transfers through joint ventures and foreign investments (Keppler & Lechner, 1997; Li & Atuahene-Gima, 2002). For example, in China, foreign investors have been offered preferential treatment (in terms of tax, land use and import/export quotas) as an incentive to establish high technology ventures within high technology science parks.

Only 19 per cent of empirical studies have focused on multiple industries, while most examined only one industrial sector. Cross-sectional studies, however, fail to consider industry-specific factors, such as different degrees of competition, marketization and cost structures, possibly resulting in bias in the overall findings. Also cross-sectional data do not reflect current

*Table 2.5 Industries sampled**

Industry	Number of samples
Manufacturing	6
Printing	2
Food processing	1
Rubber/plastics	2
Chemicals	2
Equipment	1
Textiles	2
Publishing	1
Paint	1
Metalworking and fabrication	1
Autoparts	1
Other	4
Agri-business	1
Venture capital	3
Trade	2
Services	3
Transportation	1
Construction	2
Technology	12
Electronics information	2
Computer manufacturing	1
Computer software	2
Biotechnology	1
Information technology	1
Microelectronics	1
Optical mechanical and electronic products	7
New energy and new materials	1
New pharmaceutical and bioengineering	2
Semiconductor	1
Technical Consulting	1
Technology other	2
Not specified	4

Note: * Some studies sampled multiple industries, so the number above cannot be aggregated to equal the total number of articles.

transformations that might affect the relationships specified in previous studies. It seems that longitudinal studies better capture the dynamism of constructs. Longitudinal studies have the advantage of eliminating the possibility of reverse causality among a study's variables, an unavoidable

Table 2.6 Statistical methods employed

Statistical method	Number of studies
Regression	11
ANOVA	2
Factor analysis	2
Cross-group analysis	1
Descriptive statistics	1
Structured equation modelling	1

problem associated with cross-sectional studies. Ideally, longitudinal data collection methods should involve real-time observations (Van de Ven, 1992). In such studies, researchers record the timing of events and their characteristics as the decision processes evolve and unfold over time.

Statistical Analyses

Of 43 publications, 18 are descriptive, 19 are empirical and six are purely theoretical. Statistical analysis carried out in the empirical papers was relatively simple. Eleven studies used regression analysis and two studies used ANOVA and factor analysis. More advanced techniques such as structured equation modelling have rarely been used (only in one study). There were no longitudinal studies on new ventures in emerging markets (see Table 2.6).

FUTURE DIRECTIONS AND CONCLUSION

The purpose of this chapter was to review comprehensively the development of research on new ventures in emerging markets during the period 1990–2004. This is an emerging area which has attracted increasing attention from both researchers and practitioners because of the crucial role that these new ventures can play in their domestic and global markets. This research makes several contributions to our knowledge of new ventures as well as emerging markets. First, we reviewed the major theoretical perspectives adopted by researchers during this period. Consistent with Hoskisson *et al.* (2000), our review shows the prominence of institutional theory and the resource-based view in research on new ventures in emerging markets. We also found that theories such as social capital and networking perspective and resource dependence theory play an important role in research on this area. Second, we explored in detail the methodological aspects of the studies on new ventures in emerging markets. Overall, many of the studies are exploratory, with small sample size and the use of unsophisticated statistical analyses. Also data collection mainly relied on surveys with questions and a descriptive case study. Many studies focused on a few key emerging markets, including China and Korea. This is understandable, however, in view of the tremendous physical, cultural, financial, and other difficulties encountered by researchers who are interested in new ventures in emerging markets. Third, our review showed that research on new ventures in emerging markets is very limited and sporadic. We still cannot draw any conclusions about the behaviour of new ventures in emerging markets from these studies. The findings are tentative at best, since most of them are based on one study.

There are several fertile avenues for future research. First, while previous studies have adopted several theoretical perspectives in studying new ventures in emerging markets, there are other theories which have not been paid enough attention. For example, organizational learning describes an organization's ability to experiment, improve and increase its capability, which is a critical source of competitive advantage. Because of the lack of resources and technical capabilities in emerging markets, many new ventures use alliances, joint ventures and partnerships to learn (Hitt *et al.*, 2000; Li & Atuahene-Gima, 2002). Thus more research in the future should focus on how new ventures in emerging markets learn to develop their competitive advantages and what different mechanisms these ventures follow in learning compared to their counterparts in developed markets. Also the integration of different theories is needed in future research. While a transaction cost economics perspective (including agency theory) has not been widely adopted in studies of new ventures in emerging markets, we believe that it represents an important theorctical element in our understanding of this important phenomenon. As Hoskisson *et al.* (2000:254) have noted, 'Because transaction cost economics has been primarily applied to developed market economies characterized by strong legal regimes and binding social norms, less is known about governance structures in emerging economies.' With the process of marketization and privatization in emerging markets, many new ventures face the problems of governance. For example, ownership ambiguity has become a critical factor which prohibits the growth of new technology ventures in China. Thus it is important for future researchers to investigate how new ventures deal with transaction costs and principal–agent problems under uncertain and hostile environments in emerging markets.

Most work has concentrated on applying traditional theories to the domain of emerging market entrepreneurship, yet there has been no overarching theory. McDougall and Oviatt (2000) also acknowledge this as a problem characteristic of the international entrepreneurship field. The remaining question is whether a truly unifying theory and framework can be created even though emerging markets are so distinct. Clearly, more work should be done on theory development for new ventures in emerging markets.

Our review has revealed that, if the dynamic character of new ventures in emerging markets is to be understood, more longitudinal studies are required. Since longitudinal data collection methods are more costly and time consuming, collaborative projects across national boundaries can be an important means of improving the studies on new ventures in emerging markets. Also, combining quantitative and qualitative data in emerging market research can be particularly useful in yielding novel, relevant and reliable insights (Hoskisson *et al.*, 2000:257) and the use of more advanced statistical tools will contribute toward more rigorous research efforts. In addition, future research should have a more cross-cultural orientation and hitherto neglected emerging markets should be studied.

Finally, as discussed earlier, the studies on new ventures in emerging markets have largely focused on a number of key countries, including China, Korea, Singapore and former Central and Eastern European countries. It should be noted that there are around 64 emerging economies identified in the literature and many other countries have not gained enough research attention. Thus there is a strong need for considerably broadening the research agenda to embrace developments in these countries. Also emerging markets cover a broad group of countries that continue to experience often substantial and different institutional changes. For example, emerging markets with institutional transitions in China, Russia and Eastern Europe are different from other emerging markets such as India and Brazil. Even among transition economies, they are significantly different in terms of their economic transition approaches (for example, radical versus gradual transition). In future research, the role of institutions and cultural factors should be incorporated in the examination of new ventures in emerging markets. Researchers may wish to partner with others to gain access to other countries and industries, and to aid in comparative research studies.

Finally, many of the studies made conclusions from small sample sizes. Further research should enhance generalizability through using larger samples from more varied industries. Multiple data sources and teaming with other scholars may be a way to enhance generalizability of findings. The future for research in emerging markets is still quite unclear, but promises to be a fertile ground for study.

REFERENCES

Adler, P.S. and S.W. Kwon (2002), 'Social capital: Prospects for a new concept', *Academy of Management Review*, **27** (1), 17–40.

Ahlstrom, D. and G.D. Bruton (2001), 'Learning from successful local private firms in China: Establishing legitimacy', *Academy of Management Executive*, **15** (4), 72–83.

Ahlstrom, D. and G.D. Bruton (2002), 'An institutional perspective on the role of culture in shaping strategic actions by technology-focused entrepreneurial firms in China', *Entrepreneurship Theory and Practice*, **26** (4), 53–69.

Ajzen, I. (1991), 'Theory of planned behavior', *Organizational and Human Decision Processes*, **50**, 179–211.

Barney, J. (1991), 'Firm resources and sustained competitive advantage', *Journal of Management*, **17** (1), 99–120.

Batjargal, B. and M. Liu (2004), 'Entrepreneurs' access to private equity in China: The role of social capital', *Organization Science*, **15** (2), 159–72.

Baum, J.R., E.A. Locke and K.G. Smith (2001), 'A multidimensional model of venture growth', *Academy of Management Journal*, **44** (2), 292–303.

Begley, T.M. and W.-L. Tan (2001), 'The socio-cultural environment for entrepreneurship: A comparison between East Asian and Anglo-Saxon countries', *Journal of International Business Studies*, **32** (3), 537–53.

Bruton, G.D. and D. Ahlstrom (2003), 'An institutional view of China's venture capital industry: Explaining the differences between China and the West', *Journal of Business Venturing*, **18** (2), 233–59.

Bruton, G.D. and Y. Rubanik (2002), 'Resources of the firm, Russian high-technology startups, and firm growth', *Journal of Business Venturing*, **17** (6), 553–76.

Chang, W. and I.C. MacMillan (1991), 'A review of entrepreneurial development in the People's Republic of China', *Journal of Business Venturing*, **6** (6), 375–9.

Coleman, J.S. (1988), 'Social capital in the creation of human capital', *Journal of Sociology*, **94**, S95–S120.

Cooper, A.C., G.E. Willard and C.Y. Woo (1986), 'Strategies of high performing new and small firms: A reexamination of the niche concept', *Journal of Business Venturing*, **1** (3), 247–60.

DiMaggio, P. and W. Powell (1983), 'The iron cage revisited: Institutional isomorphism and collective rationality in organizational fields', *American Sociological Review*, **48**, 147–60.

Diomande, M. (1990), 'Business creation with minimal resources: Some lessons from the African experience', *Journal of Business Venturing*, **5** (4), 191–200.

Earley, P.C. (1997), *Face, harmony, and social structure: An analysis of organizational behavior across cultures*, New York: Oxford University Press.

Eisenhardt, K.M. (1989), 'Building theories from case study research', *Academy of Management Review*, **14** (4), 532.

Eisenhardt, K.M. and C.B. Schoonhoven (1990), 'Organizational growth: Linking founding team, strategy, environment, and growth among U.S. semiconductor ventures, 1978–1988', *Administrative Science Quarterly*, **35** (3), 504–29.

Gargiulo, M. and M. Benassi (2000), 'Trapped in your own net? Network cohesion, structural holes, and the adaptations of social capital', *Organization Science*, **11** (2), 183–96.

Granovetter, M.S. (1973), 'The strength of weak ties', *American Journal of Sociology*, **78**, 1360–80.

Granovetter, M. (1985), 'Economic action and social structure: The problem of embeddedness', *American Journal of Sociology*, **91** (3), 481–510.

Gulati, R. (1998), 'Alliances and networks', *Strategic Management Journal*, **19** (4), 293–317.

Hitt, M.A., M.T. Dacin, E. Levitas, J.-L. Arregle and A. Borza (2000), 'Partner selection in emerging and developed market contexts: Resource-based and organizational learning perspectives', *Academy of Management Journal*, **43** (3) 449–67.

Holt, D.H. (1997), 'A comparative study of values among Chinese and U.S. entrepreneurs: Pragmatic convergence between contrasting cultures', *Journal of Business Venturing*, **12** (6), 483–505.

Hoskisson, R.E., L. Eden, C.M. Lau and M. Wright (2000), 'Strategy in emerging economies', *Academy of Management Journal*, **43** (3), 249–67.

Keppler, M. and M. Lechner (1997), *Emerging markets: Research, strategies and benchmarks*, New York: McGraw-Hill.

Khanna, T. and K. Palepu (1997), 'Why focused strategies may be wrong for emerging markets', *Harvard Business Review*, **75** (4), 41–8.

Krueger, N.F. and D.V. Brazeal (1994), 'Entrepreneurial potential and potential entrepreneurs', *Entrepreneurship: Theory & Practice*, **18** (3), 91–204.

Lee, C., K. Lee and J.M. Pennings (2001), 'Internal capabilities, external networks, and performance: A study on technology-based ventures', *Strategic Management Journal*, **22** (6/7), 615–46.

Lee, D.Y. and E.W.K. Tsang (2001), 'The effects of entrepreneurial personality, background and network activities on venture growth', *Journal of Management Studies*, **38** (4), 583–602.

Lee, J. and D. Miller (1999), 'People matter: Commitment to employees, strategy and performance in Korean firms', *Strategic Management Journal*, **20** (6), 579–93.

Lee, S.M. and S.I. Peterson (2000), 'Culture, entrepreneurial orientation, and global competitiveness', *Journal of World Business*, **35** (4), 401–16.

Li, H. (2001), 'How does new venture strategy matter in the environment–performance relationship?', *Journal of High Technology Management Research*, **12** (2), 183–205.

Li, H. and K. Atuahene-Gima (2001), 'Product innovation strategy and the performance of new technology ventures in China', *Academy of Management Journal*, **44** (6), 1123–34.

Li, H. and K. Atuahene-Gima (2002), 'The adoption of agency business activity, product innovation, and performance in Chinese technology ventures', *Strategic Management Journal*, **23** (6), 469–90.

Locke, E.A. and G.P. Latham (1990), 'Work motivation and satisfaction: Light at the end of the tunnel', *Psychological Science*, **1** (4), 240–46.

Lumpkin, G.T. and G.G. Dess (1996), 'Clarifying the entrepreneurial orientation construct and linking it to performance', *Academy of Management Review*, **21** (1), 135–72.

Luo, Y. and M.W. Peng (1999), 'Learning to compete in a transition economy: Experience, environment, and performance', *Journal of International Business Studies*, **30** (2) 269–95.

Luthans, F., A.D. Stajkovic and E. Ibrayeva (2000), 'Environmental and psychological challenges facing entrepreneurial development in transitional economies', *Journal of World Business*, **35** (1), 95–109.

McClelland, D.C. (1965), 'N-achievement and entrepreneurship: A longitudinal study', *Journal of Personality and Social Psychology*, **1**, 389–92.
McDougall, P. and R.B. Robinson (1990), 'New venture strategies: An empirical identification of eight "archetypes" of competitive strategies for entry', *Strategic Management Journal*, **11** (6), 447–67.
McDougall P.P. and B.M. Oviatt (2000), 'International entrepreneurship: The intersection of two paths', Guest Editor's Introduction, *Academy of Management Journal*, **43** (5), 902–8.
McDougall, P.P., J.G. Covin, R.B. Robinson and L. Herron (1994), 'The effects of industry growth and strategic breadth on new venture performance and strategy content', *Strategic Management Journal*, **15** (7), 537–57.
Meyer, J.W. and B. Rowan (1977), 'Institutionalized organizations: Formal structure as myth and ceremony', *American Journal of Sociology*, **83**, 340–63.
Nelson, R.R. and S.G. Winter (1982), 'The Schumpeterian tradeoff revisited', *American Economic Review*, **72** (1), 114–32.
North, D.C. (1990), *Institutions, institutional change, and economic performance*, Cambridge: Cambridge University Press.
Oviatt, B.M. and P.P. McDougall (1994), 'Toward a theory of international new ventures', *Journal of International Business Studies*, **25**, 45–64.
Peng, M.W. (2001), 'How entrepreneurs create wealth in transition economies', *Academy of Management Executive*, **15** (1), 95–108.
Peng, M.W. and P.S. Heath (1996), 'The growth of the firm in planned economies in transition: Institutions, organizations, and strategic choice', *Academy of Management Review*, **21** (2), 492–528.
Penrose, E.T. (1959), *The theory of the growth of the firm*, New York: John Wiley.
Pfeffer, J. and G. Salancik (1978), *The external control of organizations: A resource dependence perspective*, New York: Harper & Row.
Ray, D.M. (1994), 'The role of risk-taking in Singapore', *Journal of Business Venturing*, **9** (2), 157–77.
Scott, W.R. (1995), *Institutions and organizations*, Thousand Oaks, CA: Sage.
Shapero, A. (1982), 'Social dimensions of entrepreneurship', in C. Kent, D.L. Sexton and K.H. Vesper (eds), *Encyclopedia of entrepreneurship*, Englewood Cliffs, NY: Prentice-Hall, pp. 72–90.
Shaver, K.G. and L.R. Scott (1991), 'Person, process, choice; the psychology of new venture creation', *Entrepreneurship: Theory & Practice*, **16** (2), 23–45.
Solomon, G.T. and E.K. Winslow (1988), 'Toward a descriptive profile of the entrepreneur', *Journal of Creative Behavior*, **22**, 162–71.
Steensma, H.K., L. Marino, K.M. Weaver and P.H. Dickson, (2000), 'The influence of national culture on the formation of technology alliances by entrepreneurial firms', *Academy of Management Journal*, **43** (5), 951–73.
Stinchcombe, A.L. (1965), 'Organizations and social structure', in J.G. March (ed.), *Handbook of organizations*, Chicago: Rand McNally, pp. 142–93.
Takyi-Asiedu, S. (1993), 'Some socio-cultural factors regarding entrepreneurial activity in Sub-Saharan Africa', *Journal of Business Venturing*, **8** (2), 91–8.
Tan, J. (1996), 'Regulatory environment and strategic orientations in a transitional economy: A study of Chinese private enterprise', *Entrepreneurship: Theory & Practice*, **21** (1), 31–46.
Tsang, E.W.K. (1996), 'In search of legitimacy: The private entrepreneur in China', *Entrepreneurship: Theory & Practice*, **21** (1), 21–30.
Van de Ven, A H.G.S. (1992), 'Item homogeneity in verbal tests: A Rasch analysis

of Amthauer's verbal tests', *Educational & Psychological Measurement*, **52** (3), 623–39.

Vesper, K.H. (1980), 'New venture planning', *Journal of Business Strategy*, **1** (2), 72–4.

Wernerfelt, B. (1984), 'A resource-based view of the firm', *Strategic Management Journal*, **5** (2), 171–80.

Wilson, D., and R. Purushothaman (2003), 'Dreaming with BRICs: The path to 2050', Goldman Sachs Global Economics Paper No. 99.

Wright, M., R.E. Hoskisson, L.W. Busenitz and J. Dial (2000), 'Entrepreneurial growth through privatization', *Academy of Management Review*, **25** (3), 591–601.

Xin, K.K. and J.L. Pearce (1996), 'Guanxi: Connections as substitutes for formal institutional support', *Academy of Management Journal*, **39** (6), 1641–58.

Zahra, S.A., R.D. Ireland and M.A. Hitt (2000), 'International expansion by new venture firms: International diversity, mode of market entry, technological learning, and performance', *Academy of Management Journal*, **43**, 925–50.

Zahra, S.A., R.D. Ireland, I. Gutierrez and M.A. Hitt (2000), 'Privatization and entrepreneurial transformation: Emerging issues and a future research agenda', *Academy of Management Review*, **25** (3), 509–24.

Zhao, L. and J.D. Aram (1995), 'Networking and growth of young technology-intensive ventures in China', *Journal of Business Venturing*, **10** (5), 349–70.

3. Firm dynamics in economic transition: evidence from a Chinese science park

Quanlin Gu, Hongbin Li, Weiying Zhang and Li-An Zhou

INTRODUCTION

Seventy years ago, Schumpeter (1934) envisioned economic development as a process of 'creative destruction', in which the competition in technology innovations plays the role of a filter – picking the winners and driving out the losers. In this sense, a fundamental measure of a well-functioning and dynamic market is reflected in the extent to and pace at which the inefficient firms are wiped out in the evolutionary process of firm entry and exit.

In contrast with a mature market economy characterized by free entry and exit of firms, the traditional planning regime exhibits a dramatically different picture of firm dynamics. All the decisions regarding the birth and death of a firm were made by the government. A firm, even though financially troubled and technically inefficient, might still remain in operation through the continuing injection of state aids. Industries were monopolized by state-owned enterprises, and no private firms or joint ventures were allowed to enter.

China has started economic transition from a planning economy to a market economy and has abolished its restrictions on non-state entry in many industries since the late 1970s. As a result of the policy of encouraging entry of non-state firms, including joint ventures, product markets have become more and more competitive. The decision to exit is no longer made only by the government, and the turnover of firms takes place quite frequently.

The process of economic transition towards a market economy raises an interesting question concerning the nature of firm dynamics in China's context. Despite the increasing competitiveness of markets and enterprise reforms going on for years, one might still wonder how far the Chinese economy, after two decades of economic reforms, has moved towards a genuine market economy, an economy characterized by the firm dynamics in a Schumpeterian sense. More specifically, does market competition drive inefficient firms out of the market? Given the decisive role of government in determining the fate of a firm in the traditional regime, is there any role for government ownership in affecting survival? Did enterprise reforms have any effect on firm performance? How do firm size and age affect a firm's probability of survival?

Using a large data set from the Zhongguancun Science Park of Beijing (ZSP henceforth), we present a systematically empirical analysis of the determinants of firm survival. We first give a picture of firm evolution in the ZSP, with a focus on two major aspects: the changing mix of ownership types of firms and the patterns of entry and exit. We then employ the Cox Proportional Hazard Model to analyse the important factors affecting the hazard rate of ZSP high-tech firms. The main finding of this chapter is that both financial and productivity performance, has been an important factor in firm survival/exit in ZSP. On the other hand, we also find that, after controlling firm performance, state ownership still enjoys a survival advantage. But interestingly, enterprise reforms, such as corporatization reform, seem to lead to hardening of the budget constraint. Our evidence shows that corporatization reforms have made the exit of state-owned enterprises more sensitive to indebtedness.

Although the data set used in this study only covers one science park in China, it provides a very valuable chance to understand what is really going on with respect to the dynamics of firm evolution in China. ZSP has been the largest and the most dynamic science park in China, and it contains most of the big high-tech firms (such as Legend and Beida Founder) in China as well as some big foreign companies, such as Nokia. The ZSP's industrial output accounts for one-seventh of all science parks combined in China, and contributes to as much as two-thirds of the growth in the total industrial output of Beijing (Beijing Bureau of Statistics, 2002). Given its economic size and vital role in Chinese high technology industry, studying the determinants of the turnover of high-tech firms in this park is important in its own right.

On the other hand, ZSP has been undergoing dramatic transition since its establishment in 1988. While state-owned firms are still active in the ZSP, more and more new entrants are private firms or foreign joint-ventures, especially in recent years. This shift shares many important aspects of a general pattern of transition occurring nationwide. We believe that the results obtained from this study will shed light on the understanding of China's economic transition in general.

The rest of this chapter is organized as follows. The next section reviews the literature that relates to firm dynamics. The third section introduces the data and some basic characteristics of firm entry and exit in ZSP. The fourth section discusses briefly the method and specification of the econometric model. The fifth section reports and discusses the regression results, while the sixth section concludes.

LITERATURE REVIEW OF FIRM DYNAMICS

Classical economic theory develops its analysis on the assumption of a representative consumer/producer. In this hypothetical world, there will be no firm dynamics to be observed. This assumption turns out to be too simplistic in most cases: firms usually exhibit great differences in size, technology, productivity and other dimensions. Hence the heterogeneity across firms is an indispensable part of productivity growth, firm evolution and market dynamics.

Since the 1960s, there has been a growing literature studying both the empirical patterns and theoretical underpinnings of firm dynamics. For example, Mansfield (1962) and Du Reitz (1975) find that small firms have greater growth but higher hazard rates than large firms, which motivates Jovanovic (1982) to advance a theoretical model addressing both the heterogeneity and entry/exit of firms. In Jovanovic's model, a firm does not know its 'real' profitability when entering a market, but can learn it gradually through its realized profits in a Bayesian learning manner. A firm makes its exit decision according to the profitability that it has known. Since the 'real' profitability is assumed as fixed in this model and the learning process is just a by-product for the firm in production, this model is called the Passive Learning Model. Ericson and Pakes (1995) emphasize that, even if the firm knew its 'real' profitability, that profitability would be changed by the decisions of the other firms in the market. Hence firms will learn actively about the market condition through the outcome of its investment. Therefore this model is termed the Active Learning Model. These two learning models both explain the entry and exit of firms from the view of uncertainty. Because of the uncertainty of the demand for a new product and the opportunity cost of alternative technologies, firms will try to use different technologies and make different production decisions. This variation across firms will naturally lead to the heterogeneity in profitability or efficiency, which directly affects the firm's exit decisions. These theoretical developments provide a solid base for empirical studies to investigate the effect of efficiency factors (for example, productivity and capital intensity) on firm survival.[1]

Starting from the view expressed by Schumpeter (1934) that fixed cost is an obstacle for firms to transfer to use new technology, some economists advanced the idea that new technology is always embedded in new capital.[2] According to this idea, Mortensen and Pissarides (1994), Caballero and Hammour (1996) and Campbell (1998) build up a vintage model of technological change, which points out that productivity growth is determined by the process of new firms entering and replacing old firms. However, the learning-by-doing effect is another aspect of firm dynamics that must be given enough consideration. Jensen *et al.* (2001) make an extension that productivity growth can be decomposed into two sources: a vintage effect and a survival effect; the former is the advantage of new firms in having more advanced capital, technology and so on, and the latter is the advantage of old firms in market share, efficient scale and experience accumulation. Note that these two effects may be compounded with period effects reflecting the fluctuations of macroeconomic and market conditions. Hence it might not be surprising that the evidence of the total age (or duration) effect on the hazard rate, the combination of the above three effects, is mixed. Most studies have identified negative effects of age on exit but some others have found an inverse U-shaped effect (Audretsch and Mahmood, 1994; Honjo, 2000).

Ownership has also been studied in the literature, however, most of it is limited to the study of single-plant and multi-plant firms (Dunne *et al.*, 1989; Audretsch and Mahmood, 1994, 1995; Mata *et al.*, 1995; Disney *et al.*, 2003). Because of the coexistence of the various ownership forms in the ZSP, a more explicit analysis of the effect of ownership on firm exit can be conducted in this chapter.

While much of empirical work on firm entry and exit has so far primarily focused on developed countries, some recent studies have explored the firm dynamics in developing countries and transition economies. The exit patterns in a number of developing countries and regions, such as Chile and Colombia, have been documented in the literature (Liu and Tybout, 1996; Liu, 1993). Liedholm and Mead (1999) and Frazer (2004) have examined the determinants of firm exit in the context of Africa. Brown and Earle (2000) and Konings and Xavier (2003) study the firm exit after the privatization in the early 1990s in Russia and Slovenia, respectively.

INSTITUTIONAL BACKGROUND OF THE ZSP

Zhongguancun Science Park of Beijing[3] is the first and largest high-tech science park in China and is often called the Silicon Valley of China. It is located in Haidian District, in Northwest Beijing. Haidian is well-known for its vast intensity of research institutions and universities in China. This area is home to 232 scientific research institutions and 73 universities (including Peking University and Tsinghua University). ZSP is the leader of China's high-tech industry development, and its origin can be traced back to the late 1980s. By the end of 2003, there were more than 12 000 firms in ZSP, with a total employment of over 480 000. The gross industrial output value in 2003 was about 25 billion US dollars.

The goal of establishing ZSP for the Chinese government was to build a Chinese version of Silicon Valley. In order to achieve this ambitious goal, a series of preferential policies have been instituted to quicken the development of high-tech firms. Their preferential policies cover many aspects and are mainly manifest in the areas of tax, financing, land and foreign exchange.[4] The most prominent preferential policy is taxation. All ZSP firms will be exempted from the corporate income tax in the first three years and half of the income tax in the subsequent three years. The normal corporate income tax rate is 33 per cent in China. Furthermore, expenditures on R&D and innovation are tax-deductible. For those products listed in the *State Catalogue of New and High Tech Products*, the imported equipment, know-how and accessories enjoy tariff exemption. ZSP also set up some special funds, such as the Zhongguancun Science Park New and High Tech Industries Development Fund, to support and subsidize the development of high-tech industries. Beijing City government provided ZSP firms with certain preferential treatment in land use. The fee for the transfer of the land use right and the urban infrastructure construction fees are exempted or collected at a reduced rate.

However, a high-tech firm has to obtain the *certificate of a high-tech firm* from the Administrative Committee of Zhongguancun, which is a government agency in charge of all administrative issues in ZSP, before it is eligible to enjoy these advantages. This certificate has to be renewed annually. The certification and renewing process requires each firm to file an annual report for the Administrative Committee. The report has to disclose information about basic background, balance sheet, human resources and so on. The Committee will check the high-tech qualifications of each firm on the basis of the report. To obtain the certificate, the firm must satisfy at least the following conditions: (a) it must engage in high-tech related R&D and/or production and the revenue from these operations must reach a certain percentage of total revenue; (b) R&D expenditures must take up 3 per cent or more of the total revenue; (c) employees with college degrees or above must make up a minimum proportion of total employment. The data set used for this chapter is compiled from the annual reports by all certified firms for the period 1995–2003. On average, there are more than 5000 firms per year.

FIRM DYNAMICS IN ZSP: DATA DESCRIPTION

In contrast to Russia and Poland, where massive privatization was undertaken in the initial stage of economic transition, China took a gradualist approach to transition. Instead of privatizing state-owned firms in a big-bang manner, China made its priority encouraging entry of new non-state firms. As a result of this reform strategy, the mix of state and non-state firms has been in constant flux, with a steady declining share of the state-owned sector. This general pattern also appeared in the evolution process of ZSP over the years since its establishment. Figure 3.1 depicts the shares of different ownership types between 1995 and 2003. Four kinds of ownership types are distinguished: state-owned firms, collective firms, private firms, and foreign and joint-venture firms. We see a declining share of state and collective firms, followed by an increasing share of private and other non-state firms. It is noteworthy that state-owned firms only accounted for less than 15 per cent in 2003 while this share was more than 30 per cent in 1995. Figure 3.2 describes the pattern of ownership change from a different angle. Here we depict the proportions of various ownership types for a given entry cohort. Clearly, among the new entrants over the years between 1995 and 2003, state-owned firms have become a negligible part.

One of the enterprise reforms in China, so-called 'corporatization reform', started in the early 1990s, with the aim of transforming traditionally state-owned enterprises (SOE) into incorporated firms with limited liability, at least legally so in the initial period of the reform.[5] Since the SOEs were originally controlled by the government, which led to many problems, such as soft-budget constraint syndrome, this reform was supposed to establish well-structured corporate governance for the SOEs

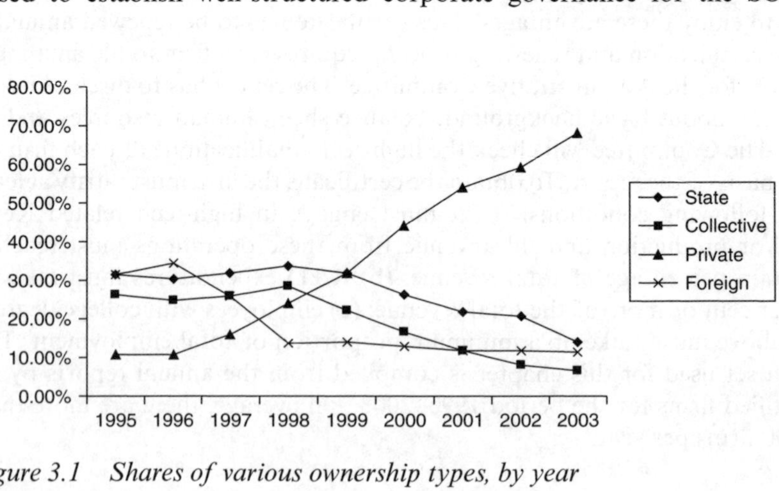

Figure 3.1 Shares of various ownership types, by year

with the monitoring of shareholders in place of that of the government. Figure 3.3 presents the proportions of corporatized firms over all state-owned enterprises in the sample during 1995–2003 by 'birth' cohort, that is, according to, the establishment year of the firm. This proportion has been increasing quite steadily from cohort 1992 to cohort 2003, reaching 90 per cent in the latter.

The constant shifts in the property rights arrangements of the state-owned firms due to corporatization and privatization reforms, plus entry of

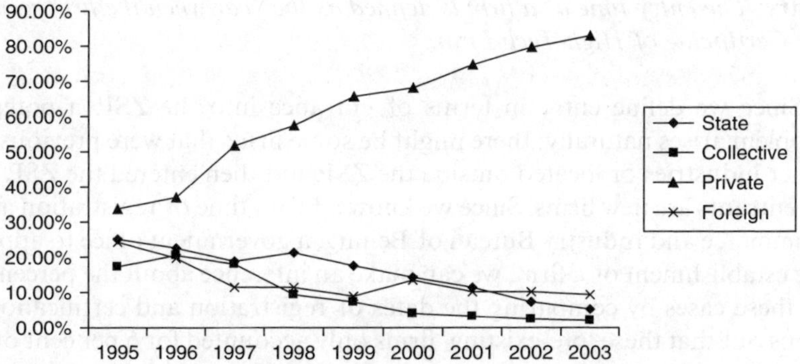

Figure 3.2 Proportions of ownership by entry cohort

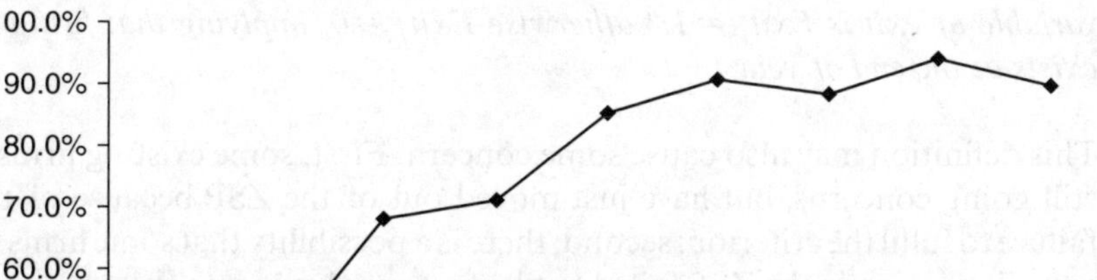

Figure 3.3 The proportion of corporatized SOEs, by birth cohort

purely private firms and joint-ventures, give rise to the coexistence of various ownership-type firms in the ZSP. The heterogeneity in ownership and the firm dynamics driven by entry and exit raise interesting questions concerning the role of ownership in determining firm survival and the effects of enterprise reforms, such as the effect of corporatization of the SOEs. We will explore these issues in the next section.

In order to give a precise understanding of the turnover of firms in the ZSP, we must first define the entry and exit of the firm.

Entry: *The entry time of a firm is defined as the year when the firm received the Certificate of High-Tech Firm.*

Since we define entry in terms of entrance into the ZSP, a potential problem arises naturally: there might be some firms that were previously in other industries or located outside the ZSP and then entered the ZSP. Not all entrants are new firms. Since we know a firm's time of registration at the Commerce and Industry Bureau of Beijing, a government office to approve the establishment of a firm, we can make an inference about the percentage of these cases by comparing the dates of registration and certification. It turns out that these 'pre-existing' firms only accounted for 5 per cent of the whole sample.

Exit: *If the data of one firm exist in year t but disappear in year t+1, the firm is considered as exiting the ZSP at some time in year t+1. In this case, the variable of exit is* $Exit_{i,t}=1$;[6] *otherwise* $Exit_{i,t}=0$*, implying that firm i still exists at the end of year t+1.*[7]

This definition may also cause some concern. First, some existing firms are still going concerns, but have just moved out of the ZSP because of their failure to fulfil the criterion; second, there is a possibility that some firms that are still running in the ZSP failed to file the annual report to the administrative committee, so it happens that some exiting firms 're-enter' the data set. However, these two concerns will not severely affect our analysis, for the following reasons: (1) since staying in the ZSP allows the firms to enjoy preferential taxes and other policies, almost all qualifying firms will want to remain whenever possible; (2) the extension of the high-tech certificate, which is contingent on the submission of an annual report, and the numerous efforts by the Committee urging each firm to file the annual report, make the number of 're-entering' firms very small. A careful examination of the data also shows that statistical errors due to reporting failures are negligible.

Based on the above definitions, Figure 3.4 depicts the general pattern of firm entry and exit in the ZSP during the period 1995–2003. In terms of

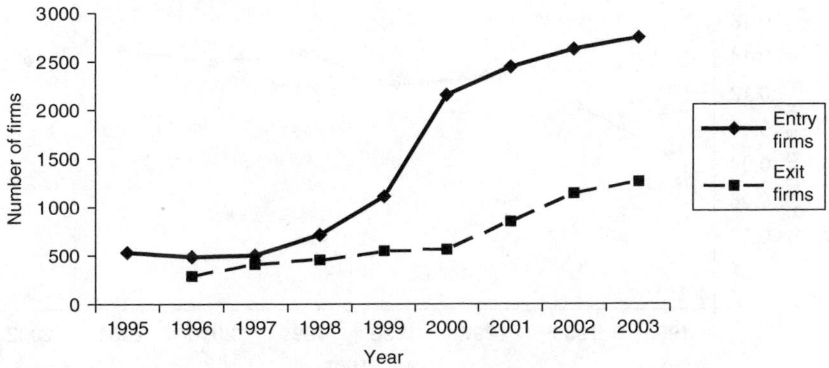

Figure 3.4 Numbers of entry and exit firms, by year (1995–2003)

entry, the first three years only saw a rather steady and relatively low inflow of new firms (only around 500 firms on average). The situation began to change from 1998. After a small increase in 1998, the new entry exhibited sharp jumps in 1999 and 2000, and hit a record high level of about 2500 firms in 2001. There are two important factors that contributed to the hikes in 1999 and 2000. First, the worldwide optimism about IT technology and the new economy in 1999, resulting from the persistent booming of new economy in the United States, propelled Chinese entrepreneurs (including some returning overseas Chinese graduates) to rush in and start new ventures in the ZSP. Second, as mentioned earlier, the Committee also issued a series of new preferential policies in 1999 to encourage entry and growth of the hi-tech firms.

The firm exit demonstrates a slightly different pattern. The number of exiting firms was rather stable, at the 500 level, until the year 2000, and then steadily going up from 2001. The increase is mainly due to the sharp increase in entrant firms and the intensified competition. Overall, the exits are less subject to fluctuations. Figure 3.5 plots the average hazard rates of firms from 1995 to 2002. The hazard rate represents the continuing firm's conditional probability of failure, which is calculated as the proportion of exiting firms in a given year $t+1$ (those firms have $EXIT_{i,t}=1$) of all the survivors by the end of year t (that is, the number of firms in year t). As a reflection of intensified competition in the ZSP over time, the hazard rate keeps rising until 2000 and then remains at that high level.

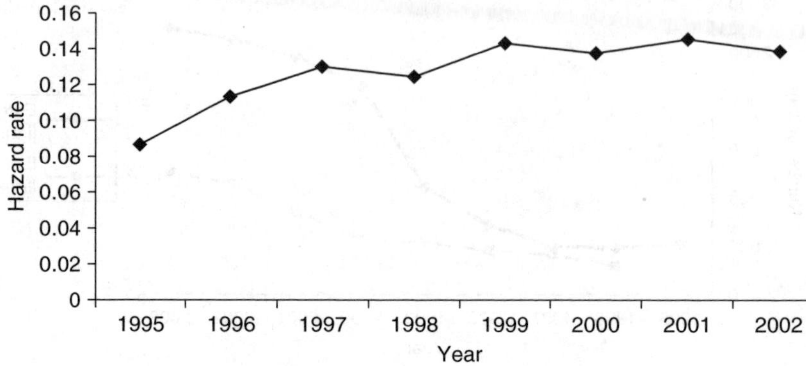

Figure 3.5 Average hazard rate, by year

METHODOLOGY AND SPECIFICATION

Usually the firm-level data do not record the exit time of all firms because some firms have not yet exited by the end of the sample period, so the exit time of some firms is unknown. This is known as the 'censoring problem' in estimation. The duration model deals with this problem by estimating the conditional probability of exit as a function of firm age and other factors. The proportional hazard model proposed by Cox (1972, 1975) is now a standard tool in survival analysis.[8]

The hazard function of the Cox model is as follows:

$$\lambda_i(t) = \lambda_0(t)\exp[x_i(t)'\beta].$$

$\lambda_0(t)$ is the baseline hazard function, which could be any non-negative function of base time or duration t. In our application, t denotes the age of the firm, not the year as commonly specified; $\lambda_0(t)$ is assumed not to be influenced by any covariates; $\lambda_i(t)$ is the hazard rate in time t, which is 'proportional' to the baseline hazard $\lambda_0(t)$. The 'proportion' is $\exp[x_i(t)'\beta]$, where x_i is the vector of covariates and β is the vector of coefficients to be estimated; it can be estimated by the partial likelihood method proposed by Cox without estimating $\lambda_0(t)$. However, $\lambda_0(t)$ can then be estimated after the estimation of β.

In the regression analysis below, the estimates of the coefficients β will be reported. The estimated coefficients β are similar to the marginal effect of a variable in the linear regression model. A negative coefficient implies that an increase in an explanatory variable is associated with a lower hazard, and therefore a higher survival possibility, while a positive coefficient implies that an increase in the explanatory variable would add to the hazard rate. The Cox model utilizes all the information in the data except for the base time t, that is, firm age in this chapter. Hence the dummy variables for the year will be included in the regression equation and the information of the age effect is contained in the baseline hazard function $\lambda_0(t)$.

The proportional hazard model implies that the hazard function is fully determined by its observed characteristics x (there is no residual term in this model). However, there might be some unobserved factors affecting firm survival, which make some firms have longer durations. If not accounted for, this so-called 'unobserved heterogeneity' (frailty) problem causes the hazard function to be biased downward. It is introduced into the duration model by multiplying the hazard by a random variable v_i, which is positive, has a mean of one, and is distributed independently of x_i. Because in this data most firms have multiple records, the v_i (frailty) is assumed to be constant across one firm's observations (Vaupel *et al.*, 1979; Hougaard, 1995; Gutierrez, 2002). The duration model with frailty can be estimated in the context of the Cox model but that is too computing-intensive to implement on such a large data set. Instead we use the parametric duration model (based on the Weibull distribution),[9] which accommodates frailty, but is less computing-intensive, to check the impact of unobserved heterogeneity. We find that the unobserved heterogeneity is negligible.[10]

The choice of variables to be included in the regression largely relies on the guidelines provided by existing theories and empirical evidence. The following six sets of factors are considered to affect firm survival/exit: (1) some basic characteristics of the firm, such as size, age and capital intensity; (2) performance and efficiency; (3) financial constraints; (4) ownership; (5) degree of diversification in production; and (6) industry-specific characteristics. A complete list of variable definitions can be found in the Appendix.

A well-documented fact in the existing literature is that large firms are more likely to survive than small firms (Ahn, 2001). A major explanation for this fact is that large firms are able to reach the output level of minimum efficient scale (MES). According to this argument, firm size, measured by log employment, is expected to have a negative effect on the hazard rate. Firms with higher capital intensities might be able to use more capital (resources) to survive the fluctuations of industry. Doms *et al.* (1995) observed lower hazard rates and higher growth rates associated with capital-intensive firms. So the effect of capital intensity (capital per employee) on the hazard rate is expected to be negative. From the average hazard rate in different years and ages presented in Figure 3.5 and 3.6, the effect of firm age on the hazard rate is expected to have an inverse-U shape.

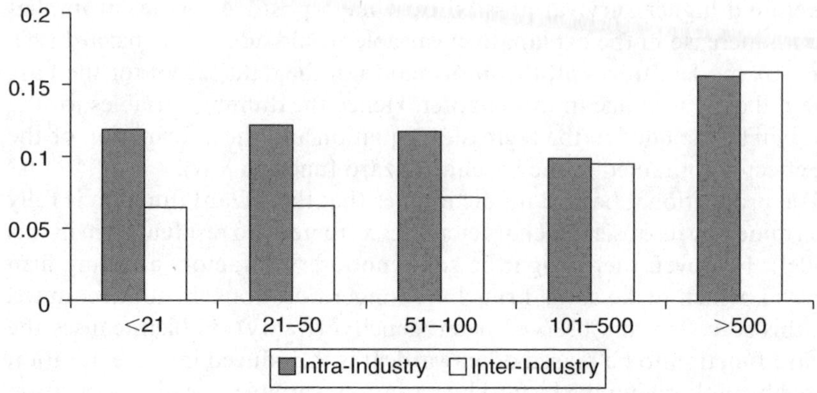

Figure 3.6 Average degrees of diversification by size classes (employment)

Thanks to the richness of our data set, we can use several variables, such as labour productivity, return on assets, R&D intensity and export, to represent various aspects of firm productivity and performance. Productivity is an important index of efficiency. Baldwin (1993), Doms *et al.* (1995) and Olley and Pakes (1996) find that TFP or labour productivity is relatively higher in surviving firms. Profitability is obviously critical in the firm's decision about exits. Here return on assets, defined as the ratio of profit before tax to total assets, will be used. R&D is found to have a favourable effect on the survival probability (for example, Kim *et al.*, 2002). The firms having more R&D expenditure might be more innovative and competitive in markets. Because the data on R&D expenditure are not available until 2000, the proportion of R&D employees is employed as a proxy for R&D intensity.[11] Whether the firm has exports can also be considered as an indicator of efficiency. Bernard and Jensen (1999) observed a positive correlation between productivity and exports. Konings and Xaviers (2003) find that those firms with previous experience in international trade prior to transition are more likely to survive in transition. We use a dummy variable – exporting firm – indicating whether the firm currently has exports. All these variables are expected to have negative effects on the hazard rate.

The financial constraint is also an important factor that affects the likelihood of survival/exit. Zingales (1998), using a data set from the American truck industry, finds that the survivors out of market competition are not only 'the fittest', but also 'the fattest'. The fitness refers to the efficiency while the fatness refers to the financial health of a firm. Konings *et al.* (2002) stress the significance of credit constraints on firms in different transition stages using a data set from four Eastern European countries. In this chapter the financial constraint is measured as the debt ratio, that is, the ratio of current liability to total assets. Firms with a higher debt ratio would have more difficulty in raising funds and would be more likely to get into the danger of bankruptcy. We will also consider the effect of debt ratio on the hazard rate of the ZSP firms.

The effect of ownership on firm survival is one of the central issues to be analysed in this study. Four dummy variables are constructed: state-owned firms, collective firms, foreign firms and firms of Hong Kong's and Taiwanese capital (HKT henceforth).[12] Hence the domestic private firms are treated as the reference group. We will also consider whether a firm affiliated to a group will exert any effect on the hazard rate (we use a dummy variable, 'Group Affiliation', to denote this). The affiliation with a firm group might provide extra protection against market hazard through the help from the group which is able to mobilize the resources of other affiliated firms. According to this hypothesis, group affiliation is expected to have a negative effect on the chance of exit.

There has been a line of literature studying the effect of diversification or specialization on firm value. While diversification has traditionally been regarded as having the advantage of risk reduction and scale optimization, Wernerfelt and Montgomery (1988) and some other recent studies find that diversification has a negative effect on the firm value, which has come to be known as the 'diversification discount'. Some evidence shows that firms that choose to diversify are significantly different in some characteristics from those that choose to specialize (Lemelin, 1982; Rondi *et al.*, 1996; Silverman, 1999). However, it is unclear how diversification or specialization affects firm survival. In this chapter we will try to shed some light on this issue.

According to the technological fields of the main product(s) produced by a firm, two variables, intraindustrial diversification (denoted by 'Intra-Industry'), and interindustrial diversification (denoted by 'Inter-Industry'), are defined as follows. If a firm produces the main product(s) in the same sub-industry, both variables are set to be zero. If its main products fall in the two or three different sub-industries within one industry,[13] 'Intra-Industry' is set equal to one or two, respectively. Likewise, if its main products are in two or three industries, 'Inter-Industry' is set equal to one or two, respectively.[14, 15]

The average degrees of diversification in different-size classes are presented in Figure 3.6. The degree of interindustry diversification clearly increases with firm size. In contrast, the intraindustry diversification does not exhibit variation for three small-size classes, and it even becomes lower for the size class between 101 and 500 employees. Only large firms with more than 500 employees choose a high degree of intraindustry diversification.

To control the heterogeneity among industries, 17 industry dummy variables are constructed and will be added into the subsequent regressions.

Table 3.1 Descriptive statistics of regression variables

Variables	Observations	Avg.	St. Dev.	Min.	Max.
Firm exit (dummy)	39 730	0.127	0.333	0	1
Firm size	39 730	45.853	234.265	1	33231
Firm size (after log)	39 730	2.979	1.076	0	10.411
Capital intensity (log)	39 730	5.118	1.407	−6.685	13.742
Firm age	39 730	4.424	3.119	1	20
Debt ratio	39 730	0.456	0.344	0	1.998
Labour productivity (log)	39 730	3.612	2.537	−5.617	12.012
Return on assets	39 730	−0.006	0.166	−1	1
R&D intensity	39 730	0.171	0.236	0	1
Export firm (dummy)	39 730	0.033	0.180	0	1
Group affil. firm (dummy)	39 730	0.073	0.260	0	1
State-owned firm (dummy)	39 730	0.272	0.445	0	1
Collective firm (dummy)	39 730	0.182	0.386	0	1
Foreign firm (dummy)	39 730	0.077	0.265	0	1
HKT firm (dummy)	39 730	0.044	0.205	0	1
Intra-industry	39 730	0.118	0.401	0	2
Inter-industry	39 730	0.069	0.281	0	2

Table 3.1 presents the descriptive statistics of the key variables used in subsequent analysis. The panel data are unbalanced owing to the continuing exit and entry of firms, and the number of observations is calculated in firm-years. In order to eliminate the effects of some abnormal values, those observations with return on assets greater than 1 or smaller than −1, or debt ratios greater than 2 are dropped, which leads to 738 observations being dropped. Firstly, note that the mean of the dummy variable 'firm exit' is 0.127, which indicates that, on average, 12.7 per cent of firms exit from the ZSP each year. The average age of the firms under observation is 4.424 years, but the oldest is 20 years. More than 27 per cent (27.2) of firms in the sample are state-owned firms, 7.7 per cent foreign firms, almost 5 per cent HKT firms. About 3 per cent of the ZSP firms have exports ('export firms' for short, henceforth). The average debt ratio is 0.424, and the return on assets is negative (−0.6 per cent) on average.

BASIC RESULTS

In this section, we will first present the basic results on the determinants of firm survival in the ZSP, and then examine the effect enterprise reforms have on the sensitivity of firm survival to financial constraint. Table 3.2

Table 3.2 Determinants of firm exit: basic results

	(1)	(2)	(3)	(4)
Firm size (log)		−0.402***	−0.494***	−0.397***
		(0.015)	(0.014)	(0.015)
Capital intensity (log)		−0.045***	−0.110***	−0.040***
		(0.010)	(0.09)	(0.010)
Debt ratio		0.093***		0.089**
		(0.036)		(0.036)
Labour productivity (log)		−0.101***		−0.100***
		(0.005)		(0.005)
Return on assets		−0.827***		−0.821***
		(0.080)		(0.077)
R&D intensity		−0.397***	−0.399***	−0.399***
		(0.062)	(0.060)	(0.061)
Group affiliation (dummy)		−0.238***	−0.181***	−0.312***
		(0.071)	(0.069)	(0.070)
Export firm (dummy)		−0.289***	−0.278***	−0.331***
		(0.115)	(0.114)	(0.110)
Intra-industry		0.246***	0.246***	0.252***
		(0.030)	(0.030)	(0.030)
Inter-industry		−0.030	−0.057	−0.034
		(0.054)	(0.053)	(0.054)
State-owned firm (dummy)	−0.611***	−0.502***	−0.490***	
	(0.041)	(0.041)	(0.041)	
Collective firm (dummy)	−0.330***	−0.521***	−0.504***	
	(0.040)	(0.041)	(0.040)	
Foreign firm (dummy)	−0.456***	−0.342***	−0.181***	
	(0.060)	(0.059)	(0.056)	
HKT firm (dummy)	−0.403***	−0.319***	−0.205***	
	(0.071)	(0.069)	(0.067)	
Observations	39 730	39 730	39 730	39 730
Pseudo-R^2	0.078	0.109	0.095	0.105

Notes: a All the regressions in this table also control seven year dummies and 17 industry dummies (not reported). The numbers in parentheses are robust standard errors. *, ** and *** denote 10 per cent, 5 per cent and 1 per cent level of significance, respectively.

b The pseudo-R^2 is calculated as W / (W + number of observations), where W is the Wald test statistic testing the hypothesis that all of the 'slope' coefficients equal zero. See Magee (1990).

presents the basic results. In all regressions, 17 industry dummies and seven year dummies are included but their estimated coefficients are not reported owing to space limitation. Column (1) only includes four dummy variables for ownership types. An interesting result emerges that, compared with the reference group – domestic private firms – all of the four ownership types enjoy survival advantage, indicated by their negative and significant estimated coefficients. In particular, the SOEs and foreign firms have the largest advantage since their coefficients are among the largest in magnitude (in terms of absolute values). As other firm characteristics are controlled, as done in columns (2) and (3), the basic result of the disadvantage associated with domestic private firms remains, but the estimates of the effect of various ownership types on exit are different from those documented in column (1). Overall, the SOEs and collective firms are less likely to exit than foreign firms and joint-ventures. These findings suggest that ownership has remained a critical determinant of firm dynamics in the ZSP and, for domestic firms, public ownership exhibits substantial advantages over private ownership in ensuring survival. A potential explanation for the significant role of ownership in determining firm survival rests in the lingering legacy of the planned economy, especially the persistent presence of government intervention and control over economy even to this day in China.

Table 3.2 also reveals that productivity and financial health significantly increases the chance of survival in the ZSP. The estimated coefficients of the proxies for firm productivity, such as labour productivity, R&D intensity and export indicator, are all negative and significant at the 1 per cent level. This result suggests that the rule of 'survival of the fittest' for a market economy partially applies in the development of the ZSP, and market forces have begun to play an important role in determining the fate of a firm. In addition to these productivity measures, a firm's financial performance is also shown to matter for its survival. Regression results in columns (2) to (4) show that the probability of exit is significantly affected by a firm's return on assets and debt ratio. As expected, we find that higher return on assets brings higher likelihood of survival while a higher debt ratio implies the worse prospect of survival in future. These findings are consistent with the evidence presented by Zingales (1998) which studies the determinants of firm survival in the American trucking industry.

As presented in columns (2) to (4) in Table 3.2, firm size and capital intensity have negative effects on the hazard rate. In order to understand the mechanism underlying this evidence, we make calculations of the profitability for different-sized firms in the sample. As shown in Table 3.3, the firm's profitability declines greatly from large to small size classes, which is not consistent with the general finding in existing literature that the small

Table 3.3 Profitability and debit ability of firms, by size classes

Size classes (by employment)	Number of obs.	Percentage of firms with positive profits	Return on assets	Newly added loan rate
1–20	23 934	28.2	−0.023	0.005
			(0.145)	(0.082)
21–50	10 657	44.1	0.005	0.014
			(0.147)	(0.092)
51–100	3 894	57.0	0.024	0.022
			(0.151)	(0.090)
101–500	2 526	68.8	0.037	0.034
			(0.149)	(0.145)
>500	376	88.0	0.049	0.057
			(0.985)	(0.152)

Notes: New-added loan rate equals newly-added loans divided by total asset. The numbers in parentheses are standard deviations.

firms tend to have a higher profit rate (at a greater variance) (Caves, 1998). Honjo (2000) stresses the importance of getting sufficient capital for a firm to avoid exit. A major problem facing small firms in China is that they are often denied access to state banks and venture capital financing has so far been very underdeveloped in China. As a result, small firms have to rely on their own profits as sources of capital accumulation. The limited ability to raise funds for small firms also posed a serious problem for hi-tech firms in the ZSP. As shown in the last column of Table 3.3, the ratio of newly-added loans to the assets in the smallest size class (employment below 20) is only 0.005, while this ratio for the largest size class (employment over 500) is 0.057, and the difference is more than ten times. Without the support of outside financing, it is very difficult for a small firm to reach the required MES of high-tech industries and therefore it is more likely to exit than larger firms.

The effect of firm age on the hazard rate can only be seen through its effect on the baseline hazard functions. The baseline hazard functions of columns (1), (2) and (4) in Table 3.2 are plotted in Figure 3.7.[16] Firm age has an inverse-U-shaped effect on the hazard rate. From the shape of the baseline hazard curves, we can see that the maximum age effect on the hazard rate is reached at around age ten when the dummy variables for ownership are controlled for (columns (1), (2) and (3)). That is, the entrants into the ZSP will encounter a prospect of increasing hazard rate over nearly ten years. Why does experience become a liability for a firm to survive?

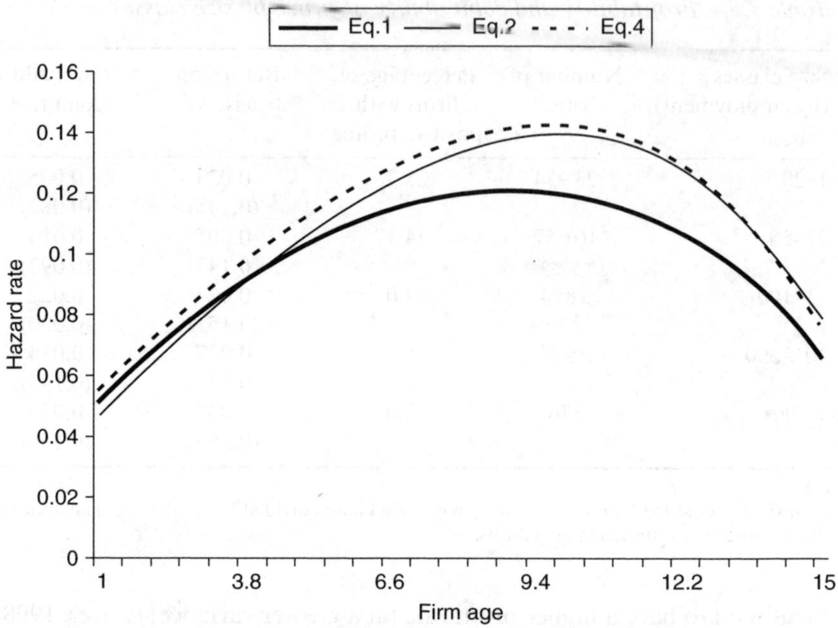

Figure 3.7 Baseline hazard functions for regression equations in Table 3.2

First we should note that the age effects identified above are a mixture of vintage effects and survival effects since we only controlled year dummies in the regressions. Thus the increasing hazard rate over early years of entry may reflect the dominant role of vintage effects. The depreciation rate of technology in high-tech industries is presumably much faster than that of other industries. Hence the vintage effect, which is manifested in the advantage of new technology and capital 'embodied' in new firms, may be greater than the 'survival effect', which refers to the advantage of learning-by-doing over time.[17] Moreover, the vintage effect may also come from the institutional aspect of the new entrants. As shown in Figures 3.2 and 3.3, the share of private firms and corporatized firms in the entry cohorts of the ZSP has been steadily increasing over time. In general these new ownership types are more efficient than old, existing SOEs.

Second, the seemingly puzzling age effects may relate to the kind of tax system in the ZSP. The ZSP implements a tax policy of waiving all income taxes in the first three years from the date of establishment and collecting only half of income taxes in the subsequent three years. The benefits of tax reduction decrease with time, which, other things being equal, increases the hazard rate accordingly.

We also have other interesting findings. First, the group-affiliated firms are found to be less likely to exit, which is consistent with our expectation. Second, the two measures of diversification have different effects on the hazard rate. The intraindustrial diversification significantly increases the hazard rate while the interindustrial diversification decreases the hazard rate, but the effect is not statistically significant.

Third, the year effects, represented by the estimates of year dummies (not reported), reveal a trend of increasing hazard rate over the time period 1995–2002. This result confirms what we found in Figure 3.5 about the intensified competition in the ZSP.

The Impact of Corporization Reform

As discussed earlier, the reforms of corporatizing traditional SOEs in the early 1990s led to the coexistence of a variety of ownership types in the ZSP. After the reforms, the SOEs consist of traditionally state-owned firms and corporatized SOEs. An interesting research question arises naturally: are these two kinds of SOEs different? More fundamentally, does corporatization reform make any difference in improving the behaviour of the SOEs? Fortunately, the information about the status of SOEs in terms of whether corporatization has been enforced is available in the data set, so it is possible to test whether this particular enterprise reform has any real impact on firm behaviour and outcomes.

Since the primary goal of initiating corporatization reforms was to establish a well-functioning governance structure and weaken the connections with the government, the corporatized SOEs are expected to have harder budget constraints than firms with traditional SOEs if the reform is successful. The main challenge is to test empirically the effect of corporatization reform on the hardening of the soft-budget constraint (SBC). The earliest and most influential treatment of SBC was made by Kornai (1979). According to his theory, a state firm in financial distress will be bailed out by the paternalistic government through various kinds of means, such as fiscal subsidy or bank loans. Walder (1995) extends the theory of SBC to understanding the rise of township–village enterprises in China's economic transition. He stresses the differential capacities in bailing out loss-making firms between the central government and local governments. The former presumably has a lot more resources available for mobilization when a massive bail-out is needed. For example, the central government has easier access to borrowing from banks and is capable of setting tax polices to finance deficits. This leads to the differential degrees of SBC syndrome between the firms which are owned by different levels of governments. Anderson *et al.* (2000) base their empirical analysis on a similar idea and find evidence on the differential expectation of state aid in times of trouble between central state firms and local ones. Our empirical test will be based on Walder (1995) and Anderson *et al.* (2000) and distinguish the SOEs into two types: central government-owned enterprises and local government-owned enterprises. About three-fourths of the SOEs in ZSP are owned by the central government, the remaining one-fourth are owned by city/county governments. For simplicity, let us call them Type I and Type II SOEs, respectively. Since Type I SOEs have easier access to government aid than Type II SOEs, they are less sensitive to the debt level while facing the hazard rate of exit. We will therefore introduce the dummy variable for corporatization and examine whether corporatization affects the sensitivity of exit to debt ratio for two types of SOEs.

Table 3.4 reports the regression results exploring the impact of corporatization reforms on the ZSP firms. We first divide the whole sample into public and non-public firms, and then divide public firms into three subsamples: Type I SOEs, Type II SOEs and collective firms. We find that corporatization itself negatively affects the hazard rate of the SOEs for both types. The key variable of interest is the interaction term between debt ratio and corporatization for two types of the SOEs since the estimated coefficients of them will reveal whether corporatization matters for the SOEs. Interestingly, for Type-I SOEs, the estimated coefficient of the interaction term is positive and significant at the 1 per cent level. Given the main effect of the debt ratio is significantly negative, that means that the debt ratio has become a more serious constraint for Type I SOEs owing to corporatization, which implies the hardening of the budget constraint. We find a similar impact of corporatization on Type II SOEs, but the significance level is much lower. The estimated coefficient of the interaction term between debt ratio and corporatization is positive for Type II SOEs. Given that the main effect of the debt ratio is positive for Type II SOEs, this indicates that corporatization makes Type II SOEs more sensitive to debt ratio although this effect is insignificant at conventional levels.

Table 3.4 also shows that the debt ratio has a positive and significant effect on the hazard rate of non-public firms while its effects on public firms are either insignificant or even negative, which essentially supports our assertion that the degree of SBC varies across different types of ownership and even across the SOEs owned by different-level governments. This result suggests that using the interaction term between debt ratio and corporatization is valid for identifying the impact of corporatization reforms on the hardening of budget constraint for the SOEs.

Table 3.4 Effects of enterprise reforms

	Public			Non-public firms
	SOEs		Collective firms	
	Type I	Type II		
Firm size (log)	−0.464***	−0.352***	−0.312***	−0.342***
	(0.045)	(0.069)	(0.037)	(0.018)
Capital intensity (log)	−0.103***	−0.032	−0.002	−0.029**
	(0.031)	(0.042)	(0.025)	(0.012)
Debt ratio	−0.328**	0.102	−0.023	0.103**
	(0.136)	(0.178)	(0.104)	(0.045)
Corporatization (dummy)	−0.465***	−0.685**		
	(0.179)	(0.285)		
Debt ratio* Corporatization	0.950***	0.437		
	(0.281)	(0.362)		
Labour productivity (log)	−0.120***	−0.123***	−0.095***	−0.086***
	(0.017)	(0.024)	(0.013)	(0.006)
Asset profit rate	−1.185***	−1.409***	−1.100***	−0.767***
	(0.377)	(0.395)	(0.207)	(0.089)
R&D intensity	−0.330**	−0.135	−0.305**	−0.613**
	(0.163)	(0.267)	(0.134)	(0.081)
Group affiliation (dummy)	−0.258**	−0.557**	−0.211	−0.242**
	(0.119)	(0.232)	(0.169)	(0.104)
Observations	6693	2024	7163	23850
Pseudo-R^2	0.094	0.131	0.107	0.101

Notes: a All the regressions in this table also include export firm, intraindustry, interindustry, seven dummy variables for year and 17 for industry. The numbers in parentheses are robust standard errors. *, ** and *** denote 10 per cent, 5 per cent and 1 per cent level of significance respectively, using a two-tail test.
b The Pseudo-R^2 is calculated as W / (W + number of observations), where W is the Wald test statistic testing the hypothesis that all of the 'slope' coefficients equal zero. See Magee (1990).

CONCLUSION

One of the central challenges in economic transition concerns how to make firm survival and growth rely on the market forces and eliminate the lingering impact of government intervention inherited from the planned economy. Although there is a large literature that investigates firm dynamics in developed and developing countries, little is known about the determinants of firm survival/exit in economic transition. Using a firm-level data set from the largest science park in China, this chapter presents a systematic analysis of the determinants of firm dynamics and the impact of enterprise reforms on the hardening of the soft-budget constraint.

We find that the ZSP has gone through a dramatic process of transition characterized by constant decline in the share of the SOEs and marked size of firm entry and exit. In the process of firm dynamics, firm productivity and financial performance have been vital for firm survival/exit in the ZSP. On the other hand, ownership still matters for survival and domestic private firms are at a substantial disadvantage compared to other ownership types. However, evidence also shows that the enterprise reform has produced the significant impact of hardening the budget constraint of the SOEs. While the positive role of firm size and financial health on survival are borne out in our data, which are consistent with existing evidence, we also present an interesting result on the inverted U-shaped age effect. According to our interpretation, the seemingly puzzling age effect may be a result of the nature of high-tech industry, economic transition and the peculiar tax system in the ZSP.

NOTES

1. A summary of empirical evidence can be found in Ahn (2001).
2. This idea originated in Solow (1960). Also see Cooper *et al.* (1999).
3. Before 1999 it was called the New-tech Industrial Development Park of Beijing.
4. A detailed list of preferential policies could be found at http://www.ZSP.gov.cn/cms/template/item_english.html?did=73&cid=73\3985.
5. Note that these reform measures concerning the SOEs were not enforced for all SOEs. Thus we can observe the coexistence of the SOEs with traditional form and those with new forms.
6. Note that, when firm i exited in period $t+1$, the notation of EXIT is $Exit_{i,t}=1$ for regression analysis. That is because the data of one year indicate the information at the end of that year.
7. One possible concern is that the exit of a firm might not be the result of bad performance or bankruptcy but be caused by mergers and acquisitions (M&A). However, such cases do not definitely lead to disappearance from the data, because the preferential policy will only be applicable to a certified ZSP firm. The new firm after M&A may still remain in the shell of the old firm that holds the certification. At any rate, such cases seldom arose in the sample period of the ZSP.
8. See Audretsch and Mahmood (1994), Mata *et al.* (1995), Honjo (2000), Kimura and Fujii (2003) and Disney *et al.* (2003) for the applications of this model in firm survival/exit analysis.
9. The functional form of the parametric proportional hazard based on Weibull distribution is $h(t)=\rho\lambda t^{\rho-1}$, which is a non-frailty hazard function. The corresponding frailty model is $h(t|\alpha)=\alpha h(t)$. The frailty α is assumed to have mean one and variance θ. Since α is unobservable, we can assume that its probability density function is $g(\alpha)$. For purposes of mathematical tractability, $g(\alpha)$ is usually chosen from either the Gamma distribution or the Inverse-Gaussian distribution. Note that the shared frailty

model treats the frailty α to be constant over time for each firm. See Gutierrez (2002) for more details.
10. The regression results are available upon request.
11. Mishel and Bernstein (1994) also used this measure.
12. It is not merely for political reasons that we distinguish HKT firms from foreign firms. Some HKT firms have more connections (say, capital, manager and so on) with Mainland China.
13. The industry and sub-industry of a product is defined by its field and sub-field of high technology.
14. By definition, it is possible that one firm's 'Intra-industry' and 'Inter-industry' both equal one only when it produces three main products, one of which is in one industry and the other two of which are in two sub-industries within another industry.
15. Because the revenue of one main product must take up at least 25 per cent of the total revenue, there are three main products at most.
16. The hazard function of column (3) is omitted because it is too close to that of column (2) in the figure.
17. Jensen *et al.* (2001) study these two effects on productivity using data of US manufacturing plants.

REFERENCES

Ahn, S. (2001), 'Firm dynamics and productivity growth: a review of micro evidence from OECD countries', OECD working paper, no.297, OECD, Paris.

Anderson, J., G. Korsun and P. Murrell (2000), 'Which enterprises (believe they) have soft budgets after mass privatization? Evidence from Mongolia', *Journal of Comparative Economics*, **28**, 219–46.

Audretsch, D. and T. Mahmood (1994), 'The rate of hazard confronting new firms and plants in US manufacturing', *Review of Industrial Organization*, **9**, 41–56.

Audretsch, D. and T. Mahmood (1995), 'New firm survival: new results using a hazard function', *Review of Economics and Statistics*, **77**, 97–103.

Baldwin, J.R. (1993), *The dynamics of industrial competition: a North American perspective*, Cambridge University Press.

Beijing Bureau of Statistics (2002), 'Firm innovations in the Zhongguancun Science Park: An international comparison', http://www.stats.gov.cn.

Bernard, A. and J.B. Jensen (1999), 'Exceptional exporter performance: cause, effects or both', *Journal of International Economics* , **47**, 3–31.

Brown, J.D. and J.S. Earle (2000), 'Competition and firm performance: lessons from Russia', Willian Davidson Institute Working Paper no. 296, University of Michigan Business School.

Caballero, R.J. and M.L. Hammour (1996), 'On the timing and efficiency of creative destruction', *Quarterly Journal of Economics*, **3** (3).

Campbell, J. (1998), 'Entry, exit, embodied technology and business cycles', *Review of Economic Dynamics*, **1**, 371–408.

Caves, R.E. (1998), 'Industrial organization and new findings on the turnover and mobility of firms', *Journal of Economic Literature*, **36** (4), 1947–82.

Cooper, R., J. Haltiwanger and L. Power (1999), 'Machine replacement and the business cycle: lumps and bumps', *American Economic Review*, **89**, 921–46.

Cox, D.R. (1972), 'Regression models and life tables', *Journal of the Royal Statistical Society*, 34.

Cox, D.R. (1975), 'Partial likelihood', *Biometrika*, **62** (2).
Disney, R., J. Haskel and Y. Heden (2003), 'Reconstructing and productivity growth in UK manufacturing', *Journal of Industrial Economics*, 2463.
Doms, M., T. Dunne and M.J. Roberts (1995), 'The role of technology use in the survival and growth of manufacturing plants', *International Journal of Industrial Organization*, **13** (4), 523–42.
Du Reitz, G. (1975), 'New firm entry in Swedish manufacturing industries during the post-war period', doctoral dissertation, Stockholm.
Dunne, T., M. Roberts and L. Samuelson (1989), 'The growth and failure of US manufacturing plants', *Quarterly Journal of Economics*, **104**, 671–98.
Ericson, R. and A. Pakes (1995), 'Markov perfect industry dynamics: a framework for empirical analysis', *Review of Economic Studies*, **62** (1).
Frazer, G. (2004), 'Used-clothing donations and apparel production in Africa', Working Paper, University of Toronto.
Gutierrez, R.G. (2002), 'Parametric frailty and shared frailty survival models', *The Stata Journal*, **2** (1), 22–44.
Honjo, Y. (2000), 'Business failure of new firms: An empirical analysis using a multiplicative hazards model', *International Journal of Industrial Organization*, **18**, 557–74.
Hougaard, P. (1995), 'Frailty models for survival data', *Life Time Analysis*, **1**, 255–73.
Jensen, J.B., R.H. McGuckin and K.J. Stiroh (2001), 'The impact of vintage and survival on productivity: evidence from cohorts of US manufacturing plants', *The Review of Economics and Statistics*, **83** (2), 323–32.
Jovanovic, B. (1982), 'Selection and evolution of industries', *Econometrica*, **50**, 649–70.
Kim, Y., S.G. Park, G.H. Wang and S.Y. Jyung (2002), 'A study on the determinants of survivorship of listed companies in Korea', *Journal of Economic Research*, 7(2).
Kimura, F. and T. Fujii (2003), 'Globalizing activities and the rate of survival: Panel data analysis on Japanese firms', NBER working paper 10067.
Konings, J. and A. Xavier (2003), 'Firm performance and selection in an emerging economy', LICOS discussion paper.
Konings, J., M. Rizov and H. Vandenbussche (2002), 'Investment and credit constraints in transition economies: micro evidence from Poland, the Czech Republic, Bulgaria and Romania', LICOS discussion paper.
Kornai, J. (1979), 'Resource-constrainted versus demand-constrainted systems', *Econometrica*, **47**, 801–20.
Lemelin, A. (1982), 'Relatedness in the patterns of interindustry diversification', *Review of Economics and Statistics*, **64**, 646–57.
Liedholm, C. and D.C. Mead (1999), *Small enterprises and economic development*, New York: Routledge.
Liu, L. (1993), 'Entry-exit, learning and productivity change: evidence from Chile', *Journal of Development Economics*, **42**, 217–42.
Liu, L. and J.R. Tybout (1996), 'Productivity growth in Chile and Colombia: the role of entry, exit and learning', in M.J. Roberts and J.R. Tybout (eds), *Industrial evolution in developing countries: Micro patterns of turnover, productivity, and market structure*, Oxford: Oxford University Press.
Magee, L. (1990), 'R2 measures based on Wald and Likelihood ratio joint significance tests', *The American Statistician*, **44** (3), 250–53.

Mansfield, E. (1962), 'Entry, Gibrat's Law, innovation, and the growth of firms', *American Economic Review*, **52**, 1023–51.
Mata, J., P. Portugal and P. Guimaraes (1995), 'The survival of new plants: start-up conditions and post-entry evolution', *International Journal of Industrial Organization*, **13**, 459–81.
Mishel, L. and J. Bernstein (1994), 'Is the technology black box empty? An empirical examination of the impact of technology on wage inequality and the employment structure', Technical Paper, Washington, DC: Economic Policy Institute.
Mortensen, D.T. and C. Pissarides (1994), 'Job creation and job destruction in the theory of unemployment', *Review of Economic Studies*, **61** (3), 397–415.
Olley, G.S. and A. Pakes (1996), 'The dynamics of productivity in the telecommunications equipment industry', *Econometrica*, **64** (6), 1263–97.
Rondi, L., A. Sembenelli and E. Ragazzi (1996), 'Determinants of diversification patterns: evidence from E.U. leading firm data', in S.W. Davis and B.R. Lyons (eds), *Industrial Organization in the E.U.*, Oxford: Clarendon Press.
Schumpeter, J.A. (1934), 'The theory of economic development', Cambridge, Massachusetts: Harvard University Press.
Silverman, B.S. (1999), 'Technological resources and the direction of corporate diversification: Toward an integration of the resource-based view and transaction costs economics', *Management Science*, **45**, 1109–24.
Solow, R.M. (1960), 'Investment and technical progress', in K. Arrow, S. Karlin and P. Suppers (eds), *Mathematical Methods in the Social Sciences*, Stanford, CA: Stanford University Press.
Vaupel, J.W., K.G. Manton and E. Stallard (1979), 'The impact of heterogeneity in individual frailty on the dynamics of mortality', *Demography*, **16** (3), 439–54.
Walder, A.G. (1995), 'Local governments as industrial firms', *American Journal of Sociology*, **101**, 263–301.
Wernerfelt, B. and C.A. Montgomery (1988), 'Tobin's q and the importance of focus in firm performance', *American Economics Review*, **78**, 246–50.
Zingales, L. (1998), 'Survival of the fittest or fattest? Exit and financing in the trucking industry', *Journal of Finance*, **LIII**, 905–38.

APPENDIX

List of Variables

Firm exit: if the data of one firm exists in year t in the data but disappears in year $t+1$, $Exit_{i,t}=1$; otherwise $Exit_{i,t}=0$.

Firm size: employment (measured in log form).

Capital intensity: assets divided by employment (measured in log form).

Firm age: current year minus the year of firm establishment.

Labour productivity: firm total revenue divided by employment.

Return on assets: firm profit divided by assets; observations whose asset profit rates are greater than 1 or smaller than -1 are dropped.

Debt ratio: ratio of current liability to assets; observations whose debt ratio are greater than 2 are dropped.

R&D intensity: number of R&D employees divided by total employment.

Export firm: a dummy variable which equals one if a firm has exports in that year and zero otherwise.

Group affiliation: a dummy variable which equals one if a firm is affiliated to a firm group and zero otherwise.

Intraindustry: a categorical variable which equals $n-1$ if the main products of a firm are in n different sub-industries but still within one industry.

Interindustry: a categorical variable which equals $n-1$ if the main products of a firm are in n different industries.

State-owned firm: a dummy variable which equals one if a firm's shares are controlled by the state and zero otherwise.

Type I SOEs: a dummy variable which equals one if a firm is controlled by the central government and zero otherwise.

Type II SOEs: a dummy variable which equals one if a firm is controlled by local governments (including city, county or township governments) and zero otherwise.

Collective firm: a dummy variable indicating whether a firm is a collective firm (including shareholding cooperative firms).

Foreign firm: a dummy variable which equals one if a firm is controlled by foreign capital and zero otherwise.

HKT firm: a dummy variable which equals one if a firm is controlled by Hong Kong or Taiwanese capital and zero otherwise.

4. Venture capital in China: high technology investing in an emerging economy*

David Ahlstrom, Kuang S. Yeh and Garry D. Bruton

INTRODUCTION

Investing in new projects and technologies by mature firms is a challenging endeavour that often fails to create lasting significant growth for firms (Jensen, 1993). Investments in entrepreneurial ventures by outsiders such as venture capitalists that often rely on new technology can, however, be even more challenging (Christensen, 2000). Venture capitalists have found ways to reduce the risk associated with new venture investment by reducing the non-systematic risk involved in investing, though much risk remains (Christensen, 2000). They do this by careful due diligence, providing management and personnel assistance, careful monitoring of the investment, and well planned exit strategies. Venture capital is well established in developed economies, such as in North America, Western Europe, Israel and parts of East Asia where the rule of law is well established and venture capitalists are able to carry out these activities (Bruton, Ahlstrom & Yeh, 2004). It is less well established in emerging markets where a lack of property rights and an enforceable rule of law reduce the systematic market risk and may add to the complexity of the venture capital process (Bruton & Ahlstrom, 2003).

* Appreciation is expressed to Katherine Ng of the Hong Kong Venture Capital Association and Professors Mike Wright and Andy Lockett of the Centre for Management Buy-out Research, Nottingham University Business School, University of Nottingham, UK for their helpful comments and discussions. The authors express thanks to Vincent Chang of Citibank and Steven S.Y. Lui of The City University of Hong Kong for their comments on an earlier version of this chapter. The work in this chapter was substantially supported by a grant from the RGC Research Grant Direct Allocation Scheme (Project no. 2070212, 2000–2001) of The Chinese University of Hong Kong, the Hong Kong Special Administrative Region. Summer research support was also provided by the Texas Christian University Tandy Center for the third author.

Investors are able to reduce the risk of new investments with the help of a clear rule of law, reliable enforcement and established institutions in the accounting profession and business community that serve to guide behaviour of the invested firm, venture capital investors and other associated parties. Institutional stability and predictability helps to create an environment that can help to control the higher risk climate of private equity and venture capital investing (Bruton, Ahlstrom & Yeh, 2004).

But how does high technology venture investing proceed in an emerging market? The institutional stability of emerging markets is not as established as in developed markets. Emerging markets are known for having volatile environments and a lack of institutions to support business and investment (Peng & Heath, 1996; Peng, 2000). In many emerging markets, intellectual property has little protection and assets can be seized and held with little recourse to the owners (Ahlstrom, Young & Nair, 2002; Peng, 2000). Emerging markets lack active markets for corporate control (Peng, Luo & Sun, 1999) and usually lack mechanisms such as established public equity markets by which investors can exit their investments (Bruton & Ahlstrom, 2003). Thus emerging markets add a great deal of risk to venture investments (Bruton & Ahlstrom, 2003; Peng, 2000). Despite these difficulties investment in emerging markets by private equity professionals has been going up. As a general rule, high risk can yield high return. This chapter will examine how venture capitalists meet these challenges and work to control the considerable institutional risk in emerging market private equity investment.

Venture capitalists clearly see a great deal of opportunity, particularly in the fast-growth emerging market of China (Bruton & Ahlstrom, 2003; Ewing, 2004). China has had one of the fastest growing economies in the world for the past two decades. In the process, enormous opportunities have been created for local and foreign investors alike. Thus our focus here is on China, although one may argue that the lessons are applicable for many emerging markets worldwide. How the specific findings on China may differ in other emerging markets will be discussed at the end of this chapter.

While there has been a rapid expansion of both foreign and domestic venture capitalists, our focus here is primarily on foreign investors. The local investors which are often referred to as venture capitalists are investment funds run by government entities whose investments are directed by concerns other than profitability. Technological advancement is seen as a key element in China's economic progress. Thus the domestic venture capital funds which are typically a Chinese governmental entity or part of the government-linked science and technology parks focus as much on the political appropriateness of their investment as on economic return. Such government-run venture capital funds act quite differently from those run by private market-oriented entities and the resulting success of their investment is considerably less. Therefore we focus here on those foreign privately controlled organizations. Experience in other domains would indicate that the private venture capital sources of funds will ultimately be far more critical to the success in the marketplace of China's technology sector.

Foreign direct investment (FDI) in China was estimated at $50 billion in 2003 (Woetzel, 2004). Just over 2 per cent of that money, or $1.3 billion, was characterized as private equity investment. To put this in perspective for emerging markets, India receives a fraction of China's FDI but attracts more private equity than China (Ewing, 2004). This may be because private equity investors in China face not only the vagaries of the market, but the relative instability and unpredictability of the institutional regime there (Bruton & Ahlstrom, 2003). China's rule of law is in its most nascent stage, and enforcement of what laws do exist is problematic and capricious (Ahlstrom, Young, Nair & Law, 2003; Lubman, 1999). Yet some investors have been quite successful, just as some private firms have successfully navigated China's institutional environment (Ahlstrom & Bruton, 2001; Woetzel, 2004). Private equity is starting to venture into previously untapped areas of China's north and west, as with Morgan Stanley's successful investment in the Inner Mongolian Dairy Mengniu. Yet, along with the excellent opportunities opening up to venture capitalists in China, there are a number of difficult challenges; this chapter examines how venture capitalists manage these challenges as they fund high tech start-ups in China.[1]

This chapter begins with a brief examination of the background of the venture capital industry in China. Next, based on interviews with venture capitalists active in China, the chapter identifies a framework by which venture capital operates in China and how this compares and contrasts with that of the West. Specifically, four main elements of private equity in China will be examined: the selection process for prospective funded high technology firms, structuring of deals and monitoring the firm, value added activities provided to the funded firm, and exit. This is followed by a discussion of implications for practice and research on venture capital and high technology ventures in China.

There are several contributions of this research. First, it expands our understanding of venture capital in China. Specifically, where much prior research has lumped public and private venture capitalists together, this research separates them and focuses specifically on privately held venture capital firms. Such firms are starting to have an important impact on investment and entrepreneurship in China, and if China's developmental experience is at all similar to that of the US, venture capital firms will play an increasingly important role in new business formation there. Second, this research examines specifically how these firms select and manage their investments relative to technology concerns. It is in this specific sector that private venture capital firms can be expected to have the biggest impact as high technology firms enter international markets. Finally, this chapter explores the institutional strategies of venture capitalists as they manage their investments and interact with various government entities in China.

VENTURE CAPITAL IN CHINA

'Venture capital' refers to privately transacted equity investments in firms, typically in small, growing firms. The deals in mature markets can include seed, expansion and pre-offering rounds of financing. In mature markets the term 'private equity' is used to cover latter stage mezzanine arrangements plus buy-outs and turnaround investments. However, in emerging markets the two terms are typically used interchangeably.

In exchange for financing venture capitalists, investors receive significant ownership claims and some control over top management. They are active investors, usually appointing directors and exercising strict oversight during the investment period. Venture capitalists bring legitimacy and credibility to their funded firms, supplying experienced management and sometimes providing key contacts in the government.

Starting in the late 1970s, the Chinese government began loosening its stultifying grip on the economy; private enterprise and investment proliferated. Starting from close to zero in 1979, the number of private enterprises zoomed past one million by the end of the century. This proliferation has helped to make the Chinese economy consistently one of the fastest growing in the world over the last two decades. However, China still lacks many standard financing mechanisms, such as efficient equity markets and debt financing for smaller firms (Ahlstrom, Bruton & Chan, 2000). As a result, there are tremendous opportunities for private equity in China.

The first venture capital firms in China entered in the early 1980s. Local governments in China encouraged and sometimes sponsored funds to invest in state-owned enterprises to bring them up to world standard. One early company was China New Technology Venture Investment, founded in 1985. Some were established firms, and others were founded solely to invest in Chinese firms, particularly established state-owned enterprises (Bruton, Dattani, Fung, Chow & Ahlstrom, 1999).

During the early 1980s, venture capital in China was limited primarily to infrastructure and property investments; new hotels and tourist facilities were especially popular (Lau, 1999). Many of these early investments fared rather poorly, and private equity professionals' interest waned (Ahlstrom,

1989). During the late 1980s, however, there was renewed interest in the Chinese market. But the goals of the government and of private equity investors were in conflict. The central government wanted high-tech and infrastructure investment, while, at the time, investors preferred to focus on lower technology industries (Ahlstrom, 1989; Lau, 1999). As a result, the entire industry was shackled and there were few venture capital investments in high-tech start-ups. Emblematic of this problem, the Shenzhen Special Economic Zone adjacent to Hong Kong was charged with attracting high-tech start-ups and venture capital seed money. Those early efforts were largely unsuccessful: the sums invested did not come close to the amounts needed, and what ventures were funded focused principally on property development and tourism that produced little or no return to many investors (Ahlstrom, 1989).[2]

In spite of the fitful start, the venture capital industry has expanded across much of Asia in recent years. Some of its fastest growth has occurred in China; currently there are over 200 private equity firms located there (Bruton *et al.*, 1999; Ewing, 2004). Many of these firms are government-linked funds, some newer ones are private and from outside of China (Ewing, 2004). Well recognized venture capital firms with a history of investing in China, such as Baring Private Equity Partners, Softbank, Hambrecht & Quist, the Walden International Investment Group, Doll Capital Management Inc., ChinaVest and IDG Technology Venture Investment have become quite active in the market (Ewing, 2004).

INSTITUTIONAL SETTING IN CHINA

Despite the growth of venture funding in China, much is still not known about the venture capital industry there. A few articles have appeared recently detailing the historical development and current state of the industry from a macro perspective (for example, Bruton *et al.*, 1999; Saywell, 1998; Ewing, 2004). This research has focused principally on mezzanine financing of established firms or on general conditions and outlook for the industry there.

Some prior research has started to examine venture capital in related settings such as those of East Asia (Bruton *et al.*, 1999; Bruton, Ahlstrom & Yeh, 2004; Lockett, Wright, Sapienza & Pruthi, 2002) or India (Wright, Lockett & Pruthi, 2002). Other research has also examined settings such as China (Bruton & Ahlstrom, 2003; Bruton *et al.*, 1999) and Southeast Asia (Bruton, Ahlstrom & Singh, 2002). This research has found that venture capital can still function in spite of the lack of formal institutions such as laws and regulations, and minimal enforcement of those laws. It also implies a highly socialized view of venture capital whereby social network ties, including those of the venture capitalist, can help to substitute for the formal institutions present in the Anglo-American system and take on more importance in the emerging market setting (Peng & Heath, 1996). The importance of the societal context was further suggested by the work of Lockett and colleagues (2002), who found that the sources of information that venture capital firms seek out and can utilize are also culturally determined and can differ from one country to another. Yet despite research illuminating such differences much research still presumes that all venture capitalists operate in highly formal institutional environments present in predominantly Anglo-American economies (see Jeng & Wells, 2000). Venture capital investors making these investments face some unusual problems with the institutional environment in conducting business in China. Successful private firms face a certain measure of hostility, which comes from two primary sources. First, there is opposition from many government officials who worry that the socialist system will be undermined. Labour 'exploitation' by investor control over the means of production violates basic Marxist principles, which still form some basis for official policy in China. This mentality insists that the smaller funded and (often) private enterprises of today could become the monopoly capitalists of tomorrow.

Second, many officials and policy makers in China have practical concerns about tolerating increased competition for state enterprises. The largest employer in China continues to be the state: in 2003, it employed half of China's 750 million workers and controlled over 50 per cent of all industrial assets there. The SOEs still dominate vital industries such as financial services, power and telecommunications (Desvaux, Wang & Xu, 2004). SOEs and state-owned enterprises (SOEs) must provide numerous employee benefits that are not required of private firms and as such receive most government financing. Many officials in SOEs oppose private firms as potential competitors – threats to their own livelihood (Steinfeld, 1998).

Consequently, government officials and other powerful individuals regularly interfere with investment in and operation of private firms in China as well as private equity investment in joint ventures with SOEs. This is manifested in the enforcement of government policies towards firms with private backing (Clarke, 1991; Tan, 1999). The assessment of taxes, licence fees, administration fees and penalties by party cadres in China can vary greatly from region to region. Officials are enthusiastic in seeking investments, but they seek to use many tactics trying to solicit or share successful people's wealth. Thus taxes, fees and penalties levied on private firms are often at punitive levels for their political and economic interests (Wank, 1996; Xin & Pearce, 1996; Gold, Guthrie & Wank, 2002).

THE STUDY

To develop this understanding of the challenges faced by venture capitalists in China, we conducted semi-structured, in-depth interviews with 26 leading participants in China's venture capital industry. These included venture capitalists in firms in China, and venture capitalists based in Hong Kong and other parts of Southeast Asia who fund firms in China. The vast majority of these firms were located in Hong Kong, which has the largest amount of venture capital for investing in Asia outside of Japan. However, these firms typically invest very little in Hong Kong. Instead they use it as a base of operations. As a result, 19 of the firms were interviewed in Hong Kong, four in China itself and three in Singapore. The Singapore firms all had significant operations in China.

The firms examined include both independent and captive venture capital firms. Venture capitalists in the United States are typically independent firms that are organized into limited partnerships. These firms only conduct private equity investing. In Europe, there are some independent venture capital firms, but the more common model is to have captive firms that are private equity departments in large multinational banks. Both types of firms were examined, but in all instances the firm was primarily privately held. For example, Transpac from Singapore does have government investment but the management of the firm is largely independent and separated from government involvement. We did not include central and provincial government investment funds. In certain parts of China, these funds supply most of the private equity capital, apart from the developmental capital coming from Hong Kong firms and outside of China. Generally, these funds invest in infrastructure and late stage private equity and restructuring. They commonly do not invest in higher growth start-up firms (Bruton *et al.*, 1999).

Extra effort was made to ensure that venture capitalists in both large and small funds were questioned to provide as wide a perspective as possible. The list of sample participants was developed through initial interviews with senior venture capitalists in the region and with the help of the Asian Venture Capital Association in Hong Kong. This was to ensure a broad coverage of leading venture capitalists and venture capital firms in the region. All of the interview subjects are responsible for funding firms and have played a role in managing some of their funded ventures.

Semi-structured interviews were conducted with the 26 venture capitalists. Using a research design known as replication logic (Eisenhardt, 1989), these semi-structured interviews were conducted to build first an understanding of how venture capital works in China and the problems that venture capitalists face. As noted by Daft and Lewin (1990), such an in-depth understanding should be developed before quantitative investigation of a topic can begin. The cases were chosen in a manner consistent with the work of Glaser and Strauss (1967), who argued that the number of cases examined should be such that incremental learning from additional cases approaches zero. Diesing (1971) argued that six cases often provide such a number; Zhao and Aram (1995) examined such a number of cases.

All but two of the interviews were conducted in English. Two of the authors speak Chinese, and at least one native speaker was present at all Chinese language interviews to take notes. Most of the interviews had at least two interviewers present to take notes. Interview transcripts were subsequently examined to ensure consistency. If any discrepancy occurred, the interview subject would be contacted for clarification. The results were summarized and venture capitalists were subsequently contacted for verification. Concepts were coded and compared among researchers.

During the interviews, the cases and insights provided by each interviewee were tested against previous findings as well as with the stylized US model of private equity, where appropriate. The summary findings were then shared with two leading venture capitalists active in China, who were not part of the study, for their comments. The concepts and explanations were then revised to reflect their further insights.

RESULTS

Laying the Groundwork

The early strategic plans of private equity investors in China were to build relationships with large, state-owned enterprises and then to ask those firms for help finding good investment opportunities among other SOEs or township and village enterprises, often in stable, well-understood industries (Lau, 1999). The private venture capitalist would bring money and managerial expertise to the table. The large state firms would source deals and provide the *guanxi*, or key relationships among business and government officials necessary for business success in China. Thus a major chemical firm like Sino-Chem would provide a venture capital fund with access to the best chemical-related funding opportunities around China. Once the fund decided to back a venture, a Sino-joint venture between itself and the large state entity that source the deal would be created (Bruton *et al.*, 1999).

However, investors have learned that good *guanxi* alone is not enough to ensure the success of funded firms. The collapse of Guangdong province's Guangdong International Trade and Investment Corporation (GITIC) surprised a large number of investors, with a number of funds losing money. This also proved to be a painful reminder that investors should not depend entirely on *guanxi* at the top. Although having connections with a central government official is helpful, there are people at a variety of levels with whom connections may be required (Ahlstrom, Bruton & Lui, 2000; Xin & Pearce, 1996). In addition, private equity investors in our sample commented that deals should not be overly dependent on political connections. This is due to the capricious nature of politics; as one private equity investor summarized, 'if your sponsor is in power you are ok, but if he or she leaves that office, you may end up with an unattractive investment with few prospects'. Venture capital firms are wise to spend time to build their own connections with officials at various governmental levels in China. It is also helpful to 'buy' *guanxi* by seeking firms and managers, but this is just one of many relevant success factors of private equity investing in China and should not be overly relied on.

Making the Initial Investment Decision

One of the most consequential difficulties for venture capitalists in any economy is determining the viability of a proposed investment. There are several key concerns for the private venture capital firm in an emerging market like China, including obtaining an accurate financial picture of the firm and conducting due diligence.

Accounting information

In China, seed technology ventures that are little more than an idea are rarely funded; investors usually fund going concerns of a later stage than in the US or Britain (Bruton & Ahlstrom, 2003). A key difficulty in making an investment decision in this environment is that China's accounting standards differ greatly from those in the developed economies (Shi & Weisert, 2002; Tang, 2000). As a result assets such as property and inventory may not be as fully valued as in more developed markets, and accounts receivable may not be realistic. China's accounting standards as practised today deviate substantially from international accounting standards (Chen, Sun & Wang, 2002).

China's accounting rules and conventions differ from international accounting standards in three major respects. First is the policy basis of the firm's goals. Many accounting mechanisms used by Chinese enterprises are aimed at managing production rather than asset valuation. For some established firms, as well as managers trained in such a system, this bent is difficult to unlearn, irrespective of changes in regulations. Second, the significantly different intended audiences for the financial reports can prove to be a challenge for private equity firms to manage. International practice identifies investors and creditors as the primary users of accounting information, whereas Chinese practice regards the internal customers (the sales managers, production managers, general managers, middle managers and so forth) as the primary users. In spite of recent changes in China's accounting regulations, owing to weak enforcement, Chinese operating statements tend to be dominated by operational data and lacking in accurate summarized financial data. As a result there is a gap between Chinese and international accounting standards earnings results despite much harmonization of accounting regulations in China with those of more developed nations (Chen, Sun & Wang, 2002). Third, terms sometimes may be defined or understood much differently in China than elsewhere, and accounting terminology can take on different meanings in different industries or regions of China, along with differing formats (Broadman, 1999).

These problems result in financial reports often being of very limited help to investors as they seek to evaluate which firms to fund. Recent attempts to align China's accounting standards with international standards will ultimately make financial reports easier for venture capitalists to interpret in a meaningful manner. But still today such accounting information on a firm seeking funding is of limited value to the venture capitalist. Venture capitalists often must hire auditors from Hong Kong to help confirm the reported accounting statements.

Due diligence

Due diligence occurs as the venture capitalist seeks to validate the information that the firm submits. This effort will include efforts at obtaining robust market information. The availability and accuracy of such information is problematic. Yet another challenge is to secure accurate engineering information for an evaluation of the high-tech venture. Venture capitalists must expect to make greater efforts in China than in the West to help aggregate the information typically required.

Management's background

A key part of due diligence is learning the background of the firm founder and other managers. In fact, the most important aspect of an initial investment evaluation in the US centres on the evaluation of the person leading the proposed investment. As legendary venture capitalist Arthur Rock commented, 'Nearly every mistake I've made has been in picking the wrong people . . . ' (Bygrave & Timmons, 1992b). This is even truer in China when a firm's financial statements may contain only marginally valuable information and the other aspects of due diligence can be problematic. Sometimes the background and connections of the top management team may be among the most valuable resources the firm possesses.

One benefit of operating in the Chinese environment over a long period is that extensive relationships and valuable personal information networks are developed. For example, university professors and individuals at various research laboratories throughout China often control key high-tech projects. These individuals, along with their students and young colleagues, constitute a foundation of high technology information that can be accessed once strong relationships are developed. In fact, many of these individuals in turn will be the ones that become a source of new ventures to fund. For venture capital firms that are new to China, locating these individuals at the research facilities and research parks and then building the necessary relationships with them can be difficult.

Negotiations

Negotiating the deal is the last part of the selection of a firm to fund and can be extremely difficult. In part this difficulty is due to the fact that Chinese business owners, government officials and venture capitalist investors often do not agree on company valuations. Chinese entrepreneurs frequently benchmark high valuations based on similar firms in the US. Venture capitalists in China, on the other hand, stress the differences and country-specific risk. Making matters worse, there is no standard methodology for valuing a company. One possible friction point is that, since national growth rates are an important component of a firm's total valuation, valuations for Chinese firms may be inflated by highly optimistic growth projections and difficult to verify economic figures (Rawski, 2002).

The negotiation process is further complicated by the fact that contracts are not viewed in China in the same absolute terms that they are viewed in the West. As one venture capitalist interviewed cautioned, 'a signed contract simply provides a green light to more negotiations when it becomes convenient'. As a result the negotiations are typically an ongoing process in funding a firm in which, while a contract may have been signed, there is far more evolution in what those terms mean and there will be greater changes in the contract than would be expected in the West. Moreover, such a process is not unique to technology-focused firms. Thus venture capitalists can expect that Chinese firms may continue to seek revisions in that contract as their relationship with the Western investors develops.

Monitoring the Relationship

Once the technology firm is funded, the venture capitalist must monitor its operations carefully. The difficulties in monitoring a firm are magnified because of the problems with the legal system identified earlier. Venture capitalists must maintain appropriate vigilance to ensure that the firm is performing as desired. Venture capitalists reported that it is important to send people into a firm to count things and to double check that things are generally going smoothly. Several venture capitalists cautioned against maintaining the same number of investments as in the US or UK because of the monitoring and assistance these firms typically need. They also stated that venture capital firms should try to place middle managers with whom they are familiar in the funded firm to help further in monitoring its operations.

Trust

The model of investing predominant among early venture funds in China was largely a 'hands off' style limited to monitoring the funded firm. Venture capitalists were not always granted a seat on the board, and financial data were often slowly and grudgingly provided. Even today, while a board seat may be offered when a venture capitalist invests in a high technology firm, the protections that a seat on the board actually offers can be limited. The corporate governance laws in China, though improving, are still rather ineffectual and generally unenforceable. Thus the ability to influence a high technology firm is crucial and the investor relies on the trust developed between the leaders of that firm and the venture capitalist.

As a result of the importance of trust, venture capital firms often work with the managers with whom they have experience since they have already built such trust. Venture capitalists interviewed were quite quick to emphasize that in China it is highly preferable to enter a deal with personnel they know, and more so than in the US.

Another type of firm that these individuals like to fund because of the greater ability to build trust are Overseas Chinese or Chinese who have studied in North America or the UK, and have returned to China to work. These individuals have Chinese language facility but at the same time understand the concept of shareholder value and the need for disclosure and monitoring by the venture capitalist. As a result the ability to build trust with them is typically easier than it is with an individual whose only experience with venture capital may be local firms that focus on political issues as much as profitability.

Recalling that in China, venture capitalists typically fund ventures that are established already to some degree, one private equity investor pointed out that it is particularly crucial to have middle managers in place that you can trust as well. 'They are needed to act as your information conduit into the function of the business' he noted. Thus, in building trust with top management it is also important for venture capitalists to have relationships with those in the lower levels in the organization.

Regionalism

Although China is described as a 'target-rich environment', it is also a country of multiple, distinct provincial and regional markets, not a market of 1.3 billion consumers as is often thought. Each region, province, or industry may require somewhat different resources. Connections with the local government officials are needed as venture capitalists seek to fund firms in new cities and provinces.

The impact of the strength of such strong regions is that economists have coined the term 'Chinese style federalism' to illustrate the strength of the provinces and other localities. This regionalism results in goods crossing provincial borders in China facing the equivalent of a 46 per cent tariff in 1997 (Poncet, 2001). Poncet adds that this tariff equivalent of 46 per cent is about what the European Union countries charged each other in the late 1990s. This means that in respect of traded goods and a number of related commercial laws, China is something closer to a loose federation of sovereign states than a unified country and has much variety in law and regulation.

The strength of such regionalism is not limited to areas like tariffs. The initial economic reforms in China were aimed at decentralizing decision making (Child & Lu, 1990). As a result, the reforms devolved authority in a number of areas from the central government to provincial and local governments, and even to powerful individuals (Naughton, 1995). Twenty years after the initial reforms, provincial governments controlled about half of the total PRC budget while the central government controlled 20 per cent (Naughton, 1995). There has yet to be a clear delineation of authority among the central, provincial and local governments. This lack of clarity in the authority structure gives local government officials an opportunity for exploitation. The result is that some local officials run their territory as if it was their own enterprise – and treat firms under their purview almost like their own property (Peng & Luo, 2000; Gilboy, 2004).

The highly local nature of commerce in China presents challenges to monitoring funded firms. Venture capitalists and entrepreneurs interviewed emphasized the importance of understanding the local setting of the firm. Laws can vary greatly from province to province, and implementation is heavily dependent upon both the views and the moods of local bureaucrats (Clarke, 1991; Tan, 1999; Gilboy, 2004).

Some private equity professionals in the West find this concept difficult to grasp because Westerners are accustomed to standardized laws and they depend heavily on their ability to conduct business by telephone, fax, mail or Internet, regardless of where the partner or funded firm is located. In China, the situation is markedly different, and this affects how venture capitalists structure deals, as one capitalist from Shanghai recalled (Ahlstrom, Bruton & Chan, 2000:6):

> I almost never fund deals more than a couple of hours' drive from my office. The reason is that the deals in China must be monitored very carefully. One Internet firm's top management recently disappeared with about $300 000 I had invested with them. Although I used to visit that firm about once a week, it was not enough: they were about 300 kilometers away – too far to monitor properly. Now I limit my investments to firms that I can visit more frequently or watch over easily, and this means [firms] in the Shanghai area.

Such problems are amplified for high-tech firms. To be profitable, high technology firms are likely to require a national market rather than rely on a single region of the country for product sales. Venture capitalists have addressed this problem by opening local offices throughout China to seek new investments and provide support to current ones expanding across the country. For example Transpac, headquartered in Singapore, was an early mover in China's private equity landscape. One of Transpac's strengths is the network of small offices near to their investments around the region they maintain – an advantage over single-office fund groups. Transpac's decentralized approach facilitates not only finding new investments outside of the traditional areas, but also investment monitoring.

Input to Managers

While managers may seldom seek strategic input from venture capitalists, the need for their guidance may be greater than in comparable situations in the West, particularly for high-tech firms. As one Hong Kong venture capitalist observed:

> Although the Chinese business owners do not want any outsiders involved, we often find we have to be involved. Perhaps the key lender asks management for help to find overseas markets. Sometimes a partner is needed to help handle the local government. The paradox is that although the owners do not want us on the board, I find that my colleagues and I must often get more involved with operational details in our funded ventures in China than we would in similar deals in the West.

The venture capitalists noted that, when there is a need for managerial input, one of the key factors in such situations is to allow managers the opportunity to maintain 'face' or respect. This can be undertaken in several ways; for example, rather than giving an ultimatum, an idea might be put forth as a suggestion. The setting where suggestions are provided is important as well – never in front of a manager's subordinates. This way, the manager can remain 'in charge' and take credit for the changes.

Relationships

Social controls have proven far more effective than economic or legal controls in countries with underdeveloped institutional structures such as in China (Guthrie, 2002; Peng, 2000). As noted before, the legal system in China is not well developed. Though its importance may be declining as institutions slowly improve (Guthrie, 1998), *guanxi* is still crucial to the success of the venture. However, these relationships are necessary not only within the firm but also to key individuals outside the funded firm in China. A key part of the value that a venture capitalist can add to its invested firms comes from such individuals outside the firm.

Many new private firms may not have extensive *guanxi* capital to call upon – these relationships develop over time. To that end, the venture capitalists can connect them with key individuals whom the firm can hire or pay a retainer to. For example, a venture capitalist may have a relationship with a town's former village head (*cun zhang*) if they have a relationship with that person and suggest to a new high technology venture that they hire them. Although the village head may even no longer hold that position, his or her influence with the people in the area will probably remain strong. People will listen and are deferential to the village head's influence; they do their best to give face and not embarrass the village head in any way. Furthermore, many employees may be from China's inland region and do not want trouble from the local people; so they too are differential to the village head, which can further aid the high technology venture. This individual likely knows everyone in the area and is accustomed to evaluating and selecting people for work teams. This makes it less likely that workers will steal assets or shirk their responsibilities on the job.

In China, the importance of hiring people with the right skills and connections cannot be overstated.[3] At times, venture capitalists may find that they must also aid the search for professional managers, sometimes helping to recruit individuals from multinationals throughout Asia and offering them a small stake in the portfolio company (Pohndorf, 1997). Outside knowledge and connections that venture capitalist bring to their funded firms in China can be especially valuable in the area of human resources.

Reconciling conflicting goals

In monitoring their investment venture capitalists need to recognize that their goals for the firm may be fundamentally different from those of the entrepreneur. Chinese firms (whether publicly or privately owned) are encouraged by the state to maximize employment and production. For investments in SOE and township and village enterprises, this can be especially troublesome, but it can also affect private firm investments. As a result, without a venture capitalist's careful oversight, firms can end up filling warehouses with unsaleable inventory or placing dozens of extra workers on the payroll.

EXIT

Investors in large companies in China expect to exit through IPOs. Presumably the firms would be listed on either the Hong Kong Stock Exchange or, in more recent years, the stock exchanges within China (Shanghai or Shenzhen). But exiting even from successful ventures has proved to be problematic. The market for corporate control is not active in China and around Asia (Peng, Luo & Sun, 1999). China's Shenzhen and Shanghai stock exchanges typically list state-owned enterprises that are politically connected. The founders' shares become non-tradeable when listed domestically, which eliminates this venue as a liquidity source.

Thus most venture capitalists investing in large firms have to look to strategic buyers as an exit or a liquidity source. This is clearly the case for almost all high technology ventures. Strategic buyers in East Asia are typically larger corporations looking to acquire new capabilities or institutional positions in China (Peng, Lee & Wang, 2005).

Many firms have discovered that, when they do exit, their returns have not been very high. Venture capitalists typically hope for annual returns of 20–25 per cent, although our sample reported that, to date, actual returns for their funds are much lower. Therefore venture capital firms have wanted to fund deals of lower risk to compensate for the lower than hoped for return. This in part explains why most venture capitalists will not fund a firm that is not already an ongoing venture. The venture capitalists in our sample added that the funding of on-going ventures with follow-on rounds of additional financing is not common in China. Thus, while in the United States there are typically multiple rounds of financing, each at a higher valuation, such multiple rounds of financing are more difficult to achieve in China.

Even when securing a strategic buyer there are other difficulties. A strategic buyer almost always will want a majority stake in the venture. However, in China, venture capitalists normally have only a minority interest in the funded firm. Thus the venture capitalist needs to convince the Chinese firm to sell part of its share and become a minority owner in the venture – a 'tough sell' to a Chinese firm that may not want to give up control. The multiple paid on the firm's revenue stream results in a return significantly lower than that traditionally associated with an IPO.

Some local venture capitalists claim that exits are getting easier. One venture capitalist stated that of the 100-plus companies in which his firm has invested, 24 were already public, 26 were sold 100 per cent to other parties, and in many other cases their minority stake went back to the majority owner. Exit mechanisms do exist in China, but the options are more constrained and complex than they are in the West.

DISCUSSION

Venture investing in China is no simple task. Understanding the local China situation is critical to success in this market. Until recently, fund managers would argue that their financial success would be quick once they exited their funded firms. However, having been active in the China market for at least five years and managing a number of investment exits, many venture capitalists interviewed affirm that the situation is very complex in China. Success cannot be achieved by simply 'cherry picking' a few of what appear to be the best investment opportunities, as was once believed.

A significant change in recent years has been the process of selecting and monitoring the investee firm. Increasingly, investors acknowledge that China is unique but they also acknowledge that they must employ methods similar to those used in the West on issues such as due diligence. Thus the methods need to be the same. However, accomplishing those levels of due diligence is much more difficult to achieve than in the West.

Venture capital investors bring far more to the high technology venture than just money. Their advice and connections and global perspective are all important to their investee firm's success. However, an investment in a Chinese firm needs to be closely and locally monitored, and venture capitalists should be prepared to be proactive in helping the funded firm in hiring workers with valuable skills and assisting in securing and forging needed connections. Although Chinese business owners have been reluctant to permit venture capitalist access to financial and strategic decisions, they increasingly recognize and accept that investors will not be passive providers of money. The most sophisticated of the Chinese firms actively seek out a broader array of input from the smorgasbord offered by their private equity investors.

IMPLICATIONS FOR RESEARCH

This research has demonstrated that there are significant differences between venture capital in China and that of a mature market like the United States. These differences arise despite the fact that many of these same firms have operations not only in China but in those same mature markets. Venture capital firms seem to be responding to the institutional differences between China and the Anglo-American venture environment (c.f. Bruton & Ahlstrom, 2003). That is, the continuing lack of a more formal institutional structure in China portends the continuing importance of *guanxi* in a variety of places in the venture capital process. Future research should explore these differences further. For example, how do such firms integrate their various unique approaches into a worldwide system of venture capital? A firm like Walden Ventures has operations not only in Asia but also in the US and Europe. How does a worldwide firm allow such variances but at the same time maintain their worldwide standards? This has implications for any professional firm that has to conform to both national (local) and professional institutional environments.

This research has also examined both independent and captive firms. The majority of the sample were independent firms, with limited examination of captive firms. One shortcoming of the research is that the sample was not large enough to differentiate between the two types of venture capital firms as regards the nature of their adaptation to the Chinese market. Future research should expand the sample and explore the impact of these different ownership structures on the performance and actions of the venture capital firms, not only in regard to high technology ventures, but in the funding of all types of ventures.

This research has specifically examined privately held venture capital firms. We argue that there is a significant difference in government-backed funds called 'venture capital funds' and those that are privately held. Such a view is consistent with the dominant literature on venture capital. However, further research is necessary to fully understand such differences. The government-sponsored investments in China can be expected to respond to local political concerns and the need for employment (Gilboy, 2004).

If government venture capital funds focus on the employment and political concerns as expected, there will be a wide range of implications not only for the performance of the venture funds themselves, but also for the actions taken by the venture capitalists. Thus there is a rich set of issues that need to be examined for domestic Chinese venture capital. The findings here for independent firms in the nation can serve as a foundation for the issues that need to be examined. A few critical areas that should especially be examined are how new ventures are chosen for funding, how such investments are monitored and how advice is provided to the firm: specifically, what advice and actions are taken by the venture capitalists when difficulties in the funded firm are identified. It can be expected in each of these areas that differences will be found from independent firms. These findings have implications not only for venture capital firms but also for the various government entities in China as they seek to encourage technology development.

IMPLICATIONS FOR PRACTICE

Today, three broad strategies seem to be emerging in China for venture capital investors. First, the SOEs are far from closing up shop in China, as they have elsewhere around the world. China is still giving a lot of attention to SOEs as they employ about half of China's workforce. Hundreds of SOEs are considering selling, restructuring or privatizing portions of their assets and trying to get a public listing. The demand for repackaged, competitive SOEs with low costs, good assets and strong political connections, and maybe even a semi-captive provincial market in a more isolated part of China, is a very attractive target for private equity firms and their minority shareholders. This category is challenging to navigate as it is dominated by political insiders in China, and is fraught with hazards, but some private equity firms have been able to do so. The central government and bureau and departments with SOEs under their purview have been testing the viability of these strategic sales and repackaging, and our sample has reported that more are likely.

Venture capital firms have also funded many well-established private firms. Private investment funds are often drawn to entrepreneurial start-ups that can take advantage of China's low cost coupled with the growing East Asian market. Private equity firms are trying to target firms that can take advantage of China's peculiar demographics, such as the one child policy and an aging population, firms in the health care sector and firms in life sciences that can take advantage of China's strong life sciences and agricultural programmes in the universities.

CONTRIBUTIONS OF RESEARCH

As noted above, this research focuses on private venture capital efforts in China. Prior research has lumped private and government venture capital firms into a single unit. The expectation is that researchers will ultimately find that they can view venture capital as a continuum. At one end will be the pure model that is widely understood from mature markets like the US. In the middle will be private firms active in China that we have detailed here. These firms have differences but also strong similarities with the actions of firms in mature markets. It is expected that the domestic firms with strong government influence will be substantially different from the Western model and will anchor the other end of the continuum. This research helps to extend the understanding of venture capital in emerging markets by suggesting models of venture capital practice that differ from more traditional approaches that have developed in the West. This can facilitate further research on understanding the role venture capital can play in the reform of state-owned and other government linked firms – a very important sector in China and certain other emerging markets.

The research also helps to establish for the first time the interaction between technology and venture capital in a transition economy. Prior research has tended to treat the investments of venture capitalists as a uniform set with little differentiation between the different types of investments. This is the first effort to differentiate and focus on one specific investment type, high technology investments.

Finally, the research expands our understanding of institutional theory. In particular, this research expanded the understanding of how different institutions affect a specific industry. It is clear from the discussion that cognitive institutions and regulatory institutions both influence the strategic choices of the firms in a transitional environment. Future research can build on such information to examine in greater detail which institutions have the greatest power in different settings.

CONCLUSION

Despite the many challenges, the private equity market in China has been improving. The standards and professionalism of the industry in the country are also improving. One symbol of this movement is the fact that, in June 2002, a group of private equity firms founded the China Venture Capital Association (www.cvca.com.hk) to promote the rights of entrepreneurs and laws on investor protection and disclosure. The new association has about 50 members, including big private equity names such as Newbridge, the Carlyle Group and Goldman Sachs. Additionally, new regulations governing venture capitalists have been promulgated that have increased the stability of the industry (Burke, 2003). Additional positive steps for venture capitalists are also occurring with the approval of Shenzhen Stock Exchange's separate board for small and medium-sized enterprises, much like the NASDAQ and Hong Kong's GEM market.

The future for China venture capital is likely to be a model much more similar to that predominant in the US or Europe than the old China mould of the past. For example, due diligence is already starting to look more like the Western template. Where once China's information on market research may have been neither available nor reliable, the sophistication of such data is expanding, and most private equity investors are taking advantage of that information. Additionally, investigation of the investment opportunity now goes much deeper than merely getting to know the promoter presenting the business concept. This trend is likely to continue, especially for the high-tech investments that are likely to be a major part of China's future.

Venture capital in China will continue to grow. This nation, with a quarter of the world's population, undergoing such rapid levels of economic growth and change, offers an assortment of investment opportunities. Additionally, as institutional investors seek to diversify their investments, an obvious move is to include China-based firms in their portfolios. The low rates of return that some firms are experiencing will discourage other investors, and there will be a period of transition as the market continues to mature. Nevertheless, private equity in China holds many interesting opportunities and challenges, and Western venture capitalists must be prepared and well versed in the nuances of the environment there if they hope to seize the opportunities present in this intriguing and complex market.

NOTES

1. For more details on the historical development of China's venture capital industry, as well as general differences at a macro level from the US model, see Bruton *et al.*, 1999.
2. Recently Shenzhen has been attracting more high technology manufacturing, based on a nascent venture capitalist industry, local university talent and a new NASDAQ-type stock market.
3. Some private equity investors halt their investigations after the first step of evaluating the entrepreneur, as they believe no other type of information from China is adequately and consistently reliable for them to make a judgment on the business plan.

REFERENCES

Ahlstrom, D. (1989), 'The Shenzhen Special Economic Zone: progress and problems', The Chinese University of Hong Kong Working Paper Series, Winter, 1–34.

Ahlstrom, D. and G.D. Bruton (2001), 'Learning from successful local private firms in China: establishing legitimacy', *Academy of Management Executive*, **15** (4), 72–83.

Ahlstrom, D., G.D. Bruton and E.S. Chan (2000), 'Venture capital in China: ground-level challenges for high technology investing', *Journal of Private Equity*, **2** (Spring), 1–10.

Ahlstrom, D., G.D. Bruton and S.S.Y. Lui (2000), 'Navigating China's changing economy: strategies for private firms, '*Business Horizons*, **43** (1), January–February, 5–15.

Ahlstrom, D., M.N. Young and A. Nair (2002), 'Deceptive managerial practices in China: strategies for foreign firms', *Business Horizons*, **45** (6), 49–59.

Ahlstrom, D., M.N. Young, A. Nair and P. Law (2003), 'Managing the institutional environment: Challenges for foreign firms in Post-WTO China', *SAM Advanced Management Journal*, **68**, 41–9.

Broadman, H.G. (1999), 'The Chinese state as corporate shareholder', *Finance & Development*, **36** (3), September, 52–5.
Bruton, G.D. and D. Ahlstrom (2003), 'An institutional view of China's venture capital industry: explaining the difference between China and the West', *Journal of Business Venturing*, **18**, 233–59.
Bruton, G.D., D. Ahlstrom and K. Singh (2002), 'The impact of institutional environment on the venture capital industry in Singapore', *Venture Capital*, **4**, 197–218.
Bruton, G.D., D. Ahlstrom and K.S. Yeh (2004), 'Understanding venture capital in East Asia: the impact of institutions on the industry today and tomorrow', *Journal of World Business*, **39**, 72–88.
Bruton, G.D., M. Dattani, M. Fung, C. Chow and D. Ahlstrom (1999), 'Private equity in China: differences and similarities with the western model', *The Journal of Private Equity*, **1**, Winter, 7–13.
Bruton, G.D., M. Dattani, M. Fung, C. Chow and D. Ahlstrom (1999), 'Private equity in China: differences and similarities with the western model', *The Journal of Private Equity*, **1**, Winter, 7–13.
Burke, M.F. (2003), 'Venture capital options expand – a bit', *The China Business Review*, July/Aug, **30** (4), 24–8.
Bygrave, W.D. and J.A. Timmons (1992a). 'Management assessment methods in venture capital: an empirical analysis of human capital valuation', cited in G.H. Smart (ed.) (1999), *Journal of Private Equity*, **2**, 29–45.
Bygrave, W.D. and J.A. Timmons (1992b), *Venture Capital at the Crossroads*, Boston, MA: Harvard University Press.
Chen, S., Z. Sun and Y. Wang (2002), 'Evidence from China on whether harmonized accounting standards harmonize accounting practices', *Accounting Horizons*, **16**, (3), 183–97.
Child, J. and Y. Lu (1990), 'Industrial decision-making under China's reform, 1985–1988', *Organization Studies*, **11**, 321–51.
Christensen, J.L. (2000), 'Venture capital and the innovation process', in U. Gupta (ed.), *Done Deals: Venture Capitalists Tell Their Stories*, Cambridge, MA: Harvard Business School Press, pp. 19–38.
Clarke, D.C. (1991), 'What's law got to do with it? Legal institutions and economic reform in China', *Pacific Basin Law Journal*, **10** (1), 1–76.
Daft, R.L. and A.Y. Lewin (1990), 'Can organization studies begin to break out of the normal science straitjacket?', *Organization Science*, 1–11.
Desvaux, G., M. Wang and D. Xu (2004), 'Spurring performance in China's state-owned enterprises', *The McKinsey Quarterly*, Special Issue, 96–105.
Diesing, P. (1971), *Patterns of Discovery in the Social Sciences*, New York: Aldine Atherton.
Eisenhardt, K.M. (1989), 'Building theory from case study research', *Academy of Management Review*, **14**, 532–50.
Ewing, R.D. (2004), 'Private equity in China: risk for reward', *The China Business Review*, **31** (4), July/Aug. 48–66.
Gilboy, G.J. (2004), 'The myth behind China's miracle', *Foreign Affairs*, **83** (4), 33–48.
Glaser, B. and A. Strauss (1967), *The Discovery of Grounded Theory*, New York: Aldine de Gruyter.
Gold, T., D. Guthrie and D. Wank (eds) (2002), *Social Connections in China: Institutions, Culture, and the Changing Nature of Guanxi*, Cambridge: Cambridge University Press.

Gupta, U. (ed.) (2000), *Done Deals: Venture Capitalists Tell Their Stories*, Cambridge, MA: Harvard Business School Press.

Guthrie, D. (1998), 'The declining significance of Guanxi in China's economic transition', *The China Quarterly*, **154**, 31–62.

Guthrie, D. (2002), 'Information asymmetries and the problem of perception: the significance of structural position in assessing the importance of Guanxi in China', in Thomas Gold, Doug Guthrie and David Wank (eds), *Social Connections in China: Institutions, Culture, and the Changing Nature of Guanxi*, New York: Cambridge University Press, pp. 37–56.

Jeng, L.A. and P.C. Wells (2000), 'The determinants of venture capital funding: evidence across countries', *Journal of Corporate Finance*, **6**, 241–89.

Jensen, M.C. (1993), 'The modern industrial revolution, exit, and the failure of internal control systems', *Journal of Finance*, July, 831–80.

Lau, D. (1999), 'Behind the Great Wall: China opens its doors to domestic and foreign venture capital funds', *Venture Capital Journal*, **1**, July, 48–50.

Lockett, A., M. Wright, H. Sapienza and S. Pruthi (2002), 'Venture capital investors, valuation and information: a comparative study of the US, Hong Kong, India and Singapore', *Venture Capital*, **4**, 237–52.

Lubman, Stanley B. (1999), *Bird in a cage: Legal reform in China after Mao*, Stanford, CA: Stanford University Press.

Naughton, B. (1995), *Growing out of the plan: Chinese economic reform, 1978–1993*, New York: Cambridge University Press.

Peng, M.W. (2000), *Business strategies in transition economies*, Thousand Oaks, CA: Sage Publication.

Peng, M.W. and P.S. Heath (1996), 'The growth of the firm in planned economies in transition: institutions, organizations, and strategic choice', *Academy of Management Review*, **21**, 492–528.

Peng, M.W. and Y. Luo (2000), 'Managerial ties and firm performance in a transition economy: the nature of a micro–macro link', *Academy of Management Journal*, **43**, 486–502.

Peng, M.W., S. Lee and D. Wang (2005), 'What determines the scope of the firm over time? A focus on institutional relatedness', *Academy of Management Review*, **30**, 622–33.

Peng, M.W., Y. Luo and L. Sun (1999), 'Firm growth via mergers and acquisitions in China', in L. Kelley and Y. Luo (eds), *China 2000: Emerging Business Issues*, Thousand Oaks, CA: Sage Publications, pp. 73–100.

Pohndorf, D.R. (1997), 'The fine art of making a success of Asian equity funds', *Global Finance*, April, 42–6.

Poncet, S. (2001), 'Is China disintegrating? The magnitude of Chinese provinces' domestic and international border effects', paper presented at the Annual Conference of the Canadian Economics Association, 31 May–3 June, Montreal.

Rawski, T.G. (2002), 'How fast is China's economy really growing?', *The China Business Review*, **29** (2), March/Apr., 40–43.

Saywell, T. (1998), 'High stakes: foreign venture capital spurs China's hi-tech start-ups', *Far Eastern Economic Review*, **161** (37), 66–7.

Shi, S. and D. Weisert (2002), 'Corporate governance with Chinese characteristics', *The China Business Review*, **29**, Sept/Oct., 40–44.

Steinfeld, E.S. (1998), *Forgiving reform in China: the fate of state-owned industry*, New York: Cambridge University Press.

Tan, J. (1999), 'The growth of entrepreneurial firms in a transitional economy: the case of a Chinese entrepreneur', *Journal of Management Inquiry*, March, 83–9.
Tang, Y. (2000), 'Bumpy road leading to internationalization: a review of accounting development in China', *Accounting Horizons*, **14** (1), 93–102.
Wank, D.L. (1996), 'The institutional process of market clientelism: Guanxi and private business in a South China city', *The China Quarterly*, **147**, 820–38.
Woetzel, J.R. (2004), 'A guide to doing business in China', *McKinsey Quarterly*, Special edn, 36–45.
Wright, M. A. Lockett and S. Pruthi (2002), Internationalization of western venture capitalists into emerging markets: risk assessment and information in India', *Small Business Economics*, **19**, 13–29.
Xin, K.R. and J.L. Pearce (1996), 'Guanxi: connections as substitutes for formal institutional support', *Academy of Management Journal*, **39**, 1641–58.
Zhao, L.M. and J.D. Aram (1995), 'Networking and growth of young technology-intensive ventures in China', *Journal of Business Venturing*, **10**, 349–70.

PART II

Environment and new venture strategies

5. How do entrepreneurs create wealth in transition economies?*

Mike W. Peng

The rise of entrepreneurship throughout the transition economies of Central and Eastern Europe (CEE), the newly independent states (NIS) of the former Soviet Union, and East Asia in the past two decades has fundamentally transformed these economies, and attracted worldwide attention. Entrepreneurs and the start-ups they found create wealth and push these economies to a higher level of competitiveness through their sheer energy, relentless strategies and sometimes controversial practices. Although there are numerous country- and region-specific studies (Audretsch, 2000; Chang & MacMillan, 1991; Puffer, 1994; Mugler, 2000; OECD, 1996), to date, there have been few attempts that shed light on the *overall* development of entrepreneurship throughout these transition economies, from Shanghai to St Petersburg. While there are many cultural, political and economic differences permeating these countries, recent publications such as those by the World Bank and OECD have grouped them together under the collective label of 'transition economies' (The World Bank, 1996). We believe that increased knowledge about the wealth creation process throughout transition economies can greatly enrich global entrepreneurship practice and research. Therefore, in this chapter, we focus on two important questions: (1) how do entrepreneurs, who in this chapter are defined as 'founder of a new business'[1] (Gartner, 1985; Low & MacMillan, 1988; Lumpkin & Dess, 1996; Spicer, McDermott & Kogut, 2000; Wright, Hoskisson, Filatotchev & Buck, 1998) create value in transition economies? (2) what can be learned from their experience?

* Previously published as Peng, M.W. (2001), 'How entrepreneurs create wealth in transition economies', *Academy of Management Executive*, **15** (1), 95–108.

THE RISE OF ENTREPRENEURSHIP

Entrepreneurship during the Socialist Era

Despite the harsh political conditions, entrepreneurship existed in virtually all of these countries *before* major transitions took place in the 1980s (Morris, 1998). Before the transition, the private sector, which had a number of peculiar labels such as the 'grey,' 'second' and 'underground' economy, was usually limited to being small, labour-intensive, and often informal. By the 1980s, most socialist governments started to loosen restrictions on the private sector, resulting in an initial wave of entrepreneurship. However, at that time, these countries were reluctant to legalize private property. The government still imposed a limit on the size of a private firm, such as seven employees in Hungary and eight in China (which was lifted in the 1990s). What is remarkable is the rapid rise of entrepreneurship in such an ambiguous environment with little protection of private property. It seems that, wherever and whenever the government had relatively few restrictions on the private sector, 'pockets' of entrepreneurship, such as those in South China, would start to develop.[2]

A Golden Era of Entrepreneurship during the Transition

After a period of slow but steady growth in the 1980s, private entrepreneurship blossomed in the 1990s. The most fundamental driving force was the removal of the yoke of communism throughout CEE and the NIS. The other underlying force was the continued deterioration of the state sector. We may regard the lure of capitalism as a 'pull' factor, and the failure of state-owned enterprises (SOEs) as a 'push' factor. Overall, a combination of the pull and push factors resulted in the abolishment of many restrictions on private firms. In turn, these transitions opened the floodgates of entrepreneurship, which rose to undermine the foundation of the socialist economy.

Since the mid-1990s, most of GDP has been contributed by the private sector throughout CEE and the NIS (for example, approximately 80 per cent in Hungary, 75 per cent in the Czech Republic and 70 per cent in Russia) (European Bank for Reconstruction and Development, 1998: 26). At the same time, the non-state sector in China has quietly but steadily become the backbone of the economy, contributing nearly 70 per cent of total industrial output. It is the growth of the private sector that has created jobs and (at least partially) compensated for the decline of SOEs. During the 1990s, in CEE, about 5 per cent of the adult working population attempted to start up a new firm or become self-employed, a figure very similar to the percentage of *nascent* entrepreneurs in the United States and

Western Europe (Aldrich, 1999). In a nutshell, the 1990s was indeed a golden era for entrepreneurial start-ups throughout transition economies.

WHO ARE THESE ENTREPRENEURS?

What drives people to be entrepreneurs has remained as an intriguing puzzle around the world, and perhaps more so in transition economies. It seems that four types of entrepreneurs have emerged: (1) farmers, (2) 'grey' individuals, (3) former cadres, and (4) professionals.

Farmers

Although private farming was eradicated in most socialist countries, Poland never nationalized its agriculture, and its private farmers owned more than 70 per cent of the land in 1987. Even in countries where private farming had not been allowed before, the loosening of government regulations spurred a great deal of private farming. However, most private farmers would not bother to register their undertaking as a 'company'. Over time, some of them started to get organized along more formal lines, and attempted to grow beyond the family holdings. While most of them remain small, some of the better-managed ones have become larger and more visible. For example, the largest private company in China during the 1990s, the Hope Group, could trace its roots to private farming (Au & Sun, 1998).

'Grey' individuals

Under state socialism, because there were very few possibilities to lawfully organize entrepreneurial ventures, unlawful ways emerged in a 'grey' economy[3]. The socialist era left a legacy of disregard of the supremacy of the law. First, there were not enough formal laws to govern economic behaviour. Second, there was little legitimacy for the laws that did exist, which were routinely ignored. Thus, despite its lack of legality, the 'grey' economy was widely tolerated and accepted by the public.

During the transition, frontier-style overnight accumulation of wealth through 'grey' activities becomes possible. Ranging from small-scale tax evasion and bribery to large-scale mafia practices, all of these activities may be in violation of some laws or regulations (Puffer & McCarthy, 1995).[4] Since the emerging legal and regulatory frameworks are underdeveloped, their enforcement leaves numerous loopholes. As a result, these 'grey' entrepreneurs emerge to take advantage of loopholes as *intermediaries* connecting individuals and organizations with economic exchanges that otherwise would not have taken place (Peng, 1998; Peng, Hill & Wang, 2000; Peng & Ilinitch, 1998).

While strictly speaking violating laws and regulations, these individuals do not necessarily belong to crime organizations, although some of them certainly do. Most of these people are 'entrepreneurs' in a classical sense: 'persons who add value by brokering the connection between others' (Burt, 1997). They blur the boundaries separating different sectors by taking advantage of the information and resource *asymmetry* across different sectors. In turn, they profit from these arbitrage opportunities. For example, they can trade foreign exchange through black markets, obtain a business licence from officials, and enforce contracts through 'security services'. Some of these 'services' are clearly of a criminal nature, such as resolving contractual disputes through assassination by the Azerbaijan mafia active in Russia, literally resulting in 'cut-throat' competition (Peng, 2000). The size of the overall 'grey' economy is, of course, very difficult to estimate. Tentative figures in the mid-1990s (Table 5.1) put the total size of the unofficial 'grey' economy at approximately 11–12 per cent of the GDP in the Czech Republic and Poland, about 30 per cent in Bulgaria and

Table 5.1 The share of the unofficial economy as a percentage of total GDP

Country (per cent of private sector share of total GDP in 1998)	Per cent of the unofficial, 'grey' economy share of total GDP						
	1989	1990	1991	1992	1993	1994	1995
Central and Eastern Europe							
Bulgaria (50%)	22.80	25.10	23.90	25.00	29.90	29.10	36.20
Czech Republic (75%)	6.00	6.70	12.90	16.90	16.90	17.60	11.30
Hungary (80%)	27.00	28.00	32.90	30.60	28.50	27.70	29.00
Poland (65%)	15.70	19.60	23.50	19.70	18.50	15.20	12.60
Romania (60%)	22.30	13.70	15.70	18.00	16.40	17.40	19.10
Newly Independent States of the Former Soviet Union							
Azerbaijan (45%)	12.00	21.90	22.70	39.20	51.20	58.00	60.60
Belarus (20%)	12.00	15.40	16.60	13.20	11.00	18.90	19.30
Estonia (70%)	12.00	19.90	26.20	25.40	24.10	25.10	11.80
Georgia (60%)	12.00	24.90	36.00	52.30	61.00	63.50	62.60
Kazakhstan (55%)	12.00	17.00	19.70	24.90	27.30	34.10	34.30
Latvia (60%)	12.00	12.80	19.00	34.30	31.00	34.20	35.30
Russia (70%)	12.00	14.70	23.50	32.80	36.70	40.30	41.60
Ukraine (55%)	12.00	16.30	25.60	33.60	38.00	45.70	48.90

Sources: European Bank for Reconstruction and Development (1998); Johnson, Kaufmann and Shleifer (1997).

Hungary, 40–50 per cent in Russia and Ukraine, and, in the extreme case, over 60 per cent in Azerbaijan and Georgia[5] (Johnson, Kaufmann & Shleifer, 1997; Sharma & Merrell, 1997).

Cadres

The term 'cadres' refers to former communist party leaders and officers. For three reasons, cadres are widely believed to benefit from the transition by becoming entrepreneurs. First, because people with more education in general are found to do better in market economies than those with less education (Cooper & Dunkelberg, 1987), cadres, who as a group are better educated than the general population, are likely to be in an advantageous position during the transition. Second, power accumulated under state socialism can be converted into assets of high value in a transition economy. During privatization, for example, strategically located cadres can take advantage of their positions by acquiring state property (for example, the 'spontaneous' privatization throughout CEE and the NIS). Third, cadres can tap into their personal networks to gain access to valuable resources from their former colleagues still in the government, manoeuvring across different sectors as intermediaries who seek rents for their services.

In one case, a former Chinese cadre, who quit his post at the State Planning Commission in 1989, operated a $120 million company by 1995. It encompassed a futures-and-commodities trading operation, a clinic to treat nearsightedness with lasers, and a collection of high-tech start-ups (Peng, 2000). One of the key reasons he was doing so well was that he had access to powerful friends and contacts in many government agencies. In another case, during the first period of major transitions in Hungary (1989–91), cadre entrepreneurs more than doubled their personal income, while non-cadre entrepreneurs and the entire population increased their income by 73 per cent and 59 per cent, respectively (Rona-Tas, 1994). These examples suggest that cadre entrepreneurs seem to have done well in transition economies.

Professionals

The term 'professional entrepreneurs' refers to entrepreneurs who previously held professional positions not directly related to the party state, such as lawyers, managers, engineers and professors. In transition economies, they enhance the technology and professionalism of private firms, which traditionally concentrate on low-tech, labour-intensive sectors such as farming, restaurants and retail shops. Professionals also increase the legitimacy of private firms. Less educated farmers and 'grey' individuals with dubious backgrounds and activities do not inspire much confidence among the public. Cadre entrepreneurs are widely viewed with suspicion and resentment by the public. Professionals, on the other hand, are better educated and have few connections with the party state or the 'grey' economy. The added legitimacy of professional-run private firms, in turn, is likely to attract more experienced professionals and recent college graduates, thus fuelling the development of these firms.

For example, in 1991, a group of defence scientists started Vimpelcom, a high-tech start-up which later became Russia's largest cellular-telecom provider and the first Russian company to earn a full listing on the New York Stock Exchange. Combining defence industry know-how, contacts in the telecom industry and the lack of Soviet-era baggage which typically plagued many privatized firms, Vimpelcom built from the ground up with Western-style management and accounting principles that made it easy to present the three years of US standard audits required to be listed in New York (Peng, 2000).

On the other hand, the development of products and services with more technology content requires more long-term investment such as R&D. Unfortunately, the general environment in transition economies, characterized by policy instability and regulatory chaos, is not conducive to such investment (Audretsch, 2000).[6] Moreover, professional entrepreneurs' lack of connections with other sectors may be a liability rather than an asset. In an environment where personal ties figure prominently, entrepreneurs without deep and strong network relations may have a lot of difficulties in getting things done.

In sum, economic transitions have provided powerful incentives for all sorts of entrepreneurs to mushroom. They tend to specialize in different fields, taking advantage of their particular strengths. Farmer entrepreneurs usually focus on food and vegetable production, farm produce distribution and low-tech manufacturing. 'Grey' individuals are likely to specialize in various intermediary services, both legal and illegal forms. Cadre entrepreneurs are known for their interest in relationship-intensive industries, such as trading, entertainment and property development. Professionals-turned-entrepreneurs tend to be more interested in knowledge- and technology-based ventures (for example, computer software, architecture design). Next, we move on to explore their major wealth-creating strategies.

MAJOR ENTREPRENEURIAL STRATEGIES

Recent research focusing on transition economies has highlighted three major entrepreneurial strategies: (1) prospecting, (2) networking, and (3) boundary blurring (Peng, 1997; Peng & Luo, 2000; Tan, 1996;

Xin & Pearce, 1996; Johnson, *et al.*, 1997; Mugler, 2000; Charap & Webster, 1993; McCarthy, Puffer & Shekshnia, 1993; Zhuplev, Konkov & Kiesner, 1998). While they are not the only viable strategies in transition economies, individually and/or in combination, they do appear to be associated with the most successful entrepreneurs in these countries. These strategies are not necessarily unique to transition economies; prospecting and networking, for example, have been widely practised by entrepreneurs elsewhere. What is interesting in transition economies is the importance of these strategies, whereby alternatives (such as acquisitions) are few. On the other hand, blurring the boundaries of public and private sectors in multiple directions (detailed later) does appear to be a unique challenge to global entrepreneurship practice and research. Therefore our discussion below starts with a most generic strategy, and moves progressively to highlight more unique strategies.

Strategy I: Prospecting

Prospectors are firms with a changing market, a focus on innovation and change, and a flexible organizational structure headed by younger, more aggressive managers, all of which are characteristic of entrepreneurial firms in transition economies (Miles & Snow, 1978). In contrast, *defenders* are firms with a narrow market, a stable customer group and an established organizational structure managed by older, more conservative executives. Compared with larger SOEs and recently privatized ex-SOEs which tend to be defenders, private firms are usually smaller and have a higher level of market orientation, agility and flexibility. Headed by younger individuals, start-ups also tend to adopt a simple organizational structure, which allows them to react quickly to opportunities. Further, start-ups have little inherited organizational baggage from the socialist era, low fixed costs and the ability to attract the most talented people.

Another way to view this strategy is to treat these start-ups as 'underdogs', which have very little choice but to adopt *guerrilla* warfare tactics. Underdog firms cannot compete against the larger and more established rivals head on. Instead, smaller firms conserve scarce resources for crucial battles. They use speed and stealth to create disruption by pre-empting competitors, and to be the first movers while forcing their competitors to be defenders or reactors. Such a quick movement often gives entrepreneurial start-ups substantial first-mover advantages by allowing them to build up a market share and increase brand name awareness in new niche markets.

Poland's Optimus Computer exemplifies such a strategy. Founded in 1988, this start-up held 35 per cent of the Polish PC market by 1995. The owner attributed his success to a 'guerilla' strategy that sought first-mover advantages in the PC market when the PC revolution was starting to gain momentum in that country in the early 1990s. Optimus thrived by always moving ahead of the competitors in terms of products and services. Specifically, while competitors imported models approaching the end of their product life cycle, Optimus provided locally assembled, low-cost PCs equipped with the latest versions of Intel chips, Samsung monitors and Microsoft operating systems (Miller, 1995).

On the other hand, this prospector (or guerrilla) strategy has at least three limitations. First, the industries that entrepreneurs enter tend to have relatively low entry barriers and less capital intensity. As a result, they often focus on labour-intensive farming, light manufacturing and small-scale services, and shy away from large-scale, technology- and capital-intensive industries (Poland's Optimus Computer is an exception in this regard). Second, while their larger rivals find it difficult to compete on speed and stealth, successful start-ups often attract a large number of other private firms to follow the first movers. Owing to the nature of these industries (including low entry barriers), the first movers are often unable to defend themselves and, consequently, sustain a competitive advantage. Therefore entrepreneurial firms have a tendency to be 'footloose', exiting existing industries or niches and searching for new opportunities elsewhere. Finally, while the prospector (or guerrilla) strategy may be viable during the initial phase of the transitions when there are a large number of unfilled niches, it is questionable whether this strategy can be pursued in the long run when the economy becomes more developed and mature.

Strategy II: Networking

Networking as a strategy is not exclusive territory reserved only for entrepreneurial start-ups in transition economies. To some extent, entrepreneurial firms around the world rely on this strategy. In transition economies, virtually every firm needs to pay attention to its networks, which are necessitated by the institutional environment during the transition. Basically, the lack of certain market-supporting institutions often leads managers and entrepreneurs to develop networks to perform certain basic functions, such as obtaining market information, interpreting regulations and enforcing contracts. In an environment where *formal* institutional constraints such as laws and regulations are weak, *informal* institutional constraints, such as those embodied in interpersonal networks, connections and ties cultivated by managers and entrepreneurs (for example, *blat* in Russia and *guanxi* in China), help at least partially to overcome the infrastructure deficiencies by facilitating economic exchanges among members (Boisot & Child, 1996; Khanna & Palepu, 1997; Khanna & Palepu, 1999; Peng & Heath, 1996).

Compared with other firms such as SOEs, privatized firms and foreign entrants, a networking strategy is perhaps more important for private start-ups. What is noteworthy about entrepreneurial networking is (1) its urgency, (2) its intensity, and (3) its impact. Private firms initially suffer from a lack of legitimacy as new organizations owing to their 'liability of age', which prompts stronger *urgency* for them rapidly to establish network ties with the environment. Specifically, they have to cultivate two sets of networks. The first one is with entrepreneurs and managers at other firms, such as suppliers, buyers and competitors, which may be useful in most economies. A second set of networks is those with government officials, which may be more specific to transition economies, because official harassment from various government officials remains a constant danger. In Russia, for example, every private company must provide 28 *quarterly* reports to the tax authorities. In China, most private firms have to pay nearly 50 kinds of different taxes. Often taxation and regulatory policies are self-contradictory, and even the most scrupulous entrepreneurs cannot be in consistent compliance. It is not surprising that entrepreneurs clearly understand the importance of having good relationships with government officials (especially those in tax bureaus). In China, for example, the impact of network linkages with officials on firm performance is found to be *more* important than those with entrepreneurs at other firms (Peng & Luo, 2000).

A second characteristic that distinguishes entrepreneurial networking is its *intensity* (Morris, 1998, Morris & Sexton, 1996). In order to ensure survival, smaller firms often have to intensify their networking activities with larger, more legitimate and more powerful players. Moreover, a large number of them are in service industries, which in general are more relationship-intensive than manufacturing industries. Further, relative to the manufacturing sector, legal frameworks in transition economies are even less developed in the service sector, thus calling for intense networking efforts.

Finally, because of the small size of these start-ups, the contributions of individual entrepreneurs' personal networks tend to have a stronger *impact* on firm performance. In comparison, the impact of similar networks cultivated by managers at larger firms may be less pronounced because of these firms' sheer size. Moreover, being private owners, entrepreneurs can directly pocket the residual income if their firms perform well, thereby providing powerful incentives for them to network through entertainment, gift giving and/or bribery.

For example, the struggles that entrepreneurs heading China's Lucky Transportation, a trucking company servicing the construction industry, went through can serve as a case in point. Several state-owned construction and trucking firms formed an informal enterprise group aiming at internal collaboration and excluding non-members. In order to grow, Lucky Transportation had to become a member of the group by cultivating personal ties between its entrepreneurs and other managers in the group as well as government officials. The entrepreneurs worked hard to be their friends, by taking them out to dinner and occasionally giving them gifts such as 'red envelopes' (containing cash)[7] (Hill, 2000; Vogl, 1998). Eventually, Lucky Transportation was accepted as a member of the group, thus enabling the start-up to achieve significant growth (over 500 per cent growth in sales during its first three-year period, 1992–95) (Peng, 1997).

On the other hand, it is important to note the limitations of a networking strategy. One common and erroneous tendency is to exaggerate the importance of personal networks. Actually, possessing effective personal networks may be 'a necessary, but not sufficient, condition for business success' (Tsang, 1998). After all, a start-up needs to deliver value-added in the marketplace by having strengths in 'traditional' areas such as product/service quality, advertising and delivery. This seems to be increasingly important in light of the drive toward more 'normal', market-based competition in these countries (Peng & Luo, 2000).

Strategy III: Boundary Blurring

Although a 'boundary blurring' strategy is closely associated with networking, two specific types of 'boundary blurring' seem to warrant separate discussions. These are the blurring (1) of boundaries separating public and private sectors, and (2) of those separating legal and illegal sectors.

Blurring the public–private boundaries

A surprisingly large number of entrepreneurial start-ups are not purely private-owned companies in a classical sense. Called 'collective enterprises', these non-state, yet non-private start-ups are especially visible in the Chinese economy, and as a group they have become the largest contributor to GDP (that is, larger than the purely private sector and the SOE sector) since the early 1990s. Specifically, they are non-SOEs subordinate to local governments and owned and operated collectively (Bruton, Lan & Lu, 2000; Luo, Tan & Shenkar, 1998). Local governments benefit from these firms because these firms not only generate jobs, but also provide income streams and tax revenues over which local governments can have discretion. In contrast to CEE, outright privatization of these firms had not occurred in China until the late 1990s. However, hidden or informal privatization has been widespread. Specifically, entrepreneurs can bid for long-term leases to control these firms. Although such lease agreements do not entitle leaseholders to formal property rights, in practice, these agreements are widely viewed by the entrepreneurial lease-holders as well as the employees and the public as de facto property rights. These public–private *hybrid* firms, therefore, represent a gradual evolution from public to private ownership. On the other hand, a large number of pure, private start-ups move in the opposite direction by choosing to register themselves as 'collective' firms in an effort to 'wear a red cap'. Given the residual antagonism towards private entrepreneurs, many entrepreneurs are concerned about renewed hostility directed towards them and the possible appropriation of their assets. For example, Lucky Transportation is such a 'collective' company that it is a private firm in disguise.

In an environment which is still institutionally unfriendly to private ownership, it makes good sense for many entrepreneurs not to advertise the private nature of their firms. Even when discriminatory policies are removed, pure, private firms are still at a great disadvantage in obtaining state-controlled resources such as bank credit. For example, frustrated by its inability to acquire credit, a highly successful private start-up (that is, non-collective) in China, Carpenter Tan, had to use advertisements in the national media in 1997 to plead to the banks (all of which were state-owned), thus stirring up a nationwide debate on why it was so hard for private firms to raise capital. This contrasts sharply with the situation in developed economies, where banks had to advertise to promote their loans. What was more ironic in this case was that Chinese banks, while refusing to support Carpenter Tan, continued to supply capital to numerous money-losing SOEs that hardly paid interest, let alone principal. It was found that although there was no discriminatory policy banning loans to private firms, bankers practised self-imposed and unfair sanctions on private firms. Specifically, while it was normal for banks not to recover anything from loans to SOEs, any loan loss associated with private firms would automatically lead to suspicion on the part of the loan officer of embezzlement and collusion with entrepreneurs. 'So why do I want to take any risk to provide loans to private firms?' one loan officer was quoted as saying (Peng, 2000).

Unfortunately, the experience of Carpenter Tan is not unique. Table 5.2 reveals a striking pattern of underfunding for non-SOEs in China: while their share in total industrial output and value-added rose to 67 per cent and 80 per cent, respectively, in 1994, their share of total bank loans remained at around 16 per cent.[8] It is not that non-SOE firms did not need bank loans; in most cases, their loan applications, such as those made by Carpenter Tan, were simply denied while banks continued to channel precious financial resources to SOEs. On the other hand, because of the influence of local governments, changing to a 'collective' status may allow private firms to gain better access to critical resources such as loans.

Table 5.2 The non-state sector in China: contributions and shares of bank financing*

	(1)	(2)	(3)
1987	40.30	51.20	17.40
1988	43.20	52.40	17.00
1989	43.90	47.50	15.70
1990	45.40	62.20	15.00
1991	47.10	56.30	14.80
1992	51.90	67.50	14.40
1993	56.90	68.70	15.80
1994	66.90	80.10	15.60

Notes: * The non-state sector covers (a) registered, pure private firms, (b) collective (public–private hybrid) firms, and (c) foreign-invested firms. In other words, it includes all non-SOEs.
(1) Percentage of industrial output.
(2) Percentage of industrial value-added.
(3) Percentage of total bank loans.

Source: Peng (2000).

Overall, while some 'collective' firms embody the evolution away from *public* ownership, other 'collective' enterprises represent a movement away from *private* ownership. Given the general movement toward clearer specification of property rights throughout transition economies and the ambiguous property rights surrounding these firms, the question becomes: can ambiguous property rights sometimes be efficient? The answer is a qualified 'yes' (Nee, 1992). Specifically, under the particular circumstances of the transition, such a 'collective hybrid' strategy may lead to the best of the two worlds. On paper at least, these firms still retain 'public' ownership, and many local governments take these firms under their wings by shielding them from harassment from other intrusive government agencies and helping them obtain needed resources. At the same time, through 'creeping' privatization, most of these firms behave more like pure, private firms. In short, the public–private hybrid represents an interesting and previously unencountered phenomenon in global entrepreneurship practice and research, and deserves further attention from practitioners, researchers and policy makers.

Blurring the legal–illegal boundaries

In some CEE and NIS countries, the blurring of the legal–illegal boundaries has reached epic proportions. Among them, Russia seems to stand out as the most 'corrupt' major economy in the world. While the true extent of the 'grey' and/or illegal economy in Russia is difficult to assess, according to one estimate, approximately 70 to 80 per cent of private companies may be paying extortion money to organized, mafia-type criminal gangs (Manev, Manolova & Yan, 1998). Outside Russia, rising organized crime has occurred in just about every transition economy. Taking advantage of the entrepreneurial boom, many crime organizations operate under the title of fully legal business firms with impeccable offices, letterheads and bank accounts. Consider Multigroup, a small start-up founded in Bulgaria in 1989. By 1996, it had bocame a giant, with 8000 employees, $1.5 billion in annual sales, and offices in a dozen countries, from Russia to the Philippines. Despite its success, public opinion in Bulgaria widely suspects Multigroup of being an efficient scheme for siphoning off public money from the communist era and laundering it to the benefit of ex-communist officials (Manev *et al.*, 1998).

It is important to note that to acknowledge the blurring of legal–illegal boundaries does not mean to celebrate it. However undesirable, the emergence of these 'grey' organizations may be a natural by-product of economic transitions. In the absence of a strong legal and regulatory regime as formal constraints, informal constraints such as rules and regulations imposed by the mafia rise to fill the vacuum as a form of 'self government' to provide some 'public' goods such as protection from thieves and contract enforcement. In many cases, the mafia seems to have more effective contract enforcement mechanisms (for example, the collection of payments and the delivery of punishment such as the 'cut-throat' method discussed earlier) than the weak court and regulatory systems. To the extent that crime organizations are able to provide better enforcement 'services' than the predatory government, there will continue to be a demand for such services (Hay & Shleifer, 1998).

While such a 'boundary blurring' strategy may be viable during the initial, chaotic phase of the transition, how sustainable this strategy is in the long run when the 'dust settles' remains to be seen. It seems that lawlessness cannot work in the long run. As transition economies establish more legislation and regulations backed by gradually credible law enforcement, these 'grey' organizations will have to confront increasing pressures for legitimization. The CEO of Bulgaria's Multigroup perhaps provided the best advice on a future strategy, called 'tail cutting': 'The lizard survives if it cuts off its tail. It's time for our [illegal] economic groups to cut off their illegal tails.'

Overall, these three entrepreneurial strategies, while analytically distinct, are not necessarily separate in practice, and are often employed concurrently by start-ups. In other words, a start-up can adopt a prospector or guerrilla strategy, while engaging in intense networking that blurs the public–private and/or legal–illegal boundaries.

WHAT CAN BE LEARNED?

The development of entrepreneurship throughout transition economies has generated important lessons for entrepreneurs in these economies, as well as foreign entrepreneurs and managers interested in these emerging markets. This section elaborates on these lessons.

The Lessons for Entrepreneurs in Transition Economies

Dealing with environmental turbulence

At the dawn of the new millennium, the political, social and economic environment in many transition economies continues to be characterized by turbulence, which is not likely to stop soon. In CEE and the NIS, the transition brought hyperinflation in the early 1990s, which was only tamed by the mid-1990s. Then came the Russian crash in 1998, which not only sparked a collapse of Russia's own financial system, but also forced countries across the region to brace themselves against the dangers of contagion. In China, although the constitution was finally amended in 1999 to catch up with reality by acknowledging the private sector's important role in the economy (*Business Week*, 1999a), the government has continued to behave unpredictably. In 1998, it banned direct marketing without any public consultation, despite nearly $200 million invested by American firms such as Amway, Avon and Mary Kay, and an estimated involvement of 20 million Chinese entrepreneurs. More recently, in 1999, the government in a similar manner announced a ban on all foreign investment in Chinese Internet content providers, most of which are entrepreneurial start-ups (*Business Week*, 1998, 1999b).

However, despite its complexity and unpredictability, environmental turbulence seems to be a major trigger, or catalyst, for entrepreneurial activity in transition economies. The more dynamic, hostile and complex the environment, the higher the level of innovation, risk taking, and proactivity among the most successful entrepreneurial firms (Morris, 1998). Three ways of dealing with environmental turbulence in transition economies can be identified.

1. *Establish alliances with larger, more legitimate and more powerful players*. This is the heart of networking and boundary-blurring strategies discussed earlier. Partners in these alliances can include not only

more established domestic firms, but also foreign entrants and certain government agencies. Tapping into the resources of these partners may help deter environmental turbulence for entrepreneurs. China's Lucky Transportation's efforts to register as a 'collective' firm and join an enterprise group serve as a case in point here. For the same reason, many private Internet start-ups in China have investment from government-run Internet providers.

2. *Take collective action to promote entrepreneurial development*. As a new organizational form, private start-ups are misunderstood by many people in certain transition economies, who associate these firms with criminal organizations. Entrepreneurs should mobilize to form industry or business owners' associations in order to lobby the new government and the media, as well as the public about the wealth creation role they play in the economy. Of course, similar to lizards cutting off their tails, 'grey' organizations with a criminal or dubious background may have to cut off their illegal tails, as advised by the CEO of Bulgaria's Multigroup, in order to advance their legitimate interests.
3. *Create linkages with established educational institutions*. Collaboration with educational institutions confers legitimacy on entrepreneurs among the future generation of employees and entrepreneurs. Entrepreneurs can also contact researchers in these institutions, whose findings may further disseminate the role of entrepreneurship in transition economies. As a result, many start-ups, after they survive the first stage, often establish linkages with educational institutions through scholarships, internships and research support. The importance of such linkages is especially noted by professional entrepreneurs, such as those running Russia's Vimpelcom.

Transforming raw entrepreneurship into strategic leadership

While the tactics for dealing with environmental turbulence above focus on strategic alliances and collective actions, recent research has suggested that smaller, entrepreneurial firms may face an inherent disadvantage when collaborating with larger and more powerful players, given the unequal balance of power (Alvarez, 1999). Entrepreneurs may have better odds for success if they can develop capabilities that allow them to stand on their own and grow the firm (Peng & Heath, 1996). One key enabler is to focus on strategic leadership, defined as the 'ability to anticipate, envision, maintain flexibility, think strategically, and work with others to initiate changes that will create a viable future for the organization' (Ireland & Hitt, 1999). Such a capability has been argued to be a major factor differentiating the winners from the losers in the new competitive landscape of the 21st century.

With little exaggeration, we can suggest that most early entrepreneurial strategies in transition economies are highly opportunistic, making the first move to fill many unfilled gaps. This is precisely the heart of a prospector or guerrilla strategy. As transitions deepen and competition becomes more saturated, a higher level of entrepreneurial capability, namely, strategic leadership, will be required to transform such raw entrepreneurship. Specifically, entrepreneurs need to focus on four areas.

1. *Develop and communicate a long-term strategic vision.* While entrepreneurs might operate without a clearly articulated strategy when the organizational size is small, developing an explicit, long-term strategic vision becomes more critical for increasingly large organizations. Such an ability seems to characterize the best performing start-ups, such as China's Hope Group and Russia's Vimpelcom. Increasingly, the need to strategize is felt among entrepreneurs interested in taking their business to a new height.
2. *Build dynamic core competencies.* The days when entrepreneurs could 'hit and run' in the early stages of the transition seem to be passing. The new competition requires sustained investment in core competencies-based strongholds which can be defended and strengthened, thus often leading to a 'deep niching' strategy for many entrepreneurial firms. Moreover, these core competencies have to be dynamic, that is, they should be continuously updated and extended. Facing gigantic multinationals targeting these economies, the built-in flexibility of the entrepreneurial firms, thanks to their small size and informal structure, may be especially helpful (Dawar & Frost, 1999). Poland's Optimus Computer can serve as a vivid case in point here.
3. *Focus on human capital.* Given their thin resource base, entrepreneurial firms must compete on *resourcefulness*, that is, the ability of doing more with less. In this regard, how to make the most out of the human capital of their employees becomes critical. Entrepreneurs should not only seek to hire, train and invest in the best talents, but also make sure that such human capital stays within the firm as it grows. One particular hurdle that many entrepreneurial firms need to overcome, such as the one confronting China's Hope Group, is the family-style management, which tends to rule out criticism of the boss and discourages creativity. Few employees aspiring to top posts will be satisfied with an organization which will not allow them a role in business strategies. As a result, the ability to motivate and retain talented employees may become a source of competitive advantage for entrepreneurial firms that are able to do so.
4. *Make effective use of new technology.* While the technological base of most firms in transition economies is obsolete, many entrepreneurial

firms are in a unique position to leapfrog by acquiring some of the latest technology in sophisticated manufacturing and services. A recent example is the numerous Internet start-ups popping up in these economies. These new start-ups change the low-tech, labour-intensive image of many entrepreneurial firms in transition economies, and push both the scope and pace of technological progress to a new level. Given the lacklustre performance of many SOEs and privatized ex-SOEs in these economies, entrepreneurial start-ups may offer these countries the best hope of catching up with the global technological race.

In sum, these lessons for current and would-be entrepreneurs in transition economies call for continuous management of environmental turbulence and fundamental transformation from raw entrepreneurship to strategic leadership. Similarly, important lessons can be drawn for foreign entrants.

The Lessons for Foreign Entrants

Up to this point, most of the interactions that foreign firms have with local firms in transition economies are with larger SOEs as joint venture partners or ex-SOEs as acquisition targets (Si & Bruton, 1999; Uhlenbruck & Castro, 1998; Yan & Gray, 1994). However, as the entrepreneurial sector becomes more established in these countries, some of these firms will become attractive partners or targets for foreign entrants. Without much research, the lessons for foreign entrepreneurs and managers interested in working with entrepreneurs in transition economies are more tentative and speculative. In general, foreign entrants seem to need to focus on three areas.

1. *Treat their entrepreneurial partners sensibly*. Even for foreign companies experienced in the region, so far, very few have dealt with smaller, entrepreneurial firms. Therefore, how to treat these entrepreneurial firms differently from SOEs remains a major task for interested foreign entrants. Some of the assumptions that foreign companies hold with regard to SOEs, such as SOEs' inefficiency and governance problems, may not be applicable to start-ups. Typically, the first step in restructuring firms in transition economies is a very costly, difficult and time-consuming process of conversion from SOEs to market players. Such a process may not be necessary when working with private firms. As a new breed, entrepreneurial ventures, which are smaller, younger and more aggressive, can be regarded as the opposite of SOEs in terms of market orientation. Therefore foreign entrants may need to 'unlearn' some of the ungeneralizable prior notions about firms in transition

economies (for example, the need to teach Marketing 101 and Finance 101), when working with entrepreneurial start-ups there. At the same time, new ventures in transition economies may also share certain similarities with SOEs and recently privatized firms, such as a short-term mentality and a lack of interest in continuous learning. Therefore foreign firms' experience in transforming SOEs, such as ABB in Poland, may be helpful (Oblej & Thomas, 1998). In particular, foreign entrants may need to restructure simultaneously both hard (structures and procedures) and soft (human resources) aspects of the acquired firms, instead of handling one aspect at a time in a piecemeal fashion (Newman, 2000).[9]

2. *Take collective action to promote mutual interests*. This is similar to the need for mobilization among domestic entrepreneurs. It seems that the stronger and the more effective the collective actions of foreign entrants, the more likely their goals, which usually include market opening and legal reforms in transition economies, will be accomplished. Recently, industry and trade associations exclusively representing *foreign* business interests in transition economies, such as the US–China Business Council and the Working Committee on Eastern Europe of the European Council for Small Business, have become increasingly visible. For example, facing the Chinese government's bans on direct marketing and Internet investment in 1998 and 1999, respectively, American direct marketing companies and Internet venture capital firms pressed their cases through US trade negotiators in China's World Trade Organization talks, and eventually obtained significant concessions from the Chinese side.
3. *Establish alliances*. The rationale behind foreign entrants' need to use alliances is similar to that of domestic entrepreneurs. However, such an alliance strategy does not necessarily lead to joint ventures. In certain knowledge-intensive industries, foreign entrants may have little choice but to establish wholly-owned subsidiaries to protect their intellectual assets. Nevertheless, wholly-owned subsidiaries can still pursue an alliance strategy with larger and more powerful players. For instance, when encountering extensive software piracy in China, Microsoft, through its wholly-owned subsidiary, chose to collaborate with the Ministry of Electronics to develop new software, instead of challenging the government head-on. Microsoft figured that once the government has a stake in the sales of legitimate Microsoft products, it may also have a strong interest in using its influence to crack down on sales of counterfeit software. In essence, Microsoft followed its entrepreneurial counterparts in China by wearing a 'red cap' in order to accomplish its goals.

Existing research suggests that, despite the difficulties, successful foreign-led turnaround and restructuring of firms in transition economies, which so far have been limited to current and ex-SOEs, is possible (Estrin & Meyer, 1998; Meyer & Moller, 1998; Oblej & Thomas, 1998). Such development is encouraging for foreign companies interested in restructuring entrepreneurial firms in transition economies, because, after all, the founding principles (such as private ownership and profit maximization) between foreign firms and entrepreneurial firms in the region have a better 'fit' than the radical contrasts between foreign firms and SOEs. Given the expected increase of foreign firms' dealings with entrepreneurial firms in transition economies in the future, we would suggest that, if some of the tentative lessons above are followed, the odds for successful interactions with entrepreneurial firms may be higher than those with SOEs.

Some Caveats

On the other hand, it is important to note that the outlook for entrepreneurship in transition economies is not always rosy. Entrepreneurship does not always create wealth. What many entrepreneurs in transition economies (such as Bulgaria's Multigroup) have done is not wealth creation but wealth appropriation (or redistribution in their favour). Given that many entrepreneurial undertakings are clouded by 'grey' activities, how entrepreneurship can be developed in an ethical and legal manner which is *sustainable* in the long run is important (Puffer & McCarthy, 1995). Although, in the short run, some 'grey' or illegal activities serve a role that is economically functional, accepting these activities as a legitimate and natural by-product of economic transitions is dangerous in the long run. While providing arbitrage profits to certain intermediaries such as former cadres, these activities may create new distortion in the economy, deter foreign investment and generate public resentment toward all entrepreneurs, legitimate or otherwise (Gray & Kaufmann, 1998).

Since individual entrepreneurs, however enlightened, may find it difficult to deter the tide of 'grey' activities, policy makers need to seriously curtail these activities by clearly delineating and enforcing the 'rules of the game' (Baumol, 1990; North, 1990). It is fair to say that most transition economies have made considerable, if not uniform, progress in establishing basic institutional frameworks. However, they have usually achieved greater progress in the *extensiveness* than in the *effectiveness* of the laws and regulations. It is important to note that the rise of private enforcement of contracts, some of them through illegal means, has emerged as a response to the failure of the state to provide and enforce its own rules. Therefore, cracking down on the illegal enforcement services of the 'grey' economy will not succeed until public law enforcement is sufficiently developed.

In addition, governments should minimize the possibility of harassment of entrepreneurs by rent-seeking officials. Instead of being viewed as a softer 'invisible hand', the government is often viewed as a greedy 'grabbing hand' in countries such as Russia (Frye & Shleifer, 1997). Facing such a predatory regime, many existing entrepreneurs may continue to be drawn into the 'grey' economy and interested only in short-term profits, and many more would-be entrepreneurs will simply give up on their ideas. Simplification of tax rules and reduction of marginal rates will draw more firms out of the unofficial 'grey' economy. They will also make it less likely for rent-seeking officials to succeed, because transparency of the rules creates little room for 'entrepreneurial' officials to manoeuvre around.

CONCLUSION

Despite different paces and results, the entrepreneurial transformation of transition economies is taking on increasing importance. How do entrepreneurs and the start-ups they found create wealth in these environments traditionally hostile to entrepreneurship? A short answer is that they accomplish this through aggressive prospector and guerrilla strategies, extensive networking and active boundary blurring. The lessons that can be learned all centre on enhancing entrepreneurial start-ups' competitive advantage and, by extension, promoting the entrepreneurial spirit of these economies. Since entrepreneurship inevitably implies a deviation from customary behaviour in any country (Brenner, 1987), entrepreneurs in transition economies are not without their share of controversies, leading to caveats about some of their practices in the new competition.

In light of voluminous country- and region-specific publications, this chapter has opted for a generalization approach. While the lessons are derived from a multinational triangulation process based on the experience of practitioners throughout these countries, the advice from officials and advisors guiding the transitions, and the findings of scholars around the world, it is important to caution against over-generalization. Obviously, every transition economy is different in its own right. The lessons for relatively more developed economies (such as Poland), while insightful, are not likely to be of *equal* importance to less developed economies (such as Vietnam) and some 'submerging' economies in the NIS (such as Belarus). For large countries such as China and Russia, regional differences within a country are enormous, thus making over-generalization dangerous. The history of economic transitions in the past two decades suggests that what transition economies need is not a set of standard lessons, recipes or packages (for example, 'big bang' or 'shock therapy'), but rather institutional and organizational experimentation to allow for the evolutionary emergence of entrepreneurship (Spicer, McDermott & Kogut, 2000; Gartner, 1985; Hisrich & Fulop, 1994/5).

In conclusion, how entrepreneurs create wealth throughout transition economies illustrates a 'creative destruction' process that Joseph Schumpeter first highlighted. In essence, start-up firms create an alternative organizational form that challenges and may eventually destroy the state sector. Although entrepreneurs are pursuing their private gains and could hardly care less about the official ideology, collectively they become participants in a great social movement whose 'invisible hand' pushes a bankrupt, socialist regime aside. If there could be only one message one has to draw from such wide-ranging transitions, we would like it to be a sense of the staggering 'creative destruction' power of entrepreneurship and competition – both for entrepreneurs that participate in the transitions and for economies that embrace these transitions.

NOTES

1. It is important to note that the emphasis of this article, private entrepreneurship at smaller start-ups, is different from *corporate* entrepreneurship, which focuses on energizing larger, more established organizations, mostly through privatization in transition economies.
2. Local governments in South China, such as those in Guangdong Province, are more accommodating and friendly to entrepreneurs. For example, during the 1980s, they often ignored central government directives preventing private businesses from hiring more than eight employees. See Chang and MacMillan (1991).
3. The value-neutral term 'grey economy' is used here for compositional simplicity. Other terms include a number of labels, such as the 'second', 'shadow', 'underground', and 'unofficial' economy.
4. While mafia practices are clearly unethical and illegal in both transition and developed economies, using personal favouritism and grease payments to get things done and ignoring 'senseless' tax laws and regulations, which tend to be considered as unethical in developed economies such as the United States, are often regarded as largely acceptable and even ethical in transition economies such as Russia. On the other hand, some unquestioned (and hence ethical) practices in the West, such as exorbitant salary differentials between top managers and employees, layoffs, and whistleblowing, are strongly considered as unethical in Russia (Puffer & McCarthy, 1995).
5. A more radical estimate suggested that the unreported, 'grey' economy in Russia may be *larger* than the official economy (Sharma and Merrell, 1997).
6. In general, small firms – regardless of locations – tend to be unwilling to invest in R&D. However, in eastern Germany, they are less than half as likely to undertake R&D as their Western counterparts in unified Germany.
7. It is important to note that although bribery was used in this case, it is *not* a practice to be recommended. Most entrepreneurs, such as those running Lucky Transportation, resent having to resort to bribery to get things done. However, they face a competitive dilemma: if they refuse to pay while competitors do, then entrepreneurs who do not pay bribes may

be disadvantaged in terms of market opportunities and resources. This dilemma is similar to the one confronting many US firms abroad, which are constrained by the Foreign Corrupt Practice Act that forbids them from making 'corrupt' payments; their global competitors, however, are not similarly constrained. German Law, for example, permitted bribery expenses abroad to be deducted from German corporate taxes until the late 1990s.

8. Note that the 'non-state' firms in Table 5.2 not only include (1) registered, pure private firms, but also (2) collective (private–public hybrid) firms and (3) foreign-invested firms. Therefore, we may safely infer from the table that the percentage of loans obtained by registered private firms is substantially smaller than the meager 15 per cent figure obtained by all three categories of non-state firms.
9. This proposition, based on the limited experience of successful turnaround of SOEs in transition economies, requires further empirical research. A dissenting view in the literature is that too much change at one time may inhibit organizational learning and, consequently, firm performance (Newman, 2000).

ACKNOWLEDGEMENT

This research draws on a larger project funded, in part, by the Center for International Business Education and Research, Center for Slavic and East European Studies, Fisher College of Business Research Committee, and Office of International Studies, The Ohio State University; Hong Kong Research Grants Council (project HKUST6174/98H/CUHK/EI16); and French Research Center on Contemporary China, Hong Kong. This chapter was first drafted when I was visiting the Center for East European Studies at Copenhagen Business School, Denmark. Portions of this work were presented at Copenhagen, INSEAD, De Montfort, Hong Kong, Fudan, Zhejiang and Ohio State Universities during fall 1999. I thank Jay Barney, Paul Beamish, John Child and Oded Shenkar for their encouragement, Sharon Alvarez, Trevor Buck, Jean-François Huchet, Philippe Lasserre, Yuan Lu, Yadong Luo, Klaus Meyer, Niels Mygind, Snejina Michailova, Torben Pederson, Ming-Jie Rui, Justin Tan, Denis Wang, Zhong-Ming Wang, Verner Worm, Ming Zeng, the two reviewers, the guest editors (especially Mike Hitt) and the editor (Sheila Puffer) for their helpful comments and discussions, and Seung Hyun Lee and Heli Wang for research assistance. Finally, I thank Simon Johnson, Bruce Kogut, Haiyang Li, Klaus Meyer, Karen Newman, Don Sexton and Andrew Spicer for promptly answering my inquiries during the revision process.

REFERENCES

Aldrich, H. (1999), *Organizations Evolving*, London: Sage, p. 75.

Alvarez, S. (1999), 'Entrepreneurial alliances: prescriptions for alliance success with larger firms', PhD dissertation, University of Colorado, Boulder.

Au, K. and L. Sun (1998), 'Hope Group: the future of private enterprises in China', *Asian Case Research Journal*, **2**, 133–48.

Audretsch, D.A. (2000), 'Entrepreneurship in Germany', in D. Sexton and H. Landstrom (eds), *The Blackwell Handbook of Entrepreneurship*, Oxford: Blackwell.

Baumol, W. (1990), 'Entrepreneurship: productive, unproductive, and destructive', *Journal of Political Economy*, **98**, 893–921.

Boisot, M. and J. Child (1996), 'From fiefs to clans and network capitalism: expanding China's emerging economic order', *Administrative Science Quarterly*, **41**, 600–628.

Brenner, R. (1987), 'National policy and entrepreneurship', *Journal of Business Venturing*, **2**, 95–101.
Bruton, G., H. Lan and Y. Lu (2000), 'China's township and village enterprises: Kelon's competitive edge', *Academy of Management Executive*, **14**, 19–27.
Burt, R. (1997), 'The contingent value of social capital', *Administrative Science Quarterly*, **42**, 339–65.
Business Week (1998), 'Ultimatum for the Avon Lady', 11 May, (Asian edn), 22.
Business Week (1999), 'China's new revolution', 27 September, 72–8.
Business Week (1999), 'Big brother and the e-revolution', 4 October, 132–42.
Chang, W. and I. MacMillan (1991), 'A review of entrepreneurial development in the People's Republic of China', *Journal of Business Venturing*, **6**, 375–9.
Charap, J. and L. Webster (1993), 'Constraints on the development of private manufacturing in St. Petersburg', *Economics of Transition*, **1**, 299–316.
Cooper, A. and W. Dunkelberg (1987), 'Entrepreneurship research: old questions, new answers, and methodological issues', *American Journal of Small Business*, **11**, 1–20.
Dawar, N. and T. Frost (1999), 'Competing with giants: survival strategies for local companies in emerging markets', *Harvard Business Review*, March–April, 119–29.
Estrin, S. and K. Meyer (1998), 'Opportunities and tripwires for foreign investors in Eastern Europe', *Thunderbird International Business Review*, **40**, 209–34.
European Bank for Reconstruction and Development (1998), *Transition Report*, London: EBRD.
Frye, T. and A. Shleifer (1997), 'The invisible hand and the grabbing hand', *American Economic Review*, **87**, 354–8.
Gartner, W. (1985), 'A conceptual framework for describing the phenomenon of new venture creation', *Academy of Management Review*, **10**, 696–706.
Gray, C. and D. Kaufmann (1998), 'Corruption and development', *Finance & Development*, March, 7–10.
Hay, J. and A. Shleifer (1998), 'Private enforcement of public laws: a theory of legal reform', *American Economic Review*, **88**, 398–403.
Hill, C. (2000), *International Business*, Boston: Irwin McGraw-Hill.
Hisrich, R. and G. Fulop (1994/5), 'The role of women entrepreneurs in Hungary's transition economy', *International Studies of Management and Organization*, **24**, 100–117.
Ireland, R.D. and M.A. Hitt (1999), 'Achieving and maintaining strategic competitiveness in the 21st century: the role of strategic leadership', *Academy of Management Executive*, **13**, 43–57.
Johnson, S., D. Kaufmann and A. Shleifer (1997), 'The unofficial economy in transition', *Brookings Papers on Economic Activity*, **2**, 159–238.
Khanna, T. and K. Palepu (1997), 'Why focused strategies may be wrong for emerging markets?', *Harvard Business Review*, July–August, 41–51.
Khanna, T. and K. Palepu (1999), 'The right way to restructure conglomerates in emerging markets', *Harvard Business Review*, July–August, 125–34.
Low, M. and I. MacMillan (1988), 'Entrepreneurship: past research and future challenges', *Journal of Management*, **14**, 139–61.
Lumpkin, G. and G. Dess (1996), 'Clarifying the entrepreneurial orientation construct and linking it to performance', *Academy of Management Review*, **21**, 135–72.
Luo, Y., J. Tan and O. Shenkar (1998), 'Strategic response to competitive pressure: the case of town and village enterprises in China', *Asia Pacific Journal of Management*, **15**, 33–50.

Manev, I., T. Manolova and A. Yan (1998), 'The governance, legality, and interdependence of firms in transforming economies', working paper, School of Management, Boston University.
McCarthy, D., S. Puffer and S. Shekshnia (1993), 'The resurgence of an entrepreneurial class in Russia', *Journal of Management Inquiry*, **2**, 125–37.
Meyer, K. and I. Moller (1998), 'Managing deep restructuring: Danish experiences in Eastern Germany', *European Management Journal*, **16**, 411–21.
Miles, R. and C. Snow (1978), *Organizational Strategy, Structure, and Process*, New York: McGraw-Hill.
Miller, K. (1995), 'Poland: Rising star over Europe', *Business Week*, 4 December, 64–70.
Morris, M. (1998), *Entrepreneurial Intensity*, Westport, CT: Quorum.
Morris, M. and D. Sexton (1996), 'The concept of entrepreneurial intensity: implications for company performance', *Journal of Business Venturing*, **36**, 5–14.
Mugler, J. (2000), 'The climate for entrepreneurship in European countries in transition', in D. Sexton and H. Landstrom (eds), *The Blackwell Handbook of Entrepreneurship*, Oxford: Blackwell.
Nee, V. (1992), 'Organizational dynamics of market transition: hybrid forms, property rights, and mixed economy in China', *Administrative Science Quarterly*, **37**, 1–27.
Newman, K. (2000), 'Organizational transformation during institutional upheaval', *Academy of Management Review*, **25**, 602–19.
North, D. (1990), *Institutions, Institutional Change, and Economic Performance*, New York: Norton.
Oblej, K. and H. Thomas (1998), 'Transforming former state-owned companies into market competitors in Poland: the ABB experience', *European Management Journal*, **16**, 390–99.
OECD (1996), *The Development of Entrepreneurship in Transition Economies*, Paris: OECD.
Peng, M.W. (1997), 'Firm growth in transition economies: three longitudinal cases from China, 1989–96', *Organization Studies*, **18**, 385–413.
Peng, M.W. (1998), *Behind the Success and Failure of U.S. Export Intermediaries*, Westport, CT: Quorum.
Peng, M.W. (2000), *Business Strategies in Transition Economies*, Thousand Oaks, CA: Sage.
Peng, M.W. and P. Heath (1996), 'The growth of the firm in planned economies in transition: institutions, organizations, and strategic choice', *Academy of Management Review*, **21**, 492–528.
Peng, M.W. and A. Ilinitch (1998), 'Export intermediary firms: a note on export development research', *Journal of International Business Studies*, **29**, 609–20.
Peng, M.W. and Y. Luo (2000), 'Managerial ties and firm performance in a transition economy: the nature of a micro–macro link', *Academy of Management Journal*, **43**, 486–501.
Peng, M.W., C. Hill and D. Wang (2000), 'Schumpeterian dynamics versus Williamsonian considerations: a test of export intermediary performance', *Journal of Management Studies*, **37**, 167–84.
Puffer, S. (1994), 'Understanding the bear: a portrait of Russian business leaders', *Academy of Management Executive*, **8**, 41–54.
Puffer, S. and D. McCarthy (1995), 'Finding the common ground in Russian and American business ethics', *California Management Review*, **37**, 29–46.

Rona-Tas, A. (1994), 'The first shall be last? Entrepreneurship and communist cadres in the transition from socialism', *American Journal of Sociology*, **100**, 40–69.

Sharma, A. and M. Merrell (1997), 'Russia's true business performance', *Journal of World Business*, **32**, 320–32.

Si, S. and G. Bruton (1999), 'Knowledge transfer in international joint venture in transition economies: the China experience', *Academy of Management Executive*, **13**, 83–90.

Spicer, A., G. McDermott and B. Kogut (2000), 'Entrepreneurship and privatization in Central Europe: the tenuous balance between destruction and creation', *Academy of Management Review*, **25**, 630–49.

Tan, J. (1996), 'Regulatory environment and strategic orientations: a study of Chinese private entrepreneurs', *Entrepreneurship Theory and Practice*, **21**, 31–41.

The World Bank (1996), *World Development Report: From Plan to Market*, Washington, DC: The World Bank.

Tsang, E. (1998), 'Can guanxi be a source of sustainable competitive advantage for doing business in China?, *Academy of Management Executive*, **12**, 64–73.

Uhlenbruck, N. and J. de. Castro (1998), 'Privatization from the acquirer's perspective: a mergers and acquisitions based framework', *Journal of Management Studies*, **35**, 619–40.

Vogl, F. (1998), 'The supply side of global bribery', *Finance & Development*, June, 30–33.

Wright, M., R. Hoskisson, I. Filatotchev and T. Buck (1998), 'Revitalizing privatized Russian enterprises', *Academy of Management Executive*, **12**, 74–85.

Xin, K. and J. Pearce (1996), 'Guanxi: good connections as substitutes for institutional support', *Academy of Management Journal*, **39**, 1641–58.

Yan, A. and B. Gray (1994), 'Bargaining power, management control, and performance in U.S.–China joint ventures', *Academy of Management Journal*, **37**, 1478–1517.

Zhuplev, A., A. Konkov and F. Kiesner (1998), 'Russian and American small business: motivations and obstacles', *European Management Journal*, **16**, 505–16.

6. External environments, strategy and high tech new ventures in China

Mingfang Li, Xudong Gao and Yajun Wu

INTRODUCTION

The march towards a true global economy is going forward at a ferocious pace (Arnold & Quelch, 1998; Bryan, Fraser, Oppenheim & Ral, 1999). Changes on the political landscape and technological frontiers have only enhanced this irreversible trend. So far, multinational enterprises (MNEs) seem to have been the dominant players in this drama put on a world stage for all to see. Certainly the MNEs enjoy a daunting array of advantages. They possess, to name a few, unmatchable resources, enviable economies of scale and scope, rich knowledge and skills acquired through their operations over the years and in many diverse worldwide locations, and ability to respond effectively to environmental changes (Roth & Kostova, 2003). Furthermore, somehow scholars favour the lucky ones. Considerable research attention has focused on global-oriented strategies (Bartlett & Ghoshal, 1998) and those for big businesses.

At the same time, however, there are many domestic companies throughout the world that long to get ahead in this competitive global game. While those domestic firms in developed economies, with considerable research insights from scholars who study entrepreneurship there, might have over the years developed strategies to deal effectively with their multinational counterparts (see, for example, Sandvig & Colakley, 1998), domestic companies in emerging economies are confronted by various unique challenges (Peng, 2000) including those of devising effective strategies to deal with their 'new kids on the block', those almighty neighbours that have just migrated from overseas thanks to the globalization of the world economy. They probably arrived not long ago or are knocking on the door as these countries gradually open up to the world. China might be a truly unique case in point.

Even in emerging market contexts, scholarly attention has so far been on how multinationals would successfully enter and compete effectively in the 'New World' (for example, Arnold & Quelch, 1998; Li, Qian, Lam & Wang,

2000) but few have paid attention to the way businesses in emerging economies should develop their strategies to survive and thrive within this new world order (Fiegenbaum & Lavie, 2000). The void is even more glaring in the area of high-tech new ventures. As frequently noted, high-tech new ventures may contribute to the competitive landscape in many ways. These include new market development, new product introduction and greater competitive vitality of the domestic scene as these ventures develop new strategies. However, because of the greater uncertainty high-tech new ventures experience, they suffer much more from the 'liability of newness' than other entrepreneurial ventures. Simply stated, precious little guidance is available to domestic firms and especially high-tech new ventures in such cases (Dawar & Frost, 1999). This study attempts to integrate several major research ideas to form a basis for recommending effective strategies for high-tech new ventures in China competing in global environments.

This research is promising. First of all, the external environment for local firms in China tends to be quite different along a number of dimensions, such as environmental turbulence and institutional development. Consequently the company strategy may emphasize *the integration and coordination* of a variety of critical dimensions that may not be readily relevant for firms in other settings. Understanding the need to develop such strategy dimensions, and appreciating the necessity to integrate and coordinate these dimensions effectively will be useful for both domestic firms and multinationals. We apply the domestic firm strategy model developed to the high-tech new ventures, present a series of in-depth case studies and speculate on the possibility of extending this model.

THEORETICAL BASES

From a theory-building perspective, we decided to present key components of our model based on a broader understanding of the Chinese competitive environment and its interaction with the world competitive landscape, and then strategic elements applicable to domestic Chinese firms in general. An understanding of the broad backdrop will be quite useful for an enhanced examination of the high-tech new ventures in the Chinese setting. As the idea of 'organizational birth imprints' (Stinchcombe, 1965) would imply, organizations reflect the broader context of their own time of birth. High-tech new ventures, those relatively young technology-intensive firms, are no exception. Understanding the nature of the time of their birth would help us understand these firms better. Second, entrepreneurs who find these organizations are often professionals and cadres (Peng, 2001) and also expatriates from Western companies. They are influenced by the existing business practices (both domestic and global) and will lead their respective firms using a combination of those practices and possibly new innovation. Furthermore, the pressure of institutional isomorphism plays a significant role in the formation of high-tech new venture strategies as they deal with the 'liability of newness' (Stinchcombe, 1965) through collaboration (Li & Atuahene-Gima, 2002; Zhao & Aram, 1995), innovation (Li & Atuahene-Gima, 2001) and institutional imitation. Finally, successful high-tech new ventures will most likely grow into competitive juggernauts and their earlier experiences will shape their future destiny as well. Thus the study of high-tech new ventures can not and should not be divorced from the future development of these firms. With a broader context in this discussion, the reader will appreciate more of the significant challenges high-tech new ventures face, and the relevance of the four-dimension model being proposed especially for these firms.

Two aspects are relevant for a theoretical framework of effective strategies for domestic companies in global environments. First of all, we need to understand the nature of the globalization or the global environment which we all must confront. Secondly, we need to review the main characteristics of the domestic environment in which these firms compete and develop their strategies. Upon reviewing those aspects we would consider the key dimensions of strategies domestic Chinese companies in the global economy would adopt. Certainly our focus lends itself to the company strategies from the emerging economies and more specifically in this case to firms in China. At a later point we will bring the chapter into a sharper focus on the high-tech new ventures by describing the even greater challenges they face.

Global Competitive Environment

Understanding the nature of the global environment constitutes a first step towards developing effective strategy recommendations for domestic firms, especially those in emerging economies, competing in global environments. Scholars believe that two powerful forces are at play in the global environment (Bartlett & Ghoshal, 1998; Devinney, Midgley & Venaik, 2000). The first force is the pressure to globalize. Scale economies and scope economies, ever-expanding R&D costs, and increasingly shortened product life cycles all point to the need for companies to seek global power and efficiency in their business activities. Global integration from political and cultural perspectives, and advances in transportation and communication technologies, further reinforce this important pressure. Also many companies have succeeded remarkably by pursuing global or world product strategy – Japanese electronic companies are but one example. To many the world is becoming one, and meeting the global challenges a quintessential recipe for corporate success. At the same time, the second force, the pressure to localize, is fighting back. Local cultural traditions persist in spite of global integration. Regional political forces are actively influencing various processes. Violent demonstrations in cities where global economic conferences are being held serve to reinforce people's concern about globalization. Technological forces that are responsible for globalization are also responsible for accommodating and possibly reinforcing localization. For example, utilizing advanced manufacturing technologies would enable firms to produce highly customized products to meet unique customer needs and tastes without sacrificing scale economies.

Clearly the opposing forces have put conflicting demands on businesses big and small, multinationals as well as domestic. Multinationals that pursued either pure global strategy (world product orientation) or multidomestic strategy (multi-local product orientation) are now facing both pressures (Bartlett & Ghoshal, 1998). Furthermore, the need to create global power and efficiency and local differentiation and responsiveness simultaneously leads companies to seek another important advantage, that of worldwide innovation and diffusion. Innovations along various aspects of businesses are critical for responding to localization pressures; at the same time, diffusing those innovations when appropriate on a greater scale will be critical for gaining and enhancing global power and efficiency. Product innovation popularized in one market may help enhance a firm's competitiveness in another market. Effective marketing strategies uncovered in one region exhibit their prowess in another. Multinational enterprises increasingly seek sources of innovation from developing countries nowadays (Engardio & Einhorn, 2005). Global power and efficiency, local sensitivity and differentiation and worldwide innovation and diffusion have become the tripod on which a successful multinational enterprise builds its success. Likewise they provide important insights to domestic firms in emerging economies. Effectively dealing with conflicting strategic elements and goals would certainly be one of those insights (Roth & Kostova, 2003).

Chinese Environments

The open systems theory correctly identifies the external environment as one central aspect that firms must address effectively in order to be successful (Dill, 1958; Emery & Trist, 1965; Thompson, 1967). In general, the organization environment literature treats the external environments using either a dimensional approach (for example, Aldrich, 1979) or a configurational approach (for example, Emery & Trist, 1965). Aldrich

(1979) specifically identified six major external environment dimensions: complexity, dynamism munificence, consensus, interdependence and turbulence. This set of dimensions was later developed into valid measurements (Dess & Beard, 1984; Harris, 2004) and widely adopted in empirical studies (Keats & Hitt, 1988; Li & Simerly, 1998; Simerly & Li, 2000).

Another equally useful approach considers the external environment from a configuration point of view. A widely accepted model was developed by Emery and Trist (1965). In their model, the authors contend that the relationship between the environment and focal organization is of great importance. Furthermore, the authors conceive four different types of external environment: placid randomized, placid clustered, disturbed reactive and, finally, turbulent field. This conception remains valid and provides useful guidance (Emery, 2000). We believe that these two approaches are closely related and in fact would like to suggest that a *turbulent field* (Emery & Trist, 1965) is one with significant complexity and dynamism, greater interdependence among external entities and limited consensus. (Note that we emphasize by 'turbulent field' the totality of the environmental characteristics where environmental turbulence is but one of the dimensions.) The six aspects identified below of the Chinese institutional environment would provide substantive support for our contention above.

Firms in China must scan, monitor and design strategies to deal with six major institutional environmental aspects. These are (1) continued transition to a market economy, (2) opening up to world stages, (3) continued development of laws and regulations governing economic activities, (4) paramount roles that various levels of governments play in economic activities, (5) geographical diversity and developmental stage disparity, and (6) rapidly developing emerging product markets. This conception of the institutional environment integrates what other researchers call 'institutional environments' and 'extrainstitutional environments' (Lewin, Long & Carroll, 1999) together under one broader category for conceptual parsimony and we will also introduce the culture as another critical aspect to consider.

The economic reform and open door policy the Chinese government adopted more than 20 years ago has produced drastic changes. The external environment the Chinese firms now deal with is more market-like. We have witnessed various developments in the product market, labour market and capital market, markets that are deemed critical from an institutional environment point of view (Khanna & Palepu, 1997; Lewin, Long & Carroll, 1999). These markets are still being developed. Changes in these markets will have significant impacts on firm behaviour. China's accession to WTO and the opening up to the world market represent another remarkable transition with far-reaching and long-lasting impacts not only on industrial structure but also on managerial practices (Zhu & Warner, 2004). Multinational companies find a new home and a new competitive base. Chinese firms began to learn to compete head-on against other players both domestically and internationally in a more market-like fashion. This external source of learning provides Chinese firms with interesting ideas and useful models of management and business practices as well (Li, 2001).

Furthermore, more and more laws and regulations are being developed and implemented. This process complements another important aspect of the institutional environment school, the enforcement of laws and regulations (Khanna & Palepu, 1997; Li & Wong, 2003). New laws and regulations require implementation and mutual adjustments, however. The adoption and adjustments involved may contribute to greater uncertainty and turbulence. At the same time the various levels of Chinese governments continue to play important roles in the strategy formulation and daily operation of businesses in China. Governmental involvements, whether through macro control or adjustment, or direct intervention, may create further complexity in the external environment. These complexities are multiplied as various intertwined governmental agencies get involved in influencing firm behaviour. Examples might include central banks, various ministries related to specific industries (for example, telecommunications), local governments, various specialized zones and local economic planning commissions, or 'developmental and reform commissions' as they are known today.

In addition, it is important to note the significant geographical heterogeneity of mainland China, and its linkages with other regions of the Greater China economy (Kao, 1993). The vast areas of China exhibit very different weather patterns, cultural and ethnic diversities, resource bases and economic activity configurations. These heterogeneities are intertwined as different regions depend on each other for resources, markets and support. Coupled with the geographical diversity is the developmental disparity. The eastern and southern regions in China are further ahead than the northeastern and western regions as far as economic development is concerned. Finally, we must also note the importance of the emerging product markets. New technologies create possibilities and uncertainties in the market as new products are introduced and new companies are formed. For high-tech new ventures in China, this institutional factor assumes paramount importance as it implies possibilities and entrepreneurial opportunities. An obvious example is the emerging markets of information technology and telecommunications products. Many Chinese firms seized the opportunity of this significant product market trend to develop new consumer goods. Examples might include cell phones, USB (Universal Serial Bus) devices, computer hardware, software and new information appliances (Lu, 2000).

Firms in China are developing their strategies in response to and in anticipation of the evolution of domestic environments. The six important environmental aspects identified above and the higher degree of interdependencies among them lead us to believe that the Chinese domestic environment for businesses should best be characterized as *a turbulent field*, as per Emery and Trist's (1965) conception. Following the principle of requisite variety (Ashby, 1952) we believe firms there need to develop strategies of complex elements to deal with these significant turbulences in the external environment. This consideration has led us to consider a variety of different elements in effective domestic firm responses to turbulent Chinese institutional environments.

In addition, Chinese culture could be a major consideration in understanding firm strategy formulation and implementation. Clearly, many well accomplished scholars over the ages have laboured to characterize Chinese culture, with probably at best mixed success, and much remains to be done. It is indeed a daunting task to capture the essence of Chinese culture in a short section (Gannon, 1994, 2001, 2004). For our purpose of understanding the broad influence of Chinese culture on firm strategic behaviour, we would like to focus on (1) long-term orientation and critical roles played by strong leaders, (2) Yin and Yang, (3) harmony, (4) middle way of thinking, (5) holism and (6) adaptability characteristics of Chinese culture *as a starting point* of understanding how the culture influences the formulation and implementation of strategies in Chinese firms.

National culture is believed to be significant in affecting firm strategic behaviour and management (Hofstede, 1980, 1994; Trompenaars & Hampden-Turner, 1998). Supplementing the research with Hofstede on national culture (Hofstede, 1980), Michael H. Bond and his colleagues (The Chinese Culture Connection, 1987) made important progress in identifying an additional dimension missing from the original Hofstede conception, or the notion of Confucian dynamism. While the original characterization of Confucian dynamism appears to link multiple Chinese culture characteristics together (cf. Fang, 2003), the research community seems to be more interested in the long-term orientation aspect (Hofstede, 1991). This dimension is characterized by an emphasis on *perseverance, persistence and sacrificing short-term gains for long-term prospects*. This concept of long-term orientation is considered an important aspect of Chinese culture. Researchers have utilized this notion in various business research as well. In authors' field research of Chinese firms we also hear time and again the importance of perseverance and persistence in competitive manoeuvre and success.

Long-term orientation is frequently associated with the important roles played by strong leaders (Zeng, 2005). The latter can be understood from two related perspectives. First, company managers learn from Chinese culture to develop skills and capabilities and become strong leaders, especially in complex and turbulent environments. Second, Chinese people educated in Chinese culture tend to respect and follow strong leaders, as this helps with the implementation of strategies formulated by these leaders. In fact, our field research shows that, behind every successful firm, there is a strong CEO. Existing studies also indicate that strong leadership is a key characteristic of successful Chinese firms (Chen, Zhao & Zhao, 2004).

The second aspect identified above is the notion of *Yin and Yang* or the acceptance in Chinese culture of opposites or polarities (Chen, 2001; Fang, 2003). Yin and Yang can be viewed as *dualism*, as it relates to all the opposite principles one may encounter. With Yin and Yang, opposites complement and contrast each other, each making up what is lacking in the other. The acceptance of opposite principles does not mean the Chinese culture simply allows two opposing ends of a dimension to exist in isolation. The notion of *harmony* emphasizes that the opposite principles should be brought together in an integrated fashion. The *middle way* of thinking in Chinese culture may provide an effective channel for harmony to take place. The middle way would help achieve meaningful integration of the two polarities (Cheng, 1992) by avoidance of the extremes, and generation of enlightenment through the practices of seeking middle grounds. This middle way of thinking is critical, so much so that Chen (2002) would suggest the middle way of thinking is an important mechanism for transcending paradoxes. Finally, *holism* constitutes another important aspect of Chinese culture that would help achieve and reinforce harmony. From a holistic perspective, any entity, be it a nation or an organization or an individual, is viewed as a complex and interactive whole composed of interrelated parts. The whole cannot exist without the parts or parts without the whole. So, in addition to achieving harmony along opposite principles via the middle way of thinking, for the whole to work effectively it is important to achieve harmony in all parts of the system, and in all interdependencies together. We may also consider the significant role of the middle way of thinking alongside holism as different pairs of opposites may also represent extremes that are to be avoided using the middle way of thinking in favour of coordination and integration.

In addition, we also wish to point out that Chinese culture remains an open culture and has a remarkable ability to accept and internalize new ideas and new thinking. Chinese culture has experienced significant acculturation and socialization through the centuries (Hon, 2003) and is continuing to do so. Chang Chih-tung, years ago, advocated the idea of 'Chinese learning as substance, and Western learning for applications' in his *Exhortation to Learning*. There are plenty of recent examples as well. Chinese firms are combining Western management practices effectively. 'Transplant' companies or those developed in the Silicon Valley find new homes in the Chinese culture-based regions and have begun to mould Western management practices with Chinese practices. In fact an economic cluster transcending geographical territories has been forming that successfully links the Silicon Valley, Hsinchu and Shanghai (Saxenian, 2002). Leading Chinese companies are wholesales transplanting Western business practices such as 'integrated product development' and human resource systems.

While the discussion of the Chinese institutional environment focuses on the turbulent field, and the need for Chinese firms to invent strategies that integrate and coordinate complex and paradoxical elements, we believe our discussion of the Chinese culture points in a different direction: the Chinese culture may be *an enabling and stabilizing force* for Chinese firms to formulate and implement such strategies of a complex and paradoxical nature.

As a first step in this conceptual exploration, this chapter identifies four strategy dimensions and related integration and coordination considerations of effective domestic firm strategy with a complex and paradoxical nature. With a focused discussion of the nature of high-tech new ventures in the Chinese environment, we then apply this preliminary model to analyse a series of mini-cases to illustrate the applicability of the model established to some high-tech new ventures and their further development into competitive giants, and also to set the stage for further discussion of the notion of strategy complexity. Clearly, we will cite briefly various global competitive environment considerations, and Chinese institutional and cultural aspects indicated above when providing substance to the basic framework presented below.

DEVELOPING A BASIC FRAMEWORK FOR STRATEGY

From a strategic point of view, two levels of analyses are necessary (Lengnick-Hall & Wolff, 1999; Porter, 1985, 1990). First, at the individual company level, the critical question is what product market a company can pursue and what competitive approaches it would adopt in order to survive and thrive within its respective environment. Secondly, looking at the level of companies as a business ecosystem, the critical question is how the community of businesses should behave in order to develop and sustain the business ecosystem. Given the focus of this chapter, effective strategies for domestic companies in global environments, these two questions should be restated as (1) what strategic responses can a domestic company adopt to compete successfully with its multinational counterparts and (2) what should be the collective behaviour of the business ecosystem within which multinationals and domestic companies coexist?

Firm-level Considerations

The model of strategy as actions (Grimm & Smith, 1997) offers a promising starting point for developing effective strategic responses for domestic firms. The authors suggest that companies may initiate different types of actions based on their respective resource endowment and market positions. Of interest to this chapter are entrepreneurial actions, Ricardian actions and deterrent actions. They are respectively based on entrepreneurial insights, unique resource advantages and powerful market positions. Entrepreneurial actions are creative combinations of ordinary resources to explore market opportunities. The unique knowledge and insights entrepreneurs have of their local markets and effective skills related to initiating such actions and implementing them successfully are critical; so are the abilities of the entrepreneurial firm in creating competitive uncertainty and exploring blind spots. To the extent that other competitors are unable to assess the efficacy of entrepreneurial actions and therefore will not initiate timely responses, an entrepreneurial firm would be able to position itself well, and begin to accumulate resources.

Ricardian actions are based on the unique resource endowment a firm would have (Barney, 1991; Grimm & Smith, 1997; Porter, 1985). From the resource-based view of the firm, those firms with unique resource endowment will be able to pursue cost leadership and differentiation strategies effectively (Porter, 1985) and delay as well as nullify potential competitive reactions from others. Deterrent actions stem from a powerful market leader position. The market leader may be able to initiate market discipline and deter other less powerful competitors from directly challenging it. Clearly, successful firms may move from one type of action to another. For example, the earlier business activities at Wal-Mart would be entrepreneurial in nature. When it was able to expand rapidly in small towns and rural America, and accumulated enough resources, it made a transition to a resource-based competitive approach. Today, as the market leader, Wal-Mart certainly employs a variety of competitive approaches.

The nature of the global environment described above and subsequent strategies adopted by multinationals provided significant insights into the challenges domestic companies face. Consideration of the Chinese institutional environments and the culture provides additional insights into the analysis, especially with the necessity and possibility to develop strategies that contain multiple and paradoxical elements. In addition, the two levels of analysis – firm and community – and the relative advantage many domestic firms might possess will lay the foundation for a proposed model of effective domestic strategy.

Most domestic companies, for a variety of reasons, do not possess rich resources to compete head-to-head with multinational enterprises. Clearly, in a relatively market-like environment, they would not enjoy powerful market positions either when faced with powerful multinationals. The relative resource and market position these firms find themselves in would therefore not allow them to pursue either Ricardian action or deterrent action effectively as markets begin to open up and these firms are no longer being protected from global competition.

Many of those, however, may possess unique knowledge of their local markets, especially when the local environment is turbulent. Their unique insights would enable them to pursue entrepreneurial actions. These insights may include knowledge of unique customer needs, and ways and means to satisfy their demands through combining various available technologies and resources. Stated formally,

P1. Domestic companies should emphasize their ability to pursue localization based on entrepreneurial insights.

Various major IT powerhouses in China, such as Lenovo, Founder and Great Wall, offer interesting insights in successful localization. These companies pursed the market opportunity of Chinese word processing, type setting and publishing, and have been able to build competitive strengths and resources because of their leadership positions in these areas (Lu, 2000). As another example, the financial crisis Asian countries endured in recent years has had a significant impact on consumer decision-making patterns (Ang, Leong & Kotler, 2000). Domestic firm managers, being immersed in the same experience as their customers, would be in a better position to sense market changes and develop effective products and services to meet changing needs.

Multinational enterprises are certainly not oblivious to the challenges and benefits of localization. (On the other hand, multinationals pursue localization with a different emphasis: they are more likely to leverage their resource and capability advantage as a main competitive weapon while developing localization efforts as they deepen their commitment in a particular market. Domestic firms, especially high-tech new ventures, may rely heavily on entrepreneurial insights and localization as their competitive weapon, and integrate that with resource and capability expansion considerations as they develop.) Domestic companies, while emphasizing localization in their strategies, need to incorporate another important consideration, what to do with resources and skills accumulated through constant pursuit of localization advantage. It might be important for these firms to consider applying these resources and skills on a greater scale and to other businesses (Sandvig & Colakley, 1998). These firms ought to consider expanding their business operations effectively, and the path for growth ought to be based on expansion of those resources and skills they accumulated. Stated formally,

P2. Domestic companies should emphasize development of those resources and skills that are applicable to other promising growth opportunity.

Business Ecosystem-level Considerations

The discussion so far has covered actions at an individual company level. It is also important to consider strategic responses at the business ecosystem level. Porter (1998) suggests that economic clusters or various entities closely related to one another and congregated within a geographic confine are important for developing competitive advantage. Complexity theory scholars also emphasize the importance of business ecosystems (Stacey, 1996). They believe that many mutually reinforcing entities present in a successful business ecosystem will help each participant to gain advantage. Applying this theoretical perspective, domestic companies should not view these powerful new neighbours as their worst nightmare. Instead they should seek to develop, jointly with the multinationals, their business ecosystem or economic cluster that will allow every participant to benefit. In addition, they should also collaborate with their competitors for mutual gains.

One way to do so stems from the first recommendation above. Instead of pursuing common and generic strategies, domestic companies should pursue innovative strategies and develop and upgrade competences based on those unique strategies. Together as a business ecosystem or an economic cluster, the whole community would then exhibit greater strategic divergence (Hamel, 1996) and achieve greater vitality. Stated formally,

P3. Domestic companies should all seek to pursue innovative strategies. Greater strategic divergence will enable individual companies to gain unique advantage and enhance the vitality of the business ecosystem as well.

The emphasis on seeking innovative strategies and strategic divergence should be complemented by the emphasis on collaboration with other players within the business ecosystem (Iansiti & Levien, 2004). Domestic companies should consider the benefits of collaboration among themselves to learn from one another, and to extend their unique advantages. Domestic companies should consider the benefits of collaboration with their multinational counterparts as well. These collaborations might not only enable firms to learn from one another but also enrich their respective strategies. Stated formally,

P4. Domestic firms should seek to expand their advantage via collaboration.

Integrative Considerations

As Table 6.1 indicates, from both an individual company level and a business ecosystem level, the set of recommendations emphasizes differentiation and integration. There clearly exists a tension between the two directions. Such challenging tasks call for not only creative and visionary managers but also active involvement of other participants of the business community, such as government agencies, consulting companies and universities.

From an individual firm level, the tension lies in creating a unique strategy while at the same time applying skills to broader applications, whether markets or products or customer types. The localization strategy would enable domestic companies to create unique advantage and lock out powerful multinational competitors. Pursuing growth strategies based on leveraging resources and skills developed through localization strategy would enable firms to gain scale advantage, important when dealing with multinationals. These arguments therefore assume that expansion and even globalization should be viable options for domestic firms.

From a business ecosystem level point of view, strategic divergence and collaboration for competition can and should be mutually reinforcing. First, creative and innovative strategies from individual companies (including multinationals) will not only contribute to each firm's success but also enhance the attractiveness and competitiveness of the business ecosystem. Collaboration among members of the business community at the same time will be beneficial for all in the ecosystem for several reasons. First of all, collaboration can help provide complementary skills and resources

Table 6.1 Effective strategies for domestic companies in global environments

	Differentiation	*Integration*
Individual firm level	(Entrepreneurial insights) or localization strategy	Resource expansion-oriented growth
Business ecosystem level	Strategic divergence	Collaboration for competition

other members might be lacking, and therefore will help create badly needed products to customers without long delay. Secondly, learning that takes place through collaboration will contribute to creative strategy making. In a sense, companies learning from their partners may be able to create even more innovative strategies than they can by exploring alone. We now identify a variety of key issues related to each of these elements, especially with reference to high-tech new ventures. Table 6.2 provides a summary, and guides the discussion.

High-Tech New Ventures

As described earlier, high-tech new ventures may share various characteristics with other domestic firms as they are developed and immersed in the same type of institutional environment and cultural context. However, they also possess unique characteristics that should be highlighted in order for us to examine the strategic elements identified earlier, and explore the integration and coordination challenges these firms may face. Being relatively young firms engaged in business activities with substantial technological content (Chandler & Hanks, 1993), these firms tend to experience greater liability of newness (Li & Atuahene-Gima, 2002) than even other entrepreneurial ventures. Peng (2001) observes that these entrepreneurs may be professionals and cadres. In addition, we note that expatriates with substantial experience in high-tech firms in the West may also be a significant source of such entrepreneurs. One immediate implication is that these high-tech venture founders created their businesses by relying on substantial entrepreneurial insights and also likely significant knowledge of implementation issues. In addition, whether culturally pushed, or institutionally pressured, these entrepreneurs may explore collaborative strategies more actively in order to propel their new venture to success (Li & Atuahene-Gima, 2002; Zhao & Aram, 1995). To what extent, however, these firms will systematically and strategically exploit their entrepreneurial insights and collaborative advantages remains a critical question. Furthermore, looking into the Chinese institutional environment, and also the lingering impact of the central planning era, we may need to assume that Chinese high-tech venture entrepreneurs might lack managerial acumen and professional focus related to resources and capability expanded growth, and strategic divergence building. There appears to be some supporting evidence that the empirical literature in high-tech new ventures appears to focus mainly on collaboration as a way to overcome liability of newness, leaving significant voids in other areas noted above. Listing the central tasks faced by these firms in implementing those strategic elements will provide a striking reminder of the significant challenges they face.

Recall that entrepreneurial actions require firms to pursue creative combinations of ordinary resources to exploit market opportunities. This calls for entrepreneurial insight and creativity. Knowledge of the local market conditions, demands and various available resources from the local environment becomes crucial. Such knowledge and insights are useless unless the entrepreneur is able to implement successfully these creative combinations. Hence knowledge, insight and risk-taking spirit, coupled with skills in implementation, will become necessary conditions for localization. Competitive uncertainty and blind spots are effective ways for these entrepreneurial action-oriented firms to avoid competitive reactions from other firms – most likely powerful multinationals – if the domestic firms are to succeed.

An immediate task beyond entrepreneurial action is resource accumulation. Using the resource-based view of firms, scholars (Brush, Greene, Hart & Haller, 2001) have proposed a pathway model of resource accumulation. This model suggests that resource accumulation goes from tangible to intangible, and from personal to organizational. While financial resources and physical resources are important, employee knowledge and skills are even more so. Furthermore, those resources and skills that are personal in nature will increase their capacity for improving organizational effectiveness through a transformation to organizational level. Those firms that are able to accumulate a rich set of resources that reinforce one another will be in an advantageous position.

Clearly, the key resource that binds these tangible and intangible resources together and enables firms to create sustainable competitive advantage is knowledge. An effective knowledge management process then becomes crucial for entrepreneurial domestic firms to transit from having a localization advantage to converting this to critical resources for long-lasting competitive advantage (Nonaka, Toyama & Konno, 2000). An enabling context for knowledge creation within the firm, or *ba* (the Japanese word for field), then becomes important.

From the perspective of resource expansion-oriented growth, companies may pursue several directions either separately or in combination. First of all, on the basis of enhanced economies of scale, companies may wish to broaden their market scope. This can be in the form of geographical expansion or a global niche player. Secondly, on the basis of enhanced economies of scope, companies may wish to utilize the existing resources in new business activities. A value chain model will help identify promising synergies. A carefully orchestrated diversification strategy would certainly be a viable option.

Strategic divergence emphasizes several key issues as well. First of all, if the business ecosystem boosts firms of different competitive orientation, the whole business ecosystem will be better off because there will be com-

Table 6.2 Domestic strategy elements and key issues

Localization	*Resource expansion-oriented growth*
Entrepreneurial insights Implementation skills Pathway of resource accumulation Knowledge management process	Scale-based growth Geographic expansion Specialized global player Scope-based growth – synergy
Complementary players Cluster synergy Reservoir of ideas, skills & resources Legitimacy and support	Domestic and MNE players Complementary functions Comprehensive Multi-party alliances Working with research institutes Working with universities
Strategic divergence	*Collaboration for competition*

plementary players to support market demands, and facilitate competition and innovation. The business ecosystem as an economic cluster would in addition help create skilled employees, facilitate information exchange and idea cross-pollination, and store knowledge and resources for all to benefit. Furthermore, business ecosystems in the form of geographically concentrated economic clusters will have the necessary power to create legitimacy for activities of its members, and support from various external institutions such as governments.

Finally, the call for collaboration for competition would entail a variety of approaches. These might include collaboration, either equity-based or non-equity-based, among domestic firms, or with multinational firms, and the scope might stretch from functional, cross-functional to comprehensive. Strategic alliances and joint ventures are notoriously difficult to manage because of differing motives of partners, potential for opportunism and high transaction costs (Parkhe, 1993). At the same time there are strategic necessities that dictate strategic alliance as a key component of effective strategy: they might include poor resource position, high risk, dispersion of skills and knowledge, and competitive pressure for time, and they are strategic benefits as well. They include knowledge and insight transfer from collaborator or competitor, resource base building (Alvarez & Barney, 2001) and innovation and creative strategy formation as a result of strategic alliances. In addition to collaborating with other competitors, domestic and multinational, Chinese firms have also utilized available resources and expertise from universities and major research institutes. This is another very attractive avenue of collaboration.

We have identified the extraordinary challenges the Chinese domestic firms, and especially new ventures, face because of the Chinese institutional nature. Separately scholars have addressed the need for a company to consider each strategic element. The significance of the proposed model is the consideration of these strategic elements together in an integrated and coordinated fashion to form a coherent strategy. As presented in Table 6.1 and Table 6.2, these strategic elements represent two extremes or polarities at, respectively, firm and business ecosystem level. While the extant literature provides detailed discussion of each, little is available regarding how these elements can be integrated in a meaningful fashion, and furthermore coordinated together as a coherent strategy. To put it succinctly, given the turbulent nature of the Chinese institutional environment, and the greater liability of newness of Chinese high-tech new ventures, these firms need to address simultaneously the localization, resource expansion-oriented growth, strategic divergence, and collaboration for competition in an integrated and coordinated fashion. While the high-tech new venture literature has focused on collaboration as an uncertainty-reduction strategy and innovation, firms need more holistic recommendation for their success. Exclusively pursuing any one strategic dimension will not result in enhanced firm growth and development.

THE CASE OF HIGH TECH NEW VENTURES

To provide insights into this conceptual model, we use a series of mini-cases of Chinese telecommunications industry firms. We anchor these cases along the basic framework proposed above. To start with we provide some background information related to the development of the Chinese telecommunications industry.

Background

The public telephone network in China was very small two decades ago. In 1980, the number of telephones in China was the same as that of the US in 1905, that of England in 1947, and that of Japan in 1958 (Zeng, 2001). During the Sixth Five Year Plan period or the '6.5' period (1980–1985), the central government decided to support the development of the telecom service industry. Since then this industry has experienced rapid growth. For example, in 1990, China had the world's 15th largest public telephone network. In 1995, the rank rose to the fifth. In 2000, China had the second largest public telephone network in the world.

Accompanying the rapid development of the telecom service industry is the rapid development of the telecom equipment industry. In the early 1980s, China relied heavily on imports for telecom equipment. For example, in 1982, China began to import digital switches (first in Fujian Province). At that time, no domestic firms had the capability to make digital switches. All the players were MNEs. The situation began to change in the late 1980s, with the development of local equipment firms. By 2000, 80 per cent of the switches used on the public telephone network were domestically produced; more than 50 per cent of the newly installed optical transmission equipment was domestically produced; localization of mobile telecommunication equipment had also made impressive progress. (Largely owing to the significant need for fixed-cost investment, and greater economies of scale, the Chinese telecommunications equipment industry mirrors its counterparts elsewhere in terms of industry concentration. A small number of firms dominate this industry.)

In an industry dominated by multinational companies, several local firms have developed strong technological capabilities, and some of them have emerged as strong competitors. For example, Huawei and ZTE are able to compete head-to-head with multinational giants such as Nortel, Lucent, Alcatel and Cisco. Datang Group, the leading firm in wireless communications, developed the key technologies that are the basis for the TD-SCDMA standard, one of the three international standards for 3G wireless communications. The leading telecom equipment companies are also making rapid progress in developing backbone routers and optical transmission systems.

The rapid development of the local telecom equipment firms in China provides preliminary support to our framework that emphasizes (a) entrepreneurial insights or localization-based strategy, (b) resource expansion-oriented growth, (c) strategic divergence and (d) collaboration for competition.

Entrepreneurial Insights or Localization

The starting point for the local firms' development clearly is their seizing any available market opportunity or *entrepreneurial insights-based strategy*. The quick expansion of the public telephone network in China since the early 1980s, as mentioned earlier, offered a lot of opportunities for telecom equipment makers, especially multinational enterprises (MNEs). At that time, no domestic firms had the capability to make digital switches. All the players were MNEs.

Although MNEs dominated the telecom equipment market, local firms also had opportunities in the low-end market, for example in the private branch exchange (PBX) market. A lot of local firms entered the business of making PBX, either by buying foreign technologies or by developing their own technologies, and both ZTE and Huawei were among these firms. For example, ZTE was founded in 1985. In its early days ZTE did not have its own products but did ordinary equipment manufacture (OEM) for some Hong Kong firms. Today ZTE is China's largest listed telecom equipment manufacturer and one of the 520 key state-owned enterprises supported by the central government. Before 1996, ZTE's development was mainly from its seizing the opportunity in the low-end market. For example, ZTE began to develop a 68-line PBX analog switch in 1986, and in June 1987 developed successfully the ZX-60 switch. In 1987, ZTE decided to develop a digital PBX; the resultant product was the ZX500 (1989). In 1990, ZTE decided to develop a digital public office switch, and introduced the ZX500A switch in December 1991. Because of its better price/feature ratio, this machine was widely adopted and helped to transform the rural telecom network in China. This also created a good growth opportunity for the company. In November 1993, ZXJ2000, another digital switch, passed the evaluation of the Ministry of Postal and Telecommunications (MPT). In that year the machine captured 18 per cent of the rural area market.

Resource Expansion-Oriented Growth

Moving from the entrepreneurial insights-based strategy, in February 1996, ZTE decided to make the following transitions: (1) from making a single product to making multiple products; (2) from focusing on the rural market to focusing on the urban market; and (3) from focusing on the domestic market to starting to explore the overseas market. ZTE's strategic transition in 1996 was based on its resource accumulation, especially technological resources. For example, during its early days, ZTE's heavy investment helped it build technological capabilities and attract excellent researchers and engineers. In fact, ZTE's large-scale digital office switch, ZXJ10, was successfully developed in 1995. After the successful development of ZXJ10, ZTE believed that it had developed the capability, especially technological capability, to make strategic transitions in its development.

With continuous technological innovation, ZTE had made impressive progress in implementing the transition planned in 1996. It had expanded its product line from mainly switches and access equipment to a mobile network, a data communication network, an optical transport network, an intelligent network and a next generation network. ZTE has also been making rapid progress in the world market. Its products have been used in over 40 countries and regions, including Cyprus, Bangladesh, Myanmar, Thailand, the USA, Russia, Egypt, Kenya, Congo, Zambia and Hong Kong.

Strategic Divergence

The fast development of the local telecom equipment firms in China also shows the importance of a healthy business ecosystem. We look at the impact of strategic divergence from both business ecosystem capacity building, and strategy learning and enriching.

In the 1980s, the domestic telecom equipment market was dominated by MNEs. Few people believed that domestic firms had the capability to develop advanced telecom equipments. For this reason, few local firms had the confidence to develop large-volume digital switches. Great Dragon is different. Its strategy was to develop its own large-volume digital switches. The HJD04 large-scale digital switch was successfully developed in 1991. This was the first major technological breakthrough made by a local firm, and offered the opportunity for domestic firms to learn that the traditional belief that it was extremely difficult for local firms to develop large-scale digital switch was wrong, and that domestic firms should be able to develop advanced technologies and products. This significantly increased the domestic firm's confidence in developing proprietary technology. We would attribute this type of impact to part of business ecosystem capacity building.

According to a manager at ZTE,

> the successful development of the HJD 04 switch (in 1991) was of critical importance for the domestic telecom equipment firms to improve their confidence in developing key technologies. Technologically it might not be that difficult to develop proprietary technology, but how could you convince the top management to invest in R&D to develop it? Now we have an example of success. Other people in China can do it, and they are not expert in telecommunication, so we should also be able to do it. This is very important for the developers and for the top management, especially for the top management. It is not easy to convince them that the risk is not that high. When ZTE decided to develop this machine (ZXJ10, the first large-scale public switch developed by ZTE in 1995), Great Dragon had developed the 04 machine. This increased our confidence. The success of the HJD04 machine helped us. And we did it in 1995.

In the 1980s, although their environment is similar, the strategies of the leading local telecom equipment firms were not the same. ZTE and Huawei employed different strategies for success. For example, ZTE has been emphasizing the importance of risk management, given that it had to compete with giant MNEs when it was very small. According to Hou Weigui, CEO of the firm,

> my two most important concerns are how to survive and how to grow. Our industry is a high risk industry . . . we (the domestic firms) have different policies dealing with the two issues. ZTE believes in growth with a conservative fiscal

> policy. This is different from using debts to support growth, for example. We think that is too risky . . . MNEs have huge market shares and monopolistic profits. This not only supports their current development but also supports their future expansion. Some experts come to the conclusion that MNEs are able to survive for several years even if they are not making money or losing money in China. Our competitors are really strong. No domestic firm can say that it is not afraid of losing money for several years. (Ye, Yang & Meng, 2000, p. 65)

Huawei adopted a different strategy, although it also had to compete with giant MNEs when it was very small. It likes to take risks and invest significantly in R&D. According to Ren Zhengfei, CEO of the firm, 'of course there are risks in innovation, but you should not stop innovating because of risks. If you do not take the risk, you would have to follow the leaders, and have to be a second class or third class firm. You would not be able to compete with MNEs, and would not be able to survive'. One example showing Huawei's willingness to take the risks in developing proprietary technology was the development of HJD48, a PBX switch. To develop this product, Huawei invested almost all its profits from selling imported products from Hong Kong in the first few years of its operation.

Strategic divergence between ZTE and Huawei not only helped lessen unhealthy competition such as price wars between them but also created opportunities for them to learn from each other various different elements of their respective strategies. Because of the emphasis on risk management, ZTE quickly moved to develop PHS (Personal Handyphone System), a product that did not attract Huawei's attention, as it was a relatively 'low' technology-intensive product. The end result is that ZTE has encountered no competition in this market, unlike the situation in the digital switch market where there are several competitors.

Strategic divergence has also created opportunities for local firms to learn from one another. One example was the CDMA market. In this market, ZTE and Huawei once had very different strategies. ZTE chose to develop products based on relatively mature technologies, while Huawei decided to develop products based on new and more advanced technologies. The result is that the service providers decided to use ZTE's product. Huawei learned an important lesson from this. It is critical to develop products that are 'believed' to be more reliable (by using mature technologies, for example).

Collaboration for Competition

Finally, we examine the importance of a healthy business ecosystem by looking at another important contributor to the fast development of domestic telecom equipment firms, the vibrant and intensive collaborations within the business ecosystem. Our first example is the collaboration between the local telecom equipment firms and the research institutes and universities. We first look at the case of ZTE. When ZTE was founded in 1985, its technological capability was limited. In the second half of 1987, ZTE decided to develop a PBX digital switch. The problem was that ZTE only had three or four people who majored in switch technology or computer science and technology. To strengthen its R&D capability, ZTE collaborated with the Beijing Institute of Postal and Telecommunications. In November 1989, ZX500 was successfully developed. ZTE claimed that this was the first domestically developed digital switch.

ZTE institutionalized the practice of collaborating with research institutes and universities mainly by setting up internal research institutes close to research institutes and universities. This has proved to be a very effective measure. For example, ZXJ10 was developed by the Nanjing Institute, established in September 1993. This product was one of the key large-scale digital switches developed domestically. It also helped to lay the foundation for ZTE's transformation from a low-end product maker to a major telecom equipment firm.

Datang also has a lot of collaborations with research institutes and universities. For example, Datang Group is the first domestic firm that is able to develop backbone routers, by collaborating with the University of National Defense (UND). UND is very strong in computer science and engineering. It developed the 'Yinhe' high-performance computer in the 1980s, with a speed of 100m fps. In the mid-1990s it accomplished another major achievement in developing a high-performance computer reaching a speed of more than 10 billion fps. Datang and UND started to develop DTYH-ISR9108, the first domestically developed high-end backbone router, in July 1999. A team of more than 40 people was formed to carry out the research and development work. The team was led by professor Lu Xicheng, the Dean of the School of Computer Science and Technology at UND and a fellow of the Chinese Academy of Engineering. The development team made a series of technological breakthroughs in areas such as distributed system structure, a network operating system and high-speed network exchange and data transmission. In March 2001, DTYH-ISR9108 passed the evaluation by experts from the Chinese Academy of Science, the Chinese Academy of Engineering and other organizations. In December of the same year, the machine passed the evaluation by the Ministry of Information Technology and got permission for network application.

In addition to working with universities and research institutes, domestic telecom equipment firms have developed a new understanding about their competitive relationship, and collaborated with one another. Domestic telecom equipment firms believe that they need to survive as a whole rather than by beating each other. According to Hou Weigui, CEO of ZTE,

> Minister Wu Jichuan summarizes the technological breakthrough in digital switches by the domestic firms as a *group* breakthrough, this is very precise . . . in fact, if only one firm made that breakthrough, it would be hard for it to compete with the MNEs independently, the history of the domestic telecom equipment firms would have to rewrite, because there were huge gaps between our overall strengths and that of the MNEs. Today, it is very important for us to recall this history; because we are trying to make breakthroughs in wireless, we should not ignore the role of breakthrough as a whole. (Ye, Yang & Meng, 2000, p. 62)

Because of the new understanding, domestic firms have developed mutual supportive relationships and close collaborations in various areas. For example, when Huawei was not selected as one of the CDMA equipment suppliers to China Unicom, ZTE showed its support of Huawei by pointing out the benefits of having more than one domestic firm among the suppliers. ZTE and Huawei are also working closely with Datang on the commercialization of TD-SCDMA, a 3G international standard developed by Datang. Domestic firms also have collaborations with multinational companies. For example, Datang is working closely with firms such as Alcatel, Motorola, Samsung, LG, TI, Nortel and Siemens to commercialize its TD-SCDMA technology. Huawei is working with 3Com in its development of the US market.

A Synthesis

This set of mini-cases illustrates that Chinese firms may be able to develop and implement effective strategies to survive and succeed in the increasingly competitive market environment. We have focused on entrepreneurial insight-based product market strategy, resource expansion-oriented growth strategy, and reliance on a diversity-oriented business ecosystem that not only provides supports and sources of learning but also offers opportunities to collaborate so all can compete more effectively. It is important to note that, at times, firms pursue different elements of the strategies identified sequentially, for example from entrepreneurial insights-based strategy to resource expansion-oriented growth strategy, and at other times they utilize multiple strategy elements together, for example resource expansion-oriented strategy together with collaboration for competition. In addition, some of these firms clearly were able to combine all four strategic elements successfully. For example, ZTE has been able to continue exploration of entrepreneurial insights since 1996 after it started to make a strategic transformation to enter new product areas (CDMA system, PHS, mobile phone), new market areas and new geographical territories (globalization). Furthermore, ZTE contributed to the overall health of the business ecosystem by purposefully adopting different strategies, as discussed previously. As also discussed before, these strategic elements represent two polarities in terms of differentiation and integration on the one hand, and two levels of thinking respectively at the firm level and business ecosystem level on the other. We believe that the turbulent institutional environment in China contributes to the necessity for this type of response, and the Chinese culture may appear to enable firms to pursue this type of response. Our synthesis focuses more on the cultural side.

The national culture element seems to have played a critical role in the leading local telecom equipment firms' rapid growth. We found that all the leading local telecom equipment firms are long-term oriented, and all have a strong leader. For example, instead of focusing on making money in the short run, Huawei makes it clear that profit maximization is not the key goal for Huawei. The key goal is to be a leading telecom equipment firm in the world. Datang, one of the world leaders in 3G technology, is going so far as to sacrifice its technology advantages in products such as ATM, and is betting its future on its TD-SCDMA technology. ZTE is similar. While most local firms are making mobile phones by buying technology or even OEM, ZTE is determined to develop it own core technologies.

These leading local telecom equipment firms also have great and strong leaders. For example, at ZTE, Hou Weigui, CEO of the company, has been able to insist on the initiating and carrying out of many new technology and products development projects (PHS, for example) when many people, especially experienced engineers and top-level managers, are opposing these projects. It turns out that Hou has been correct in most of the cases.

Our study also shows that the strong leaders of the leading local telecom equipment firms are strong because they have learned a lot from the Chinese culture. For example, a common characteristic of these leaders is that they know how to attract and keep highly talented people. These leaders respect the highly talented people, while many local firms are treating their employees only as 'labours'. These leaders also create a lot of opportunities for the highly talented people to be promoted quickly. These leaders also offer a considerably higher salary (several times higher than the average salary) to the highly talented people, while many local firms are trying to pay as little as possible to their employees. These leaders also understand the important influence of their behaviour on the employees. For example, Ren Zhengfei, CEO of Huawei, only holds a small percentage of Huawei's shares, although he is the founder of the company and is able to hold more shares. Hou Weigui, CEO of ZTE, refuses to stay in luxury and expensive hotels when he is travelling.

Of course, individual strong leaders and long-term orientations alone would not accomplish the complex and challenging task of integrating opposing polarities in paradoxes successfully to enable a firm to gain competitive advantage. To further understand the impact of Chinese culture, we need to look at the employees as well as systems and procedures institutionalized in an organization. We believe that employees also play a key role. In a culture where Yin and Yang are believed to represent the opposing ends of a unified phenomenon, and where the middle way of thinking as well as the focus on harmony prevail, employees are able to appreciate and internalize inherent contradictions represented by the joint consideration of entrepreneurial insights-based strategy with resource expansion-oriented growth and by the joint consideration of strategic diversity and collaboration for competition. Huawei, for example, through the Huawei Basic Law (a formalized statement of Huawei's managerial philosophies and principles), emphasized the importance of combining people orientation with procedure/system orientation (*Fa*), and that of harmony (Chen, Zhao & Zhao, 2004). Thus long-term orientations together with a strong leader function effectively with employees conditioned in the Chinese cultural environment, and supported by an institutionalized managerial philosophy (represented by Huawei's Basic Law in this case) to formulate and implement complex strategies effectively.

PRACTICAL IMPLICATIONS

By reviewing the nature of the global competitive environment, we identified clearly powerful opponents on the competitive landscape. While most Chinese firms are competing and developing in the domestic area first, they are not immune to the international competition as MNEs are making China not only as their markets but also bases for R&D, manufacturing and marketing. By examining the nature of the Chinese environments from both an institutional perspective and a culture perspective, we identified a necessity for Chinese firms to develop and implement strategies that may contain complex and often paradoxical elements. We believe this approach is not only desirable but also feasible because of the unique characteristics of the Chinese culture. Chinese high-tech new ventures possess not only shared characteristics with other domestic firms but also unique features thanks to their newness and technology intensity, plus the influences of the founders.

Clearly this set of recommendations would call for Chinese high-tech new ventures to focus on both firm-level thinking and ecosystem-level thinking, and to emphasize both on individuality and commonality. Public policy recommendations are clear as well. Governmental agencies might be in a unique position to help these new venture firms to identify and crystallize hidden opportunities and to explore those opportunities. They should also encourage the development of unique competitive strategies so these firms can best utilize these opportunities and at the same time enrich and expand the capacity of the business ecosystem. Also governmental agencies and especially research arms as well as institutions of higher education might be in a unique position to help firms collaborate effectively.

As indicated earlier, governments play a role in Chinese firm development in many ways. One of the prominent approaches in recent years is the establishment of High-tech Zones or Parks. There are at least 50 such high-tech parks or zones. It would be useful to expand our model of ecosystem-level competitive dynamics when considering the impact of governmental policies in such parks. More specifically we have focused primarily on competitors within each business ecosystem. It would make excellent sense to consider how these high-tech zones or parks collaborate and compete with one another on a higher level, and also examine how governmental policies in different zones will affect the development of one particular zone and the zones overall together.

THEORETICAL CONTRIBUTIONS AND FUTURE EXTENSIONS

Dynamic and complex environments dictate dynamic and complex strategic responses. We identified four major elements of successful domestic strategies and applied those to analyse a series of mini-cases of Chinese telecom firms. We would like to emphasize first of all that a key contribution of this chapter is the recognition that, given the turbulent Chinese institutional environment, Chinese firms need to develop strategies that integrate and coordinate complex and paradoxical elements. There is clearly also a need to ask whether those four elements will be sufficient. In fact, in this concluding note, we would also like to formulate a formal definition for the notion of strategic complexity, as follows. *Strategic complexity represents the totality of a coherent set of individually effectively integrated and collectively well-coordinated organizational paradoxes*. As proposed, this concept is dynamic in nature. The paradoxes included in a complex strategy may depend on company growth, and external environment conditions such as the evolution of the market conditions. Confronted by a turbulent field that characterizes the Chinese institutional environment, high-tech new ventures must develop such strategic complexity. This concept goes beyond the existing model presented in this chapter since it only focuses on entrepreneurial insights and resources expansion-oriented growth at the firm level, and strategic divergence and collaboration for competition at the ecosystem level. We envision a firm that has developed fully the strategic complexity to integrate and coordinate well several more such paradoxes. They may include such paradoxes as public and private ownerships (Meyer & Lu, 2005), personalism (Redding, 1993, 1995; Silin, 1976; Yeung, 2000) and professionalism (Chandler, 1977, 1990) based management, resource-building and institutional management (Li & Wong, 2003), and exploration and exploitation (Benner & Tushman, 2003; March, 1991). Of course, clearly, we must recognize that these are but examples of paradoxes and not all firms with sufficiently developed strategy complexity may embrace all these complexities; and, furthermore, the relative importance of each paradox may evolve over time as well.

The multifaceted strategy and the strategic complexity notions proposed in this chapter clearly beg another question. What conditions are necessary for high-tech new ventures to formulate and implement such strategies successfully? We believe the first condition is the turbulent institutional environment because of the unique institutional characteristics identified earlier. The second condition is the Chinese culture as an *enabling and stabilizing force*. The Chinese culture's tendency to integrate paradoxes in a harmonious fashion, and to encourage middle way thinking, may allow firms to devise various paradoxes and yet manage them effectively holistically. Furthermore, the long-term orientation would emphasize persistence, perseverance and sacrificing short-term gains for a long-term future. Coupled with an adaptability to new ideas, and 'alien' thinking, the Chinese culture may enable firms to experiment with still more paradoxes.

In the authors' ongoing research we have witnessed first hand that Chinese firms are indeed effectively managing these complex and paradoxical strategic elements. Public ownership still exercises pervasive influences on firm development, while at the same time modern management techniques have found a welcoming home in many companies. Central elements of the Chinese culture become mixed with 'foreign thinking' to form unique corporate cultures. Young expatriates returned from abroad work side-by-side with cadres and professionals turned entrepreneurs. Resources and capabilities developed from the state-owned enterprises and governmental agencies are utilized together with market-oriented competitive moves. Developmental zones that mix listed companies and governmental agencies with intricate ties are working to advance entrepreneurial activities. Group companies are formed to support high-tech new ventures, and the latter in turn inject vitality into their parents. Obviously, this study has barely touched the tip of a giant iceberg. This is fertile ground for promising and enlightening research, and we have only been able to address a limited number of issues in this chapter.

Moving back from our field research to focus on the theoretical notion of strategic complexity, we could not help but wonder. What paradoxes should be included in a sufficiently complex strategy? How should those paradoxes be integrated and coordinated for firms to gain competitive advantage? Will the set of paradoxes evolve over time and, if so, how? And how does each paradox increase or decrease in importance relative to other paradoxes? These are questions to be answered in future research. Precisely defining the construct of interest is a good first step; developing related theoretical foundations to answer challenging questions will be the next step; and, finally, we must also be able to demonstrate that indeed strategic complexity will enable firms to compete more effectively.

If, as observed, multinational enterprises are full of paradoxes (Roth & Kostova, 2003), and if, as suggested in this chapter, Chinese firms are in a unique position to integrate and coordinate paradoxes in an effective manner, they may be able to gain competitive ground not only in their home territories but also in their future efforts in a broader competitive context, the global arena. As Mark Twain said, 'Supposing is good, but finding out is better.' Significant challenges are still ahead of us. Empirical studies examining what strategic elements identified in our proposed four-element model Chinese high-tech ventures are using and if they are necessary and sufficient strategic tools to enable firm competitiveness and success would go a long way in extending the model. Furthermore, careful conceptual efforts and empirical attempts are necessary to establish strategic complexity as a useful concept in strategic management research.

REFERENCES

Aldrich, H.E. (1979), *Organizations and Environments*, Englewood Cliffs, NJ: Prentice-Hall.

Alvarez, S.A. and J.B. Barney (2001), 'How entrepreneurial firms can benefit from alliances with large partners', *The Academy of Management Executive*, **15** (1), 139–48.

Ang, S.H., S.M. Leong, and P. Kotler (2000), 'The Asian apocalypse: crisis marketing for consumers and businesses', *Long Range Planning*, **33** (1), 97–119.

Arnold, D.J. and J.A. Quelch (1998), 'New strategies in emerging markets', *Sloan Management Review*, **40** (1), 7–20.

Ashby, W.R. (1952), *Design for a brain*, London: Chapman & Hall.

Barney, J. (1991), 'Firm resources and sustained competitive advantage', *Journal of Management*, **17**, 99–120.

Bartlett, C.A. and S. Ghoshal (1998), *Managing across Borders: The Transnational Solution* 2nd edn, London: Random House Business Books.

Benner, M.J. and M.L. Tushman (2003), 'Exploitation, exploration, and process management: the productivity dilemma revisited', *Academy of Management Review*, **28**, 238–56.

Brush, C.G., P.G. Greene, M.M. Hart and H.S. Haller (2001), 'From initial idea to unique advantage: the entrepreneurial challenge of constructing a resource base', *Academy of Management Executive*, **15** (1), 64–78.

Bryan, L.L., J. Fraser, J. Oppenheim and W. Ral (1999), *Race for the World: Strategies to build a great global firm*, Boston, MA: Harvard Business School Press.
Chandler, A.D. (1977), *The Visible Hand: The managerial revolution in American business*, Cambridge, MA: Harvard University.
Chandler, A.D. (1990), *Scale and Scope: The dynamics of industrial capitalism*, Cambridge, MA: The Belknap Press of Harvard University Press.
Chandler, G.N. and S.H. Hanks (1993), 'Measuring the performance of emerging businesses: a validation study', *Journal of Business Venturing*, **8**, 391–408.
Chen, C., S. Zhao and H. Zhao (2004), *The Tao of leading firms* (in Chinese), Beijing, China: CITIC Publishing House.
Chen, M.-J. (2001), *Inside Chinese Business: A guide for managers worldwide*, Boston, MA: Harvard Business School Press.
Chen, M.-J. (2002), 'Transcending paradox: the Chinese "middle way" perspective', *Asia Pacific Journal of Management*, **19**, 179–99.
Cheng, C.-Y. (1992). 'The "C" theory: a Chinese philosophical approach to management and decision-making', *Journal of Chinese Philosophy*, **19**, 125–53.
Dawar, N. and T. Frost (1999), 'Competing with giants: survival strategies for local companies in emerging markets', *Harvard Business Review*, **77** (2), 119–29.
Dess, G.G. and D.W. Beard (1984), 'Dimensions of organizational task environments', *Administrative Science Quarterly*, **29**, 52–73.
Devinney, T.M., D.F. Midgley and S. Venaik (2000), 'The optimal performance of the global firm: formalizing and extending the integration-responsiveness framework', *Organization Science*, **11**, 674–95.
Dill, W.R. (1958), 'Environment as an influence on managerial autonomy', *Administrative Science Quarterly*, **2**, 409–43.
Emery, F.T. and E.L. Trist (1965), 'The causal texture of organizational environments', *Human Relations*, **18**, 21–31.
Emery, M. (2000), 'The current version of Emery's open systems theory', *Systemic Practice and Action Research*, **13**, 623–43.
Engardio, P. and B. Einhorn (2005), 'Outsourcing innovation', *Business Week*, 21 March, 84.
Fang, T. (2003), 'A critique of Hofstede's fifth national culture dimension', *International Journal of Cross Cultural Management*, **3**, 347–68.
Fiegenbaum, A. and D. Lavie (2000), 'Strategic management of MNCs' entry into foreign markets: experience of Israel in the 1990s', *European Management Journal*, **18** (1), 93–105.
Gannon, M.J. (1994), *Understanding Global Cultures: Metaphorical journeys through 17 countries*, Thousand Oaks, CA: Sage Publications.
Gannon, M.J. (2001), *Understanding Global Cultures: Metaphorical journeys through 23 nations*, 2nd edn, Thousand Oaks CA: Sage Publications.
Gannon, M.J. (2004), *Understanding Global Cultures: Metaphorical journeys through 28 nations, clusters of nations, and continents*, 3rd edn, Thousand Oaks, CA: Sage Publications.
Grimm, C.M. and K.G. Smith (1997), *Strategy as Action: Industry rivalry and coordination*, Cincinnati, OH: South-Western College Pub.
Hamel, G. (1996), 'Strategy as revolution', *Harvard Business Review*, **74** (4), 69–82.
Harris, R.D. (2004), 'Organizational task environments: an evaluation of convergent and discriminant validity', *Journal of Management Studies*, **41**, 857–82.
Hofstede, G.H. (1980), *Culture's Consequences: International differences in work-related values*, Beverly Hills, CA: Sage Publications.

Hofstede, G.H. (1991), *Cultures and Organizations: Software of the mind*, London: McGraw-Hill.

Hofstede, G.H. (1994), *Uncommon Sense about Organizations: Cases, studies, and field observations*, Thousand Oaks, CA: Sage Publications.

Hon, T.-K. (2003), 'National essence, national learning, and culture: historical writings in Guocui Xuebao, Xueheng, and Guoxue Jikan', *Historiography East and West*, **1**, 242–86.

Iansiti, M. and R. Levien (2004), 'Strategy as ecology', *Harvard Business Review*, **82** (3), 68–78.

Kao, J. (1993), 'The worldwide web of Chinese business', *Harvard Business Review*, **71**(2), 24–36.

Keats, B.W. and M.A. Hitt (1988), 'A causal model of linkages among environmental dimensions, macro organizational characteristics, and performance', *Academy of Management Journal*, **31**, 570–98.

Khanna, T. and K. Palepu (1997), 'Why focused strategies may be wrong for emerging markets', *Harvard Business Review*, **75** (4), 41–51.

Lengnick-Hall, C.A. and J.A. Wolff (1999), 'Similarities and contradictions in the core logic of three strategy research streams', *Strategic Management Journal*, **20**, 1109–32.

Lewin, A.Y., C.P. Long and T.N. Carroll (1999), 'The coevolution of new organizational forms', *Organization Science*, **10**, 535–50.

Li, H. and K. Atuahene-Gima (2001), 'Product innovation strategy and the performance of new technology ventures in China', *Academy of Management Journal*, **44**, 1123–34.

Li, H. and K. Atuahene-Gima (2002), 'The adoption of agency business activity, product innovation, and performance in Chinese technology ventures', *Strategic Management Journal*, **23**, 469–90.

Li, J., G. Qian, K. Lam and D. Wang (2000), 'Breaking into China – Strategic considerations for multinational corporations', *Long Range Planning*, **33**, 673–87.

Li, M. and R.L. Simerly (1998), 'The moderating effect of environmental dynamism on the ownership and performance relationship', *Strategic Management Journal*, **19**, 169–79.

Li, M. (2001), 'Firm internationalization and economic performance: a conceptual synthesis and an empirical assessment of the Chinese experience', *Advances in International Marketing*, **11**, 179–96.

Li, M. and Y.Y. Wong (2003), 'Diversification and economic performance: an empirical assessment of Chinese firms', *Asia Pacific Journal of Management*, **20**, 243–65.

Lu, Q. (2000), *China's Leap into the Information Age: Innovation and organization in the computer industry*, Oxford: Oxford University Press.

March, J.G. (1991), 'Exploration and exploitation in organizational learning', *Organization Science*, **2**, 71–87.

Meyer, M.W. and X. Lu (2005), 'Managing indefinite boundaries: the strategy and structure of a Chinese business firm', *Management Organization Review*, **1**, 57–86.

Nonaka, I., R. Toyama and N. Konno (2000), 'SECI, Ba and leadership: a unified model of dynamic knowledge creation', *Long Range Planning*, **33**, 5–34.

Parkhe, A. (1993), '"Messy" research, methodological predispositions, and theory development in international joint ventures', *Academy of Management Review*, **18**, 227–68.

Peng, M.W. (2000), *Business Strategies in Transition Economies*, Thousand Oaks, CA: Sage Publications.

Peng, M.W. (2001), 'How entrepreneurs create wealth in transition economies', *The Academy of Management Executive*, **15**, 95–108.
Porter, M.E. (1985), *Competitive Advantage. Creating and sustaining superior performance*, New York: The Free Press.
Porter, M.E. (1990), *The Competitive Advantage of Nations*, New York: The Free Press.
Porter, M.E. (1998), 'Clusters and the new economics of competition', *Harvard Business Review*, **76** (6), 77–90.
Redding, S.G. (1993), *The Spirit of Chinese Capitalism*, New York: W. de Gruyter.
Redding, S.G. (1995), 'Overseas Chinese networks: understanding the enigma', *Long Range Planning*, **28**, 61–9.
Roth, K. and T. Kostova (2003), *'The use of the multinational corporation as a research context'*, *Journal of Management*, **29**, 883–902.
Sandvig, J.C. and L. Colakley (1998), 'Best practices in small firm diversification', *Business Horizons*, **41** (3), 33–40.
Saxenian, A. (2002), 'Brain circulation', *The Brookings Review*, **20** (1), 28–31.
Silin, R.H. (1976), *Leadership and Values: The organization of large-scale Taiwanese enterprises*, Cambridge, MA: Harvard University Press.
Simerly, R.L. and M. Li (2000), 'Environmental dynamism, capital structure and firm performance: a conceptual integration and an empirical test', *Strategic Management Journal*, **21**, 31–49.
Stacey, R.D. (1996), 'Management and the science of complexity: if organizational life is nonlinear, can business strategies prevail?', *Research–Technology Management*, **39** (3), 8–10.
Stinchcombe, A.L. (1965), 'Social structure and organizations', in J.G. March (ed.), *Handbook of Organizations*, Chicago, IL: Rand McNally, pp. 153–93.
The Chinese Culture Connection (1987), 'Chinese values and the search for culture-free dimensions of culture', *Journal of Cross-Cultural Psychology*, **18**, 143–64.
Thompson, J.D. (1967), *Organizations in Action*, New York: McGraw-Hill.
Trompenaars, A. and C. Hampden-Turner (1998), *Riding the Waves of Culture: Understanding cultural diversity in global business*, 2nd edn, New York: McGraw-Hill.
Ye, R., R.Yang and M. Meng (2000), *Building Digital China*, Beijing: The Publishing House of the School of the Central Committee of the CCP.
Yeung, H.W.-C. (2000), 'Limits to the growth of family-owned business? The case of Chinese transnational corporations from Hong Kong', *Family Business Review*, **13**, 55–70.
Zeng, J. (2001), *Introduction to the Development of the Telecommunication Industry*, Beijing: The Beijing Postal and Telecommunications University Press.
Zeng, S. (2005), *The Chinese Style Management* (in Chinese), Beijing, China: China Social Sciences Press.
Zhao, L. and J.D. Aram (1995), 'Networking and growth of young technology-intensive ventures in China', *Journal of Business Venturing*, **10**, 349–70.
Zhu, Y. and M. Warner (2004), 'Changing patterns of human resource management in contemporary China: WTO accession and enterprise responses', *Industrial Relations Journal*, **35**, 311–28.

7. Entrepreneurial strategy making and performance in China's new technology ventures: the contingency effect of environments and firm competences*

Haiyang Li, Yan Zhang and Tsang-Sing Chan

INTRODUCTION

It is widely recognized that today's managers face an environment characterized by rapid technological changes and customer demand changes. To cope with such challenges, both academic scholars and the popular press have suggested that entrepreneurial strategy making will be critical for organizational success (*Business Week*, 2002; Covin and Slevin, 1989; Dess, Lumpkin and Covin, 1997). Entrepreneurial strategy making, also referred to as 'entrepreneurial orientation' (Lumpkin and Dess, 1996) or 'entrepreneurial posture' (Covin and Slevin, 1989), is an extension of the concept of entrepreneurship from the individual level to the organizational level (Lee, Lee and Pennings, 2001). It reflects the extent to which a firm is committed to risk taking, innovation and proactiveness in developing and implementing its strategies (Miller, 1983).

Despite its appealing nature, the extant literature has consistently revealed that the proposed positive relationship between entrepreneurial strategy making and performance does not exist (Covin and Slevin, 1989; Dess *et al.*, 1997) or at best is very weak (Lee *et al.*, 2001). Instead, scholars have found that this relationship is contingent on contextual variables including environmental hostility (Covin and Slevin, 1989), firm performance (Covin and Slevin, 1991), and competitive strategies (Dess *et al.*, 1997). To extend this line of research, in this study we draw upon the

* Reprinted from *Journal of High Technology Management Research*, **16** (1), H. Li, Y. Zhang and T-S. Chan, 'Entrepreneurial strategy making and performance in China's new technology ventures – the contingency effect of environments and firm competences', pp. 37–57, © 2005, with permission from Elsevier.

contingency theory (Burns and Stalker, 1961; Lawrence and Lorsch, 1967; Venkatraman and Prescott, 1990) and test the contingency value of entrepreneurial strategy making by examining both external environments (that is, environmental uncertainty and industry growth) and internal firm competences (that is, technology and marketing competences) as contingency variables. This represents a response to the call by Lumpkin and Dess (1996) which suggests examining both environmental and organizational factors as contextual variables in studying the performance implication of entrepreneurial strategy making. Our key argument is that, as entrepreneurial strategy making reflects a firm's risk taking, innovation and proactiveness (Miller, 1983), its value to firm performance should depend on the extent to which external environments provide potential opportunities for the firm to explore via this type of strategy making, as well as the extent to which internal competences enable the firm to capture opportunities when implementing this type of strategy making.

In addition, previous studies in this area are mainly limited to firms operating in Western developed markets. There is a paucity of research on the contingency value of entrepreneurial strategy making in other institutional settings. In this study, we will examine this issue in the context of new technology ventures in China, the largest and the fastest growing emerging economy in the world. Emerging economies refer to 'low-income, rapid growth countries using economic liberalization as their primary engine of growth' (Hoskisson, Eden, Lau and Wright, 2000, p. 249). China's emerging economy involves significant institutional transitions and is characterized by highly uncertain and dynamic markets. Reforms in the direction of liberalization and marketization in China introduce fundamental and comprehensive changes to 'the formal and informal rules of the game that affect organizations as players' (Peng, 2003, p. 275), which encourage entrepreneurial behaviour. For example, during China's economic transition, over 20 000 technology-based new ventures have been founded across the nation in the past 15 years. Although it can be argued that the establishment of new ventures itself is entrepreneurial, new ventures differ in their levels of entrepreneurial strategy making (Lee *et al.*, 2001). More important, how new ventures can benefit from entrepreneurial strategy making is an important yet unexplored issue. Compared with established firms in China, new ventures have little heritage from the traditional planning system. Instead, they were founded after China's economic transition started about two decades ago. In addition, these ventures are relatively more vulnerable to environmental conditions and they tend to have relatively limited resources including both marketing and technological competences (Li and Atuahene-Gima, 2001; Stinchcombe, 1965). Also it has been argued that the complex and regulatory nature of China's transition environment is characterized by discouraging future-oriented and risk-related decisions (Luo and Park, 2001; Tan and Litschert, 1994). Therefore, technology-based new ventures provide a unique context to test the contingency value of entrepreneurial strategy making in China's emerging economy.

The remainder of this chapter is organized as the follows. First, after reviewing the literature on the relationship between entrepreneurial strategy making and firm performance, we develop our theory and research hypotheses. Then we describe the method used to test these hypotheses. Finally, we present the results of the study and discuss their implications.

THEORETICAL BACKGROUND AND RESEARCH HYPOTHESES

Entrepreneurial Strategy Making and Firm Performance

Strategy making is an organization-level process that encompasses the range of activities firms engage in to formulate and enact their strategic missions and goals. These activities include analysis, planning, decision making and strategy implementation (Hart, 1992). Prior studies have suggested that there are different types of strategy-making postures or modes. For example, Miller (1983) identified three strategy making postures including rationality, interaction and assertiveness. Hart (1992) proposed an integrative framework of strategy-making processes that includes five distinctive modes of strategy making: command, symbolic, rational, transactive and generative.

For a long time, entrepreneurial strategy making has been explicitly or inexplicitly subsumed by other strategy-making postures (see, for example, Bourgeois and Brodwin, 1984; Hart, 1992). One exception is Mintzberg's (1973) study which suggests that entrepreneurial mode (referring to opportunity seeking, risk taking and decisive action catalysed by a strong leader) is an independent strategy-making posture. More recently, Dess *et al.* (1997) empirically demonstrate that entrepreneurial strategy making is an independent posture for strategy making that an organization may exhibit. Entrepreneurial strategy making captures the organizational processes, methods and styles of firms to develop and implement their strategic decisions. It is consistent with 'the type of organization-wide entrepreneurial strategy-making processes described in previous research' (Dess *et al.*, 1997, p. 677). For example, Miller and Friesen (1978) posited that entrepreneurial firms are adaptive and innovative. Miles and Snow (1978) suggested that prospectors are risk taking.

Miller (1983, p. 771) has defined an entrepreneurial firm as one that 'engages in product market innovation, undertakes somewhat risky ventures, and is *first* to come up with proactive innovations, beating competitors to the punch' (original emphasis). Thus, conceptually, there are three components of entrepreneurial strategy making: innovativeness, risk-taking propensity and proactiveness. These components have been validated and widely used in the literature (Covin and Slevin, 1989; Lee *et al.*, 2001; Lumpkin and Dess, 1996). Specifically, the component of innovativeness taps a firm's propensity to engage in novelty, experimentation and R&D activities that may result in new products or technological processes (Lumpkin and Dess, 1996). Through innovations, firms differentiate themselves from others and achieve competitive advantages. The component of risk-taking propensity refers to 'the degree to which managers are willing to make large and risky resource commitments – i.e., those which have a reasonable chance of costly failures' (Miller and Friesen, 1978, p. 923). Entrepreneurial firms tend to conduct risk taking behaviours (for example, making large resource commitments) in the interest of obtaining high returns by seizing opportunities in the marketplace (Lumpkin and Dess, 1996). The component of proactiveness refers to a firm's propensity for anticipating and acting on future needs by seeking new opportunities (Lumpkin and Dess, 1996; Miller and Friesen, 1978). Proactiveness is a crucial organizational process because it entails a forward-looking perspective.

Entrepreneurial strategy making has been argued to be positively related to firm performance because it can help firms to gain from first-mover advantages and capitalize on emerging opportunities (Covin and Slevin, 1991; Zahra and Covin, 1995). For example, Zahra and Covin (1995) argue that firms with entrepreneurial strategy making can target premium market segments, charge high prices and skim the market ahead of their competitors. These firms monitor market changes and respond quickly, thus taking advantage of emerging opportunities. Lee, Lee and Pennings (2001, p. 617) even claim that entrepreneurial posture can be regarded as an organizational resource that provides sustainable competitive advantages because it is embedded in organizational routines, is intangible and is dispersed among organization members.

Despite its appealing nature, researchers have questioned the effectiveness of entrepreneurial strategy making across different conditions. For example, Covin and Slevin (1989) did not find a significant direct relationship between entrepreneurial strategy making and performance of small firms. Instead, they found that entrepreneurial strategy making can lead to higher financial performance in hostile environments. They argued that entrepreneurial strategy making in hostile environments can help the firm to gain or maintain competitive advantages aggressively. In another study,

Dess *et al.* (1997) investigated how the impact of entrepreneurial strategy making on firm performance is contingent on competitive strategies and found some evidence that cost leadership and entrepreneurial strategy making interact to have a positive impact on the overall company performance. In brief, the extant literature indicates that the impact of entrepreneurial strategy making on performance is not direct but contingent. It is suggested that an essential key to understanding the effectiveness of entrepreneurial strategy making is 'to analyze the context in which it occurs' (Dess *et al.*, 1997, p. 691). This suggestion is consistent with the contingency theory, which posits that the congruence or fit among key variables, such as environment, resources and strategy, is critical for obtaining optimal performance (Burns and Stalker, 1961; Venkatraman and Prescott, 1990). In the present study, we will examine both external environments and internal firm competences as the contingency variables within the context of China's new technology ventures.

Entrepreneurial Strategy Making in China's New Technology Ventures

China started its economic transition following its economic reform in 1979. Before economic reform, China's economy was governed by a centrally planned system in which market competition was absent, industries were protected, and strategizing was not necessary. Since its economic reform, China's economy has been switching from a centrally planned system to a market-oriented system. China's economic transition brings tremendous opportunities for firms. According to Nee (1989, pp. 666–7), with the emergence of the free market in China, administrative redistributors lose power and firms have increasing discretion over the terms of exchange of their goods and services. More importantly, markets provide powerful incentives for firms because these firms can retain a great share of the surplus through market transactions. Reforms in the direction of marketization result in the change of opportunity structures centred on the marketplace, thus giving rise to entrepreneurship. Consequently, the market-oriented economic transition has greatly stimulated industry growth and created opportunities attractive not only to Chinese domestic firms but also to foreign multinational corporations.

However, economic transition is a process and it takes a long time to be completed. When the traditional planning system has been substantially weakened while the new market system has not been well developed, the market environments become complex and uncertain (Luo and Park, 2001). For example, Shenkar (1990) argued that the environments in China are characterized by ambiguous property rights, incomplete transmission of information, and uncertain and inconsistent government policies. Also, given poorly developed financial markets and weak institutions for distribution of capital, capital has generally low availability and high cost in China (Li and Atuahene-Gima, 2002).

For new technology ventures, entrepreneurial strategy making tends to offer innovative products and compete by being the first market entrant and by experimenting with emerging trends and technologies. It helps the ventures to monitor market changes and respond quickly in order to take advantage of emerging opportunities. However, entrepreneurial strategy making also bears high costs and risks to respond effectively to institutional and economic changes in the Chinese markets. When the market environments become more uncertain and complex, new technology ventures may become hesitant to commit heavy resources to conducting product innovations or initiating highly risky R&D projects. Also entrepreneurial strategy making involves risk taking and getting into new product markets, which could result in a high level of managerial inefficiency by underutilizing or misutilizing physical, financial and human resources (Luo and Park, 2001). Note that new technology ventures have resource limitations (Stinchcombe, 1965). Thus entrepreneurial strategy making may be unnecessarily costly and risky for these ventures in China.

Given the costs and benefits of entrepreneurial strategy making, its value for firm performance in China's emerging economy depends on whether the ventures can perceive opportunities in the environment and whether they are able to capture the perceived opportunities, based on their resources and capabilities. In the following sections, we will examine how external environments and internal firm competences in China's emerging economy can moderate the relationship between entrepreneurial strategy making and firm performance, and develop our research hypotheses.

The Moderating role of External Environments

Environmental factors as important contingency variables have been widely acknowledged in the literature. New ventures manage to mitigate the effects of external forces in order to maintain or enhance their performance. They collect information on external environments and interpret the information prior to making strategic decisions. Prescott (1986) examined the role of environments as moderators of the relationship between strategy and performance. He suggested that the strategy variables were chosen as predictors because the field of strategic management focuses on managerial choices within environmental contexts. In examining strategy making in small firms, Covin and Slevin (1989), for example, found that entrepreneurial strategy making has a differential impact on firm performance in hostile as opposed to benign environments. In this chapter, we examine two important dimensions of environments in China's emerging economy: environmental uncertainty and industry growth.

Environmental uncertainty refers to the degree of change and unpredictability of a market environment (Miller, 1983). In uncertain environments, which change rapidly, managers have great difficulties in getting access to the information they need, and they have difficulties in understanding, believing and acting upon information collected. It could be argued that, in highly uncertain environments, an entrepreneurial attitude may lead to firms achieving a very poor performance. The opposite may also be true. In highly uncertain environments, bases for competitive advantage, industry structure and product performance standards are generally short-lived or in a constant state of flux. Thus new ventures can achieve product market superiority through competitive aggressiveness which distances the ventures from their industry rivals (Dess *et al.*, 1997). Indeed, McGrath and MacMillan (2000, p. 1) claim that uncertainty can be used to a firm's benefit if the firm creates and employs an entrepreneurial mindset as is the case of new ventures in this study. Empirically, Khandwalla (1987) found that firms competing in dynamic environments can successfully cope with uncertain conditions through risk taking, innovative behaviour, and proactive strategies.

The environments faced by Chinese new ventures become dynamic and uncertain through China's economic transition. With market-oriented reform, market power and competition play an increasingly important role in the whole economic system. This requires new ventures to learn quickly and be aggressive in market competition. Also, through economic reform, customer needs and preferences become more sophisticated and unpredictable. In such environments, passive or non-agressive behaviours often lead to deteriorating performance because bases for competitive advantage and product performance standards are in a constant state of flux (Karagozoglu and Brown, 1988). Instead, new ventures need to be aggressive and risk taking to achieve their competitive advantages in the sophisticated and unpredictable environments. Thus we propose the following hypothesis.

Hypothesis 1: Environmental uncertainty will moderate the relationship between entrepreneurial strategy making and firm performance in China. In a highly uncertain environment, new ventures with a stronger emphasis on entrepreneurial strategy making will have higher performance.

Industry growth refers to the growth rates of the industry in which a firm competes and it captures the richness of opportunities in the industry (McDougall *et al.*, 1994; Porter, 1980). In the literature, industry growth has been used to reflect the level of environmental munificence (Lumpkin and Dess, 1996). Porter (1980), for example, suggests that high industry growth means that firms entering a particular industry would not provoke strong retaliation by incumbent firms. Peng and Luo (2000) argue that a fast-going industry might offer most of its firms an opportunity to expand.

Although high-growth industries offer opportunities, the window to exploit these opportunities is typically limited because, as new entrants are attracted to these industries, increasing competition will soon reduce excess profits (Porter, 1980). Thus firms' entrepreneurial strategy making will largely determine how much of the 'pie', and how fast, they can capture in competing with their rivals. Particularly in China, its economic growth and potential market opportunities have attracted many foreign firms to get into this market. China has become the largest recipient of foreign direct investment (FDI) in the world and its annual inflows of FDI increased from US$1.91 billion in 1983 to US$41 billion in 2000 (UNCTAD, 2001). It appears that new ventures with a high level of entrepreneurial strategy making should have a better chance to capture growing opportunities in high-growth industries through developing new products and proactively searching for new business opportunities. In addition, Lumpkin and Dess (1996, p. 158) argue that firms competing in munificent environments will generate additional slack resources because of a relatively higher level of profits. Such slack resources can be used to facilitate experimentation with new strategies and practices, thus enhancing the effectiveness of entrepreneurial strategy making. Nonetheless, we propose the following.

Hypothesis 2: Industry growth will moderate the relationship between entrepreneurial strategy making and performance: in a high-growth industry, new ventures with stronger entrepreneurial strategy making will have higher performance.

The Moderating Role of Internal Firm Competences

The effectiveness of entrepreneurial strategy making will also depend on internal firm competences. The reason is that entrepreneurial strategy making can be inferred from the firm's willingness to incur large resource commitments to uncertain and novel business (Lumpkin and Dess, 1996). Thus it is highly risky and has strong resource requirements. This conjecture is consistent with the resource-based view, which argues that firm resources and capabilities are critical for the effectiveness of strategic management and firm performance (Barney, 1991).

As discussed earlier, entrepreneurial strategy making includes three components: risk taking, innovativeness and proactiveness. We argue that new ventures which engage in endeavours associated with these components require the support of both technological competences and marketing competences. Technological competences refer to a venture's ability to develop substantial technological resources and use them in the development of new products (Moorman and Slotegraaf, 1999). Marketing competences indicate a venture's abilities in organization-wide information generation and dissemination and an appropriate response related to current and future customer needs and competitive situations (Moorman and Slotegraaf, 1999). These competences can provide a resource basis for new ventures to implement entrepreneurial strategic decisions successfully. This is particularly true for new ventures in China's emerging market. Compared to their counterparts in Western economies, new ventures in China's emerging economy have relatively weak technological competences and market competences (Hitt *et al.*, 2000). Thus new ventures which have strong technological competences and marketing competences are more likely to implement entrepreneurial strategy making successfully and achieve better performance.

To benefit from risk taking, innovativeness and proactiveness in strategy making, new ventures need strong technological competences. As this type of strategy making enables new ventures to identify new business opportunities, technological competences are essential for the ventures to capture these opportunities by developing high-quality new products and/or pioneering technologies. If a venture is equipped with strong technological competences and adopts entrepreneurial strategy making, it can achieve first-mover advantage in competition. More importantly, this first-mover advantage is not easy for competitors to mimic. The reason is that technological competences are tacit and enjoy a tight appropriability regime under which firms are almost assured of translating their innovation into market value for some period of time (Teece, 1986). In contrast, without strong technological competences, a venture's first-mover advantages gained from entrepreneurial strategy making is not sustainable. Competitors are very likely to come up with superior technologies and products to capture new business opportunities identified by the focal firm. In this sense, without strong technological competences, new ventures will largely take the risks associated with entrepreneurial strategy making, but have little chance of taking advantage of the benefits associated with it.

Marketing competences are also important for new ventures to benefit from entrepreneurial strategy making. As noted earlier, entrepreneurial strategy making involves monitoring market changes, responding quickly and taking advantage of emerging opportunities. Strong marketing competences enable the ventures to better analyse and understand unmet customer needs so they can forge new market segments. Strong marketing competences also enable the ventures to avoid commitments in wrong directions. Thus, with strong marketing competences, new ventures can identify the 'right' business opportunities emerging in the process of entrepreneurial strategy making. In addition, strong marketing competences are also important for successfully commercializing new products in the market to capture new business opportunities. For example, Royal Crown Cola developed Diet Rite, the first diet cola. However, Coca-Cola and PepsiCo took over the market because of their strong marketing competencies.

Prior research has noted that there exists a technology gap between firms in emerging markets and firms in developed markets (Hitt *et al.*, 2000). Firms in China's emerging economy have relatively less sophisticated technologies to compete with their counterparts in the West. Also, in emerging economy firms, marketing competences are often not well developed. Constrained for a long time by the legacy of Marxist ideology and government plans, Chinese firms usually have little marketing experience and have little incentive to understand and satisfy customer needs. They also have little exposure to modern marketing concepts, techniques and processes. When technological competencies and market competences are relatively lacking in Chinese firms, those which have developed these competences are more likely to build up competitive advantage. These competencies may enhance the firms' ability to benefit from their entrepreneurial strategy making. On the basis of this discussion, we propose the following hypotheses.

Hypothesis 3: Technological competences will moderate the relationship between entrepreneurial strategy making and firm performance: in new ventures with great technological competences, stronger entrepreneurial strategy making will be associated with higher firm performance.

Hypothesis 4: Marketing competences will moderate the relationship between entrepreneurial strategy making and firm performance: in new ventures with great marketing competences, stronger entrepreneurial strategy making will be associated with higher firm performance.

METHODOLOGY

Sample and Data Collection

We gathered the data from the population of new technology ventures in Beijing High Technology Experimental Zone (BHEZ) in China. BHEZ is one of the most developed high technology industry zones in China. The core area of the Zone is Zhongguancun in Haidian district. There are

68 universities and 213 research institutes within the zone. Over the past decade, this zone has achieved significant growth in terms of number of ventures and industry sales. In 2000, for example, revenue in the zone was US$18 billion with industrial production of US$10.8 billion. Consistent with the accepted definition of a new venture, all sampled firms were eight years old or younger (see McDougall *et al.*, 1994). We drew on a sample of 300 firms from the list of firms compiled by the BHEZ Office. We sent a letter to the general managers of these ventures, explaining the purpose of the study and inviting their participation.

Prior research has noted that in most emerging economies such as China the lack of reliable archival data and inadequate postal systems make the use of archival and mail survey research methods difficult (Li and Atuahene-Gima, 2001). We collected the data by using an on-site structured interview, whereby a trained interviewer scheduled appointments, presented the key informants with a survey questionnaire, answered general questions and collected the completed questionnaire. This approach appears to be the key to gaining access to the right respondents, to ensure the correct use and understanding of the terms, and to gain a better response. We obtained data from 184 firms for an effective participation rate of 61.3 per cent (184/300). The average age of these new ventures was 4.83 years (s.d. = 2.3 years). We found no statistically significant differences in terms of venture size and age between the responding and non-responding ventures.

Measurement and Validation

Pre-existing measures were identified where possible and adapted on the basis of the nature of the phenomena under study. All the multi-item constructs were measured, based on a 5-point Likert scale (that is, 1 for a low score and 5 for a high score). We averaged the items to create the scores for these constructs. Measuring performance of new ventures has been a challenging task for entrepreneurship scholars. For the purpose of this study, we chose to use self-report performance measures, for several reasons. First, objective financial performance measures were unavailable for most of our sample because these ventures kept their objective financial data confidential and they were reluctant to divulge this information. As McDougall *et al.* (1994) note, however, more complete performance information could be obtained using subjective measures. Second, financial performance figures are often very biased for new technology ventures since they may devote their available resources to R&D and marketing for growth goals rather than for profits (Balkin and Gomez-Mejia, 1987). Third, because of the cross-sectional nature of the sample, objective performance measures of the firms are unavailable for comparisons. Finally, it has been indicated that managerial self-reportings have a strong correlation with internally objective performance measures (Dess and Robinson, 1984). New venture studies have provided substantial evidence supporting the reliability and validity of self-reported performance measures (Li and Atuahene-Gima, 2001). In this study, we used multiple measures to assess *new venture performance*. We asked the respondents to rate how successfully their ventures have achieved the following goals over the past three years, or since its inception if the venture was less than three years old: return on investment, return on sales, profit growth, return on assets, and overall efficiency of operations sales growth, market share growth, cash flow from market operations and firm's overall reputation (1 = very unsuccessfully and 5 = very successfully). The Cronbach α for this variable is 0.88.

Following Miller (1983) and Covin and Slevin (1991), we used six items to measure the extent to which the firms conducted *entrepreneurial strategy making* (1 = not at all and 5 = to a great extent). Two items measured each of the three components: innovativeness, risk-taking propensity and proactiveness. The two items measuring innovativeness tap the extent to which the venture favours experimentation and original approaches to problem solving and designs its own unique new processes and methods of production. The two items measuring risk-taking propensity tap the extent to which the venture was risk taking in decision making and the venture had a strong proclivity towards high-risk projects with chances of very high returns. The two items measuring proactiveness tap the extent to which the venture is quick to seize opportunity and tries to be the first mover in the target market. Results of factor analysis show that these six items load on a single factor (one item for proactiveness was deleted because of low reliability) ($\alpha = 0.75$).

Environmental uncertainty ($\alpha = 0.60$) was measured by four items adapted from Miller (1983). We asked the informants to indicate their agreement with each of the four statements as it applies to the venture's principle industry: (1) the competitive intensity has been very high and uncertain; (2) severe price competition has been a characteristic of my industry; (3) our firm must change its marketing practices frequently to keep up with the market and competitors, and (4) the rate at which products or services become obsolete has dramatically increased (1 = strongly disagree and 5 = strongly agree). *Industry growth* ($\alpha = 0.75$) was measured by three items drawn from McDougall *et al.* (1994). The items include the following: (1) there is high growth in demand in this industry; (2) this industry offers many attractive opportunities for future growth; and (3) growth opportunities in this industry are abundant (1 = strongly disagree and 5 = strongly agree). The use of managerial perceptions of environment has been supported by a number of studies based on the relevance of such perceptions to the formulation of strategy (Downey, Hellriegel and Slocum,

1975), as well as their accuracy with respect to objective measures of environmental conditions (Dess and Robinson, 1984).

Technological competence ($\alpha = 0.82$) was measured by four items newly developed for this study. These items include (1) our venture uses sophisticated technologies in new product development; (2) our new products are always at the state of the art of the technology; (3) our firm has invested a large amount in R&D; and (4) our R&D team are competent in innovation development (1 = strongly disagree and 5 = strongly agree). *Marketing competence* ($\alpha = 0.80$) was measured by six items newly developed for this study: (1) we are able to monitor constantly our level of commitment and orientation to serving customer current and future needs; (2) we have routine or regular measures of the changes in customer demand and competitor behaviour; (3) our advertising and promotion resources are more than adequate; (4) we use sophisticated techniques to do market research; (5) we have a very strong marketing team in this industry; and (6) we poll end-users frequently to assess the quality of our products and services (1 = strongly disagree and 5 = strongly agree).

We controlled for several factors in order to account for alternative explanations, including venture size, venture age, venture ownership and industry type (Li and Atuahene-Gima, 2001; Zahra, 1991). Venture *size* was measured by the natural log of the number of full-time employees; Venture *age* was measured by the number of years the new venture has been in existence (8 years or less); Venture *ownership* was measured by asking the respondents to indicate the nature of the current ownership of the venture. There were four different types of ownership: state/collectively-owned, joint share, privately owned and foreign invested firms (including international joint ventures). Accordingly, we created three ownership dummy variables by using foreign invested ventures as the base group. *Industry type* was also controlled for. Following the standard of industry classification adopted by the Administration Commission of the Park, we classified the industries of these ventures into five types: electronic information industry, new energy and new material industry, new pharmaceutical and biotechnology industry, integrated optical industry and others (for example, scientific instruments and aerospace). We created four industry dummy variables by using the others type as the base group.

Our reliance on perceptual measures for both dependent and independent variables raises a legitimate concern about the potential problem of common method variance. In addition to the procedures mentioned previously, we checked this problem further with Harman's one-factor test, as suggested by Podsakoff and Organ (1986). To test for this potential threat to validity, we subjected all the measures to a factor analysis. If common method variance was a serious problem, a single factor would have emerged from the factor analysis or one general factor would account for most of the variance. Our test yielded several factors with eigenvalues greater than one and no single factor was dominant, suggesting common method variance is not a problem in the data.

We further assessed unidimensionality, reliability and validity. First, we subjected each construct separately to principal component factor analysis. In each case, one factor emerged with an eigenvalue greater than one, which provides support for the unidimensionality of the scales. Second, we calculated the Cronbach alpha for each construct. The constructs have high reliability, with all but one having alphas over 0.70. Third, to examine the convergent and discriminant validity of the measures, we conducted confirmatory factor analysis. Given the sample size restrictions, we divided the constructs into three sub-models of theoretically related groups (c.f. Bentler and Chou, 1987): new venture performance and entrepreneurial strategy making ($\chi^2 = 117.21$, $p = 0.00$; GFI = 0.91, CFI = 0.95, NNFI = 0.94, RMSEA = 0.07); environmental uncertainty and industry growth ($\chi^2 = 12.94$, $p = 0.11$; GFI = 0.98, CFI = 0.97, NNFI = 0.94, RMSEA = 0.05); technological competence and marketing competences ($\chi^2 = 79.44$, $p = 0.00$; GFI = 0.93, CFI = 0.98, NNFI = 0.97, RMSEA = 0.06). All items loaded on their respective constructs, and each loading was large and significant at the 0.01 level (see the Appendix). These results indicate convergent and discriminant validity of the scales.

DATA ANALYSES AND RESULTS

Table 7.1 provides means, standard deviations and correlations between variables examined in this study. We used hierarchically moderated regression analysis to test the hypotheses. To reduce the potential problem of multicollinearity, both independent and moderator variables were mean-centred prior to the creation of interaction terms (Aiken and West, 1991). We computed the variance inflation factor (VIF) for all variables in the study. Results suggested no need for concern with respect to multicollinearity.

Regression results are reported in Table 7.2. Model 1 in Table 7.2 included control variables, moderating variables and entrepreneurial strategy making. In Model 2, the interaction terms between entrepreneurial strategy making and environmental conditions were added to Model 1, which explained 2 per cent more variance of new venture performance ($\Delta F = 2.39$, $p < 0.10$). In Model 3, the interaction terms between entrepreneurial strategy making and firm competences were added to Model 1, explaining 3 per cent more variance of new venture performance ($\Delta F = 2.74$, $p < 0.05$). In Model 4, all the interaction variables were put together, explaining 5 per cent more variance of new venture performance relative to Model 1. These significant R-square changes suggest the existence of interaction effects (Aiken and West, 1991). For the sake of simplicity, we used results in Model 4, which has the full model specification, to test our hypotheses.

Our results show that the main effect of entrepreneurial strategy making on new venture performance is positive but statistically insignificant across all the models. Hypothesis 1 suggests that, in a highly uncertain environment, a stronger emphasis on entrepreneurial strategy making will be positively related to firm performance. This hypothesis is supported ($b=0.32$, $p<0.05$). To facilitate interpretation, we plotted the interaction effect in Figure 7.1. To create this figure, all variables in Model 3 except entrepreneurial strategy making and environmental uncertainty were constrained to the mean. The variables of entrepreneurial strategy making and environmental uncertainty took values of one standard deviation below and above mean. As shown in Figure 7.1, the relationship between entrepreneurial strategy making and new venture performance is positive when environmental uncertainty is high, but negative when it is low. However, our results do not support Hypothesis 2, which proposes that, in high-growth industries, stronger entrepreneurial strategy making will be positively related to firm performance ($b=0.12$, n.s.).

Hypothesis 3 regarding the moderating role of technological competences is not supported ($b=-0.01$, n.s.). Hypothesis 4 states that, in new ventures with great marketing competences, stronger entrepreneurial strategy making will be associated with higher firm performance. Our results support this hypothesis ($b=0.60$, $p<0.05$). As shown in Figure 7.2, the relationship between entrepreneurial strategy making and new venture performance is positive when marketing competence is high, but it is negative when marketing competence is low.

Among control variables, we found that state/collectively-owned ventures and share-owned ventures have a lower performance than foreign-invested ventures (including international joint ventures) ($b=-0.55$, $p<-0.001$ and $b=-0.26$, $p<0.10$, respectively). Consistent with Li and Atuahene-Gima's (2002) findings, it appears that new ventures which can gain support from foreign investors can perform better. In addition, industry growth ($b=0.12$, $p<0.10$), technological competence ($b=0.21$, $p<0.05$) and marketing competence ($b=0.21$, $p<0.05$) are all positively related to new venture performance. As theoretically expected, when new ventures are in a growth industry with greater opportunities and when the ventures have stronger technological competence and marketing competence, they are more likely to build up competitive advantages and achieve better firm performance. Our results, however, show that environmental uncertainty has no significant relationship with new venture performance. Considering the

Table 7.1 Means, standard deviations and correlations

Variables	Mean	S.D.	1	2	3	4	5	6	7	8	9	10	11	12	13	14
1. New venture performance	3.53	0.62	—													
2. Entre-preneurial strategy making	3.13	0.39	0.24**	—												
3. Environ-mental uncertainty	3.88	0.68	−0.15	−0.12	—											
4. Industry growth	3.85	0.76	0.47**	0.21**	0.06	—										
5. Technological competence	3.56	0.72	0.40**	0.32**	0.06	0.42**	—									
6. Marketing competence	3.99	0.68	0.45**	0.22**	−0.10	0.32**	0.44**	—								
7. Venture size	160.98	200.16	0.04	0.10	−0.02	0.05	0.05	−0.09								
8. Venture age	4.83	2.03	−0.01	0.02	0.10	0.10	−0.03	−0.06	0.25**	—						
9. State/ collective ownership	0.30	0.45	−0.27**	0.01	−0.07	−0.09	−0.08	−0.07	0.10	0.18*	—					
10. Share ownership	0.30	0.46	0.12	0.06	0.07	0.12	0.20**	0.08	−0.06	−0.05	−0.43**	—				
11. Private ownership	0.18	0.39	−0.03	−0.09	0.06	−0.06	−0.09	−0.08	−0.24**	0.04	−0.31**	−0.31**	—			

12. Electronic information industry	0.51	0.50	−0.03	−0.01	−0.02	0.02	−0.00	0.00	−0.19*	0.04	0.09	0.03	0.05	—		
13. New energy/ material industry	0.13	0.33	0.03	0.09	0.14	0.01	0.04	−0.03	0.20**	0.11	0.01	0.04	−0.05	−0.38**	—	
14. New pharma-ceutical industry	0.10	0.31	0.03	0.08	0.02	0.03	0.14	0.09	0.08	−0.08	−0.11	−0.03	−0.06	−0.35**	−0.13	—
15. Integrated optical industry	0.17	0.38	−0.08	−0.13	−0.02	−0.06	−0.09	−0.12	0.04	−0.05	−0.07	−0.01	0.10	−0.45**	−0.17*	−0.16

Note: Significance level: * $p < 0.05$, ** $p < 0.01$; N = 184.

Table 7.2 Results of regression analyses for entrepreneurial strategy making and new venture performance

Variables	New venture performance			
	Model 1 b	Model 2 b	Model 3 b	Model 4 b
Constant	1.18†	1.38*	1.29*	1.46*
Venture size	0.02	0.03	0.02	0.03
Venture age	0.03	0.02	0.03	0.03
State/collective ownership	−0.55***	−0.52***	−0.59***	−0.55***
Share ownership	−0.29*	−0.24†	−0.31*	−0.26†
Private ownership	−0.15	−0.09	−0.18	−0.16
Electronic information industry	−0.02	−0.06	−0.06	−0.06
New energy/material industry	−0.04	−0.05	−0.06	0.06
New pharmaceutical industry	−0.23	−0.25	−0.26	−0.26
Integrated optical industry	−0.08	−0.10	−0.07	−0.07
Environmental uncertainty	−0.07	−0.09	−0.07	−0.09
Industry growth	0.13†	0.13†	0.13†	0.12†
Technology competence	0.23**	0.21*	0.22*	0.21*
Marketing competence	0.25**	0.26**	0.22**	0.21*
Entrepreneurial strategy making (ESM)	0.11	0.08	0.13	0.12
Interaction terms				
ESM * Environmental uncertainty		0.26*		0.32*
ESM * Industry growth		0.21		0.12
ESM * Technological competence			0.07	−0.01
ESM * Marketing competence			0.51*	0.60*
R^2	0.44	0.46	0.47	0.49
Adjusted R^2	0.37	0.38	0.39	0.40
F-value	5.84***	5.47***	5.62***	5.40***
ΔR^2 relative to Model 1		0.02	0.03	0.05
ΔF relative to Model 1		2.39†	2.74*	2.60*

Note: N = 184; significance level: † $p < 0.10$; * $p < 0.05$; ** $p < 0.01$; *** $p < 0.001$.

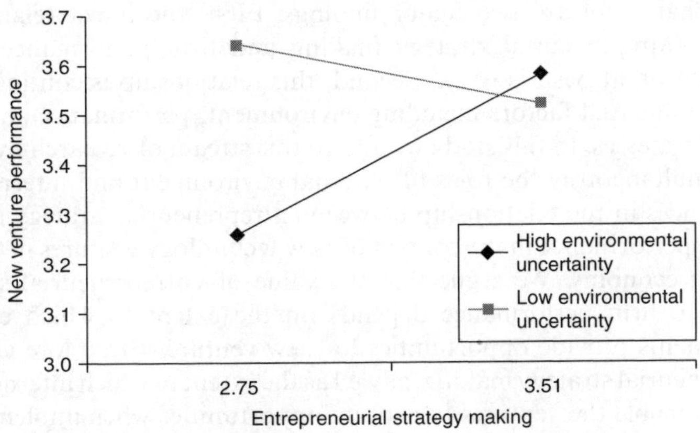

Figure 7.1 The moderating role of environmental uncertainty

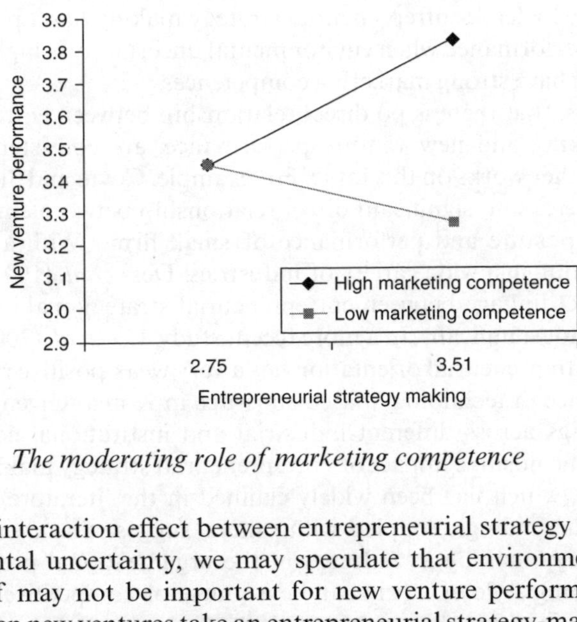

Figure 7.2 The moderating role of marketing competence

significant interaction effect between entrepreneurial strategy making and environmental uncertainty, we may speculate that environmental uncertainty itself may not be important for new venture performance, but it matters when new ventures take an entrepreneurial strategy-making model.

DISCUSSION AND CONCLUSIONS

The important role of entrepreneurial strategy making in firm performance has been extensively studied in the literature. Our review of the literature reveals that there are two major findings. First, the direct relationship between entrepreneurial strategy making and firm performance is not significant or, at best, is weak. Second, this relationship is contingent on several contextual factors including environment, performance and competitive strategies. In this study we add to this stream of research by examining simultaneously the roles of external environment and internal firm competences in the relationship between entrepreneurial strategy making and firm performance in the context of new technology ventures in China's emerging economy. We argue that the value of entrepreneurial strategy making to firm performance depends on the extent to which external environments provide opportunities for new ventures to explore via their entrepreneurial strategy making, as well as the extent to which internal competences enable the ventures to capture opportunities when implementing entrepreneurial strategy making. With a sample of technology-based new ventures in China, our results support the contingency view and suggest that entrepreneurial strategy making may not have a direct relationship to firm performance. Instead, entrepreneurial strategy making has a positive relationship to performance when environmental uncertainty is high and when new ventures have strong marketing competences.

Our results, that there is no direct relationship between entrepreneurial strategy making and new venture performance, are consistent with the findings of other works on this topic. For example, Covin and Slevin (1989) show that there is no significant direct relationship between entrepreneurial strategic posture and performance of small firms. With a sample of firms competing in a wide variety of industries, Dess *et al.* (1997) also find that the direct linkage between entrepreneurial strategy making and performance is not significant. In a more recent study, Lee *et al.* (2001) demonstrate that entrepreneurial orientation has a very weak positive relationship to performance in technology-based start-ups in Korea. Given these consistent findings across different industrial and institutional contexts, we believe that the positive impact of entrepreneurial strategy making on firm performance, which has been widely claimed in the literature, is at most marginal. Our findings, coupled with others, suggest that even entrepreneurial strategy making as reflected in the organizational processes and decision-making style of a firm can be a source of competitive advantage (Lee *et al.*, 2001; Lumpkin and Dess, 1996); its value largely depends on the context in which it occurs. This is consistent with the contingent resource-based view which posits that a firm's resources or capability to provide sustainable competitive advantage often lie in their combination with other organizational skills and capabilities and deployment in the appropriate environment (Eisenhardt and Martin, 2000; Teece, Pisano and Shuen, 1997).

Prior research has noted that, in China's emerging economy, characterized by market distortion and limited demands for new and innovative products, firms are more defensive in strategic decision making rather than risk-taking and innovative (Luo and Park, 2001; Tan and Litschert, 1994). However, our study tells a different story. Our findings imply that Chinese firms, particularly new technology ventures, can pay off with entrepreneurial strategy making. Specifically, our results show that, when the level of environmental uncertainty is perceived as high, and when the firm has strong marketing competencies, it is more likely to benefit from its entrepreneurial strategy making. Our study sends a clear message to practitioners that even if the Chinese transitional environment is characterized by discouraging risk-related decisions (Tan and Litschert, 1994), firms can benefit from entrepreneurial strategy making, depending on what they perceive from the environment and what they are able to do.

We examined the moderating role of two environmental dimensions: environmental uncertainty and industry growth. As expected, our findings indicate that entrepreneurial strategy making has a significantly positive relationship to new venture performance when the level of environmental uncertainty is high. Contrary to our hypothesis, however, we find that industry growth does not play a moderating role in this relationship. These findings suggest that environmental uncertainty and industry growth have a differential impact on the effectiveness of entrepreneurial strategy making in Chinese new ventures. It appears that high-growth industries provide tremendous opportunities for new ventures competing in the industries. In such an environment, new ventures achieve a better performance by taking the external opportunities rather than making entrepreneurial strategy. As Covin and Slevin (1989) argued, in benign environments (characterized by munificence and richness in investment and marketing opportunities), the relationship between entrepreneurial strategy making and firm performance may be weaker and possibly negative because in such environments firms 'are not typically forced to engage in uncertain, resource-consuming endeavors in order to maintain viability' (Covin and Slevin, 1989, p. 77). In contrast, environmental uncertainty provides both opportunities and threats to new ventures. The extent to which new ventures can benefit from the opportunities and attenuate the threats depends significantly on their entrepreneurial strategy making. It is likely that new ventures with a high level of entrepreneurial strategy making can take advantage of the opportunities in an uncertain environment through their innovativeness, risk taking and proactiveness, while those with a low level of entrepreneurial strategy making cannot. Using the metaphor of the theory of Darwinian evolution, most ventures (as species) can survive and succeed in 'good' environments (for example, high growth environments) while, in 'bad' environments (for example, high uncertain environments), only those who can adapt to them can survive and succeed. Clearly, entrepreneurial strategy making is an important way for new ventures to adapt to the uncertain environment and to pursue opportunities.

These findings contribute to our understanding of the role of environmental conditions in the relationship between entrepreneurial strategy making and firm performance. Covin and Slevin (1989), for example, show that environmental hostility is an important environmental factor which may moderate the linkage between entrepreneurial strategy making and firm performance. They conceptualized environmental hostility based on a hostile–benign continuum. They defined hostile environments as those 'characterized by precarious industry settings, intense competition, harsh, overwhelming business climates, and the relative lack of exploitable opportunities. Non-hostile or benign environments, on the other hand, provide a safe setting for business operations due to overall level of munificence and richness in investment and marketing opportunities' (p. 75). Clearly, this definition of hostility includes elements of both uncertainty and munificence. Our study has advanced the literature by examining environmental uncertainty and industry growth (as indicators of environmental munificence) separately. Our findings suggest that these two dimensions of environmental conditions play different roles in this relationship.

Our study extends the literature by examining internal firm competences as important contingency variables. The results suggest that entrepreneurial strategy making has a positive relationship to new venture performance when marketing competences are high. However, our results suggest that technological competences do not moderate this relationship. These results suggest that marketing competences are complementary assets for entrepreneurial strategy making. As noted earlier, marketing competences represent a venture's abilities to generate and disseminate information related to current and future customer needs and competitive situations. As new ventures have better knowledge about customer demands and competitive situations, they are more able to identify and capture new business opportunities through their innovative, risk-taking and proactive behaviours. Note that marketing resources and skills are relatively less available in China. Thus those new ventures with marketing competencies are more likely to develop a differential advantage in enabling them to carry out proficiently their entrepreneurial strategy making. These findings, therefore, highlight the importance of firm competences in enhancing the effectiveness of entrepreneurial strategy making.

The results of this study urge managers to pay special attention to the context requirements of successful entrepreneurial strategy making. Our findings suggest that the effectiveness of entrepreneurial strategy making is context-specific and depends on the level of environmental uncertainty and/or the firm's marketing competences. Therefore, when making entrepreneurial strategy decisions, managers should consider the contextual variables which may limit or enhance the effectiveness of this strategy-making mode. In other words, managers who ignore the contextual situations may suffer rather than benefit from entrepreneurial strategy making.

Limitations and Further Research

The interpretations of our findings must be viewed in light of the limitations inherent to the research. Consistent with most survey research, our results relied primarily on subjective evaluation of the environment and competence variables. Hence they are subject to issues related to common method variance, though we took several actions in data collection to improve reliability and validity of retrospective reporting. However, we believe that issues related to common method variance should not be serious in our data analysis, because there is no theoretical reason to expect an interaction with common method variance. As concluded by Evans (1985), correlated error from the use of similar methods to collect data on criterion and predictor variables cannot create spurious interactions. Nonetheless, a longitudinal design, cross-validation of the findings, and more sources of data would enable us to assess further the hypothesized relationships.

The study's cross-sectional design is also a limitation. The cross-sectional data used in this study do not allow for causal interpretations among the variables. In addition, the environmental conditions and new ventures' competences are not stable but change over time. Our results, however, do not capture the dynamics of the way in which the change in the environment and firm competences may affect the effectiveness of entrepreneurial strategy making. Future research could investigate the dynamic relationship between the environment, firm competences and entrepreneurial strategy making by using longitudinal data.

The sample of the study could be biased because of our focus on new technology ventures. Since the firms included are new, they have a more marked entrepreneurial attitude; they are technology-based firms and therefore they have a high level of technological competence. However, even in such a group of firms with a more marked entrepreneurial attitude, we still observe the contingency effects of entrepreneurial strategy making on firm performance, which makes our findings more robust. We tend to agree that the non-significant results regarding the interaction term between entrepreneurial strategy making and technological competence may be that most of the firms have a high level of technological competence. Future studies should explore how our findings can be generalized to other types of firms.

Finally, this empirical study was conducted in a specific region in China and the sample is limited to new technology ventures in BHEZ. Given the sampling criterion, the sample was heterogeneous regarding factors such as venture size, industry and ownership. Variation in the sample has the potential to increase the generalizability of the findings. Nonetheless, more research is needed to ensure that findings of this study can be extrapolated for the rest of the country. For example, currently, there are 53 high technology development zones (including the BHEZ) in China. These zones were established at different times and in different locations. Thus it will be interesting to explore how our findings may vary for new technology ventures from other zones in different locations in the future.

Lee, Lee and Pennings (2001, p. 617) have claimed that, though the positive relationship between entrepreneurial strategy making and performance has been proposed, 'the strong theoretical frame has not been combined with effective empirical work'. We have advanced the literature by focusing on new ventures' entrepreneurial strategy making in China and providing insights into the environmental and organizational factors that affect the value of entrepreneurial strategy making. We hope that our results described here will facilitate further development of this important but neglected area of understanding. In particular, as China's economy is shifting towards a market-based economy and its increasingly important role in the world, more research about entrepreneurial strategy making in this emerging market has become more significant both for theory and for practice.

REFERENCES

Aiken, L.S. and S.G. West (1991), *Multiple Regression: Testing and Interpreting Interactions*, Newbury Park, CA: Sage Publications.

Balkin, D.B. and L.R. Gomez-Mejia (1987), 'Toward a contingency theory of compensation strategy', *Strategic Management Journal*, **8** (2), 169–82.

Barney, J.B. (1991). 'Firm resources and sustained competitive advantage', *Journal of Management*, **17**, 99–120.

Bentler, P.M. and C.P. Chou (1987), 'Practical issues in structural modeling', *Sociological Methods and Research*, **16**, 78–117.

Bourgeois, L. and D. Brodwin (1984), 'Strategic implementation: five approaches to an elusive phenomenon', *Strategic Management Journal*, **5** (3), 241–64.

Burns, T. and G. Stalker (1961), *The Management of Innovation*, London: Tavistock.

Business Week (2002), Michelin, September, 23–6.

Covin, J.G. and D.P. Slevin (1989), 'Strategic management of small firms in hostile and benign environments', *Strategic Management Journal*, **10** (1), 75–87.

Covin, J.G. and D.P. Slevin (1991), 'A conceptual model of entrepreneurship as firm behavior', *Entrepreneurship Theory and Practice*, **16** (1), 7–25.

Dess, G.G. and R.B. Robinson, (1984), 'Measuring organizational performance in the absence of objective measures', *Strategic Management Journal*, **5** (3), 265–73.

Dess, G.G., G.T. Lumpkin and J.G. Covin (1997), 'Entrepreneurial strategy making and firm performance: tests of contingency and configurational models', *Strategic Management Journal*, **18** (9), 677–95.

Downey, H., D. Hellriegel and J. Slocum (1975), 'Environmental uncertainty: the construct and its application', *Administrative Science Quarterly*, **20**, 613–29.

Eisenhardt, K.M. and J.A. Martin (2000), 'Dynamic capabilities: what are they?', *Strategic Management Journal*, **21** (10/11), 1105–21.

Evans, M.G. (1985), 'A Monte Carlo study of the effects of correlated method variance in moderated multiple regression analysis', *Organizational Behavior and Human Decision Processes*, **36** (3), 305–23.

Hart. S. (1992), 'An integrative framework for strategy making processes', *Academy of Management Review*, **17**, 327–51.

Hitt, M.A., M.T. Dacin, E. Levitas, J.-L. Arregle and A. Borza (2000), 'Partner selection in emerging and developed market contexts: resource-based and organizational learning perspectives', *Academy of Management Journal*, **43**, 449–67.

Hoskisson, R.E., L. Eden, C.M. Lau and M. Wright (2000), 'Strategy in emerging economies', *Academy of Management Journal*, **43**, 249–67.

Karagozoglu, N. and W.B. Brown (1988), 'Adaptive responses by conservative and entrepreneurial firms', *Journal of Product Innovation Management*, **5**, 269–81.

Khandwalla, P.N. (1987), 'Generators of pioneering innovative management: some Indian Evidence', *Organization Studies*, **8**, 39–59.

Lawrence, P. and J. Lorsch (1967), *Organization and Environment*, Cambridge, MA: Harvard University Press.

Lee, C., K. Lee and J.M. Pennings (2001), 'Internal capabilities, external networks, and performance: a study of technology-based ventures', *Strategic Management Journal*, **22**, 615–40.

Li, H. and K. Atuahene-Gima (2001), 'Product innovation strategy and the performance of new technology ventures in China', *Academy of Management Journal*, **44** (6), 1123–34.

Li, H. and K. Atuahene-Gima (2002), 'The adoption of agency business activity, product innovation, and performance in Chinese technology ventures', *Strategic Management Journal*, **23**, 469–90.

Lumpkin, G.T. and G.G. Dess (1996), 'Clarifying the entrepreneurial orientation construct and linking it to performance', *Academy of Management Review*, **21** (1), 135–72.

Luo, Y. and S.H. Park (2001), 'Strategic alignment and performance of market-seeking MNCs in China', *Strategic Management Journal*, **22**, 141–55.

McDougall, P.P., J.G. Covin, R.B. Robinson and L. Herron (1994), 'The effects of industry growth and strategic breadth on new venture performance and strategy content', *Strategic Management Journal*, **15** (7), 537–54.

McGrath, R.G. and I. MacMillan (2000), *The Entrepreneurial Mindset*, Boston, MA: Harvard Business School Press.

Miles, R.E. and C.C. Snow (1978), *Organizational Strategy, Structure, and Process*, New York: McGraw-Hill Book Co.

Miller, D. (1983), 'The correlates of entrepreneurship in three types of firms', *Management Science*, **29** (7), 770–91.

Miller, D. and P. Friesen (1978), 'Archetypes of strategy formulation', *Management Science*, **24**, 921–33.

Mintzberg, H. (1973), 'Strategy-making in three modes', *California Management Review*, Winter, 44–53.
Moorman, C. and R.J. Slotegraaf (1999), 'The contingency value of complementary capabilities in product development', *Journal of Marketing Research*, **36**, 239–57.
Nee, V. (1989), 'A theory of market transition: from redistribution to markets in state socialism', *American Sociological Review*, **54**, 663–81.
Peng, M.W. (2003), 'Institutional transitions and strategic choices', *Academy of Management Review*, **28** (2), 275–96.
Peng, M.W. and Y. Luo (2000), 'Managerial ties and firm performance in a transition economy: the nature of a micro–macro link', *Academy of Management Journal*, **43** (3), 486–501.
Podsakoff, P.M. and D. Organ (1986), 'Self-reports in organizational research: problems and prospects', *Journal of Management*, **12**, 531–43.
Porter, M.E. (1980), *Competitive Strategy: Techniques for Analyzing Industries and Competitors*, New York: Free Press.
Prescott, J.E. (1986), 'Environments as moderators of the relationship between strategy and performance', *Academy of Management Journal*, **29** (2), 329–46.
Shenkar, O. (1990), 'International joint ventures' problems in China: risks and remedies', *Long Range Planning*, **23** (3), 82–90.
Stinchcombe, A.L. (1965), 'Organizations and social structure', in J.G. March (ed.), *Handbook of Organizations*, Chicago, IL: Rand McNally, pp. 142–93.
Tan, J.J. and R.J. Litschert (1994), 'Environment–strategy relationship and its performance implications: an empirical study of the Chinese electronics industry', *Strategic Management Journal*, **15**, 1–20.
Teece, D.J. (1986), 'Profiting from technological innovation: implications for integration, collaboration, licensing and public policy', *Research Policy*, **15**, 785–805.
Teece, D.J., G. Pisano and A.A. Shuen (1997), 'Dynamic capabilities and strategic management', *Strategic Management Journal*, **18**, 504–34.
UNCTAD, (2001), *World Investment Report*, New York: United Nations.
Venkatraman, N. and J.E. Prescott (1990), 'Environment–strategy coalignment: an empirical test of its performance implications', *Strategic Management Journal*, **11** (1), 1–23.
Zahra, S.A. (1991), 'Predictors and financial outcomes of corporate entrepreneurship: an exploratory study', *Journal of Business Venturing*, **6** (4), 259–85.
Zahra, S.A. and J.G. Covin (1995), 'Contextual influences on the corporate entrepreneurship–performance relationship: a longitudinal analysis', *Journal of Business Venturing*, **10** (1), 43–58.

APPENDIX

Table 7A.1 Construct measurement and confirmatory factor analysis by LISREL[a]

Constructs in Model 1	Standardized loading	t-value
New venture performance		
Indicate how successfully your venture has achieved these goals in the last three years.		
1. Return on investment	0.74	10.50
2. Return on sales	0.54	7.58
3. Profit growth	0.53	6.90
4. Return on assets	0.69	10.27
5. Overall efficiency of operations	0.80	11.51
6. Sales growth	0.58	8.28
7. Market share growth	0.59	8.40
8. Cash flow from market operations	0.57	7.47
9. Firm's overall reputation	0.60	9.03
Entrepreneurial strategy making		
Indicate the extent of agreement about how well the statements describe the actual norms in your venture.		
1. The top managers of my venture favour experimentation and original approaches to problem solving.	0.57	7.26
2. The top managers of my venture prefer to study a problem thoroughly before deploying resources to solve it. ®	0.54	6.62
3. My venture prefers to design its own unique new processes and methods of production.	0.59	7.52
4. My firm is quick to seize opportunities that we think will give us a good pay-off.	0.67	8.72
5. My firm has a strong proclivity for high-risk projects with chances of very high returns.**		
6. When confronting decision-making situations involving uncertainty, my firm typically adopts a cautious, 'wait-and-see' posture in order to minimize the probability of making costly decisions. ®	0.61	7.72
χ^2= 117.21, p = 0.00; GFI = 0.91, CFI = 0.95, NNFI = 0.94, RMSEA = 0.07		

Table 7A.1 (continued)

Constructs in Model 2	Standardized loading	t-value
Environmental uncertainty		
1. The competitive intensity has been very high and uncertain.	0.64	8.15
2. Severe price competition has been a characteristic of my industry.	0.43	3.61
3. Our firm must change its marketing practices frequently to keep up with the market and competitors.	0.42	4.62
4. The rate at which products or services become obsolete has dramatically increased.	0.85	10.76
Industry growth		
Rate the degree to which each of these statements describes your principal industry over the last three years:		
1. There has been high growth in demand in this industry.	0.65	8.25
2. This industry offered many attractive opportunities for future growth.	0.85	10.42
3. Growth opportunities in this industry have been abundant.	0.63	8.10
$\chi^2 = 12.94$, p = 0.11; GFI = 0.98, CFI = 0.97, NNFI = 0.94, RMSEA = 0.05		

Constructs in Model 3	Standardized loading	t-value
Technological competence		
1. Our venture uses sophisticated technologies in new product development.	0.73	10.39
2. Our new products are always at the state of the art of the technology.	0.70	9.67
3. Our firm has invested a large amount of sales in R&D.	0.78	11.36
4. Our R&D team is competent in innovation development.	0.67	9.40
Marketing competence		
1. We are able to constantly monitor our level of commitment and orientation to serving customer current and future needs.	0.57	7.43
2. We have routine or regular measures of the changes in customer demand and competitive behaviour.	0.69	8.74
3. Our advertising and promotion resources are more than adequate.	0.67	8.59
4. We use sophisticated techniques to do market research.	0.63	8.06
5. We have a very strong marketing team in this industry.	0.66	8.35
6. We poll end-users frequently to assess the quality of our products and services.	0.57	7.43
$\chi^2 = 79.44$, p = 0.00, GFI = 0.93, CFI = 0.98, NNFI = 0.97, RMSEA = 0.06		

Notes:

[a] For new ventures that were less than three years old the response time frame was since their founding.

® The item was reverse-coded.

** Items were deleted because of low item-to-total correlations.

8. Innovation capabilities and the rise of the local telecom equipment firms in China

Xudong Gao and Weiqing Gao

INTRODUCTION

The existing literature on technological catch-up examines the technology strategies of local firms in developing countries from a resource-based perspective (Barney, 1986; Dierickx and Cool, 1989; Prahalad & Hamel, 1990; Penrose, 1959; Peteraf, 1993; Teece, Pisano & Shuen, 1997; Wernerfelt, 1984), and studies local firms' technology strategies by comparing the technological resources of local firms with those of multinational enterprises (MNEs). The basic argument is that local firms should choose to compete against MNEs by developing strong manufacturing capabilities, because local firms do not have proprietary technologies, while MNEs do, and only when there are windows of opportunity should local firms try to compete against MNEs by developing innovation capabilities and proprietary technologies (Amsden, 2001; Amsden & Chu, 2003; Kim, 1997; Lall, 1982; Lee & Lim, 1998; Hobday, 1990; Westphal, Kim & Dahlman, 1985).

Although the argument that local firms should choose to compete against MNEs by developing strong manufacturing capabilities has strong empirical support, there are accumulating studies showing that local firms can compete with MNEs by developing innovation capabilities and proprietary technologies by themselves (Cusumano, 1985; Gao, 2003; Lee & Lim, 1998; Lu, 2000; Lu & Lazonick, 2001; Perez & Soete, 1988; Shen, 1999; Wang, 1999; Zhang, 2000), and it is important to explore the possibilities for local firms to compete against MNEs by developing innovation capabilities and proprietary technologies (Gao, Zhang & Liu, forthcoming). In this chapter, we explore this issue by applying two concepts: 'barriers to appropriability' and 'opportunities for improvement'.

We define 'barriers to appropriability' in terms of the difficulties MNEs have in using their superior technological resources in a host country's market. Barriers to appropriability arise when there are factors that constrain MNEs' ability to utilize fully their superior technological resources to compete with local firms. We define 'opportunities for improvement' as opportunities for local firms to improve their technological resources. Opportunities for improvement arise when there are factors helping local firms to develop innovation capabilities and proprietary technologies.

We have argued that local firms might be able to compete against MNEs by developing innovation capabilities and proprietary technologies if there are opportunities for improvement. In this study, we illustrated this argument by using case studies of Chinese telecom equipment firms.

Our study offers new insights into the analysis of local firms' technology strategies. For example, our study indicates that local firms should apply the concept of barriers to appropriability and the concept of opportunities for improvement in choosing their technology strategies. Our study also indicates that, in the era of globalization, it is hard for local firms to compete against MNEs by developing strong manufacturing capabilities, but it might be possible for local firms to compete against MNEs by developing innovation capabilities and proprietary technologies. Also our study indicates that too strong competition from MNEs could be harmful for local firms' developing innovation capabilities and proprietary technologies, and controlled competition and public ownership could play positive roles.

This chapter is organized as follows. We first introduce the concepts of barriers to appropriability and opportunities for improvement. We then discuss the possibilities and conditions for local firms to compete with MNEs by developing innovation capabilities and proprietary technologies. Third, we use cases from the telecom equipment industry to illustrate our argument that local firms might be able to compete against MNEs by developing innovation capabilities and proprietary technologies if there are opportunities for improvement. We conclude with a discussion of opportunities for additional research.

CONDITIONS FOR CHOOSING THE STRATEGY OF BUILDING INNOVATION CAPABILITIES

Barriers to Appropriability

The existing literature has long recognized that there exist 'liabilities of foreignness' (that is, firms setting up operations abroad face certain unavoidable costs that firms operating in their home environment do not) and that MNEs' superior technological resources will not necessarily lead to competitive advantage when they globalize their businesses (Buckley & Casson,

1976; Hymer, 1976; Lou, 2000; Zaheer & Mosakowski, 1997). In this chapter we introduce the concept of barriers to appropriability to further examine the difficulties MNEs have in using their superior technological resources to compete with local firms. We identify five kinds of barriers to appropriability: regulatory, information, resource, coordination and commitment.

Although there is overlap between the concepts of barriers to appropriability and liabilities of foreignness, the differences are crucial. The concept of liabilities of foreignness mainly covers two kinds of barriers to appropriability: regulatory and information. The other three kinds of appropriability barriers (resource, coordination and commitment barriers) are not really covered by the concept of liabilities of foreignness. These three types of barriers to appropriability mainly arise from factors such as competition between MNEs themselves, the huge size of MNEs, and MNEs' low commitment to the specific host country market. These factors are not necessarily specific to operating in a foreign market. They could also influence MNEs' utilization of technological resources in regional markets of their home countries. We find it necessary to introduce the broader concept of barriers to appropriability in order to provide a better understanding of the difficulties facing MNEs when they compete with local firms in a foreign market.

We now examine in detail the five kinds of barriers to appropriability. *Regulatory barriers* (host or home country government regulations) are the most obvious barriers preventing MNEs from using their superior technological resources in host country markets (Dunning, 1992; Hymer, 1976). The development of the Japanese automobile industry shows that regulatory barriers prevented US and European automobile companies from using their superior technological resources in the Japanese market. Leading US and European automobile firms had more advanced technologies than Japanese automobile firms, both before World War II and at its end, yet superior technological resources did not create competitive advantages for the US and European automobile firms. Before World War II, Japanese government regulations did not allow foreigners to be the major shareholders in Japanese automobile firms. Mainly because of this policy, Ford and GM, the two dominant leaders of the Japanese automobile industry before World War II, had to leave Japan. After World War II, the Japanese government basically prohibited the import of small cars until the early 1970s, when local firms had developed international competitiveness (Cusumano, 1985).

Information barriers arise from MNEs' relative unfamiliarity with the unique characteristics of a host country's business environment. These characteristics include customers' tastes, business customs, supporting industries, the culture and the legal system (Buckley & Casson, 1976; Hymer, 1976; Lou, 2000; Zaheer & Mosakowski, 1997). Because of their unfamiliarity with customers' tastes, it is hard for MNEs to customize their products to meet local market demand. Similarly, MNEs will find it difficult to work with wholesalers and retailers if they are not familiar with the business customs of the host country. Lack of familiarity with material or parts suppliers will also make it difficult for MNEs to choose the most appropriate products and processes.

Resource barriers arise from resource constraints, especially financial resource constraints, that face MNEs when they try to employ their superior technological resources in a host country. Although a commonly accepted view in China is that MNEs have the advantage in financial resources, in many cases this is not true. In fact, many MNEs face strong competition, mainly from other MNEs (Lester, 1998; Mowery & Nelson, 1999; Womack *et al.*, 1991). The result is that many MNEs often have low profit margins and have to operate on tight budgets. On the other hand, local firms actually might enjoy higher profit margins. For example, according to the State Commission of Economy and Trade, the 512 key local enterprises in China had much higher average profit margins in 2001 than the Fortune 500. The profit/sales ratio for the former is 6 per cent compared to 2.8 per cent for the latter. This indicates that MNEs are not necessarily strong financially, and it might not be easy for MNEs to find the financial resources to invest in a host country, even though they have superior technological resources.

Coordination barriers arise from the complexity of coordinating activities within MNEs (Bartlett & Ghoshal, 1989; Carroll, 1993; Chandler, 1962; Doz, 1986; Dunning, 1992; Freeland, 1996; Studer-Noguez, 2002). For example, Chandler (1962) has shown that a firm needs to create new organizational structures in order to manage its geographic expansion and diversification effectively. Freeland (1996) further shows that it is not easy to make a multidivisional structure effective. He points out that GM intentionally violated the axioms of efficient organization to create managerial consent. Doz (1986) also shows that coordinating business units around the world is a complex and costly process for an MNE. Because of the great complexity of coordination, it is not easy for MNEs to transfer superior technological resources from one country to another.

Commitment barriers come from MNEs' low commitment to a host country market. For many MNEs, the home country market is of strategic importance (Bartlett & Ghoshal, 1989; Studer-Noguez, 2002). Home-country markets might be the key sources of revenue and profits. Also superior technological resources might be developed mainly in the home country. In contrast, because a host country might be of peripheral importance in contributing to revenue and profits, or creating superior technological resources, MNEs might have little commitment to the host country's market. This is supported by empirical studies (Amsden, 2001; Amsden & Chu, 2003; Zhang, 2000). Low commitment to the host country's markets makes it hard for MNEs to employ their superior technological resources effectively.

Opportunities for Improvement

Although the existing literature on technological catch-up emphasizes the huge gaps in technological capabilities between local firms and MNEs, and the difficulties of local firms developing technological capabilities, there is evidence showing that local firms can compete with MNEs by developing new technologies, reinventing mature technologies, or transferring technologies. We introduce the concept of opportunities for improvement to explore further the potential for local firms to improve their technological resources.

We first look at four kinds of opportunities for improvement in the development of emerging technologies: learning, cultural, incentive and organizational opportunities. *Learning opportunities* arise because entry barriers are low when a new technology is emerging. A key feature of emerging technology is that the entry barriers are not high: this offers opportunities for many firms to try many different ways of developing the technology (Perez & Soete, 1988; Tushman & Rosenkopf, 1992; Utterback, 1994). For this reason, when technologies are just emerging, local firms might be able to develop equivalent or even better technologies than those developed by MNEs (Gao, 2003; Hobday, 1990; Lee & Lim, 1998; Perez & Soete, 1988; Wang, 1999).

Cultural opportunities are the corporate culture advantages that local firms might have over MNEs in perceiving and developing emerging technologies. Some studies have shown that a leading organization tends to develop the NIH (not invented here) syndrome, believing that it has a monopoly on knowledge in its field and fails to consider seriously the possibility that other organizations might produce important new ideas (Cohen & Levinthal, 1990; Hamel, Doz & Prahalad, 1989; Katz and Allen, 1982). A typical example is GE. In the late 1970s, GE's Major Appliance Business Group (MABG) did not believe that Matsushita and other Japanese firms could develop better refrigerator compressors, even when GE's Canadian subsidiary was about to buy Matsushita's products (Magazine & Patinkin, 1989). Given that local firms are followers and are thus less established, they should be less subject to an NIH syndrome. Accordingly, local firms might be able to take advantage of MNEs' less favourable culture to perceive and develop emerging new technologies.

Incentive opportunities come from the greater incentives that local firms might have, compared to MNEs, to develop and use emerging technologies. Many studies have shown that incumbent firms are reluctant to develop and use emerging technologies because these technologies might cannibalize their existing businesses (Foster, 1986; Henderson, 1993). Local firms, as followers, should be less worried about the cannibalization effects and should have greater incentives to develop and adopt new technologies. Local firms can take advantage of MNEs' lack of interest in developing and adopting emerging technologies.

Organizational opportunities arise from the greater structural flexibility of local firms. Many studies have shown that incumbent firms' existing organizational structures and procedures have developed to support existing strategies and operations. It is therefore difficult for incumbent firms to change these structures and procedures and allocate necessary resources to support the development and utilization of emerging technologies, especially when the emerging technologies are competence-destroying (Christensen & Rosenbloom, 1995; Clark, 1985; Henderson & Clark, 1990; Tushman & Rosenkopf, 1992), when the technologies are not well aligned with current corporate strategies (Burgelman, 2002), or when the technologies fall out of the existing value network (Christensen & Bower, 1996). Given that local firms are followers, they should be less locked into existing organizational routines and existing value networks, and should be in a more advantageous position to develop and use emerging technologies.

Now we turn to the discussion of opportunities for improvement in mature technologies. Here we identify two kinds of opportunities for improvement in mature technologies: reinvention and adaptation opportunities. *Reinvention opportunities* are the possibilities for local firms to reinvent mature product and process technologies. For leading firms such as MNEs, further investment in mature product and process technologies usually leads to diminishing returns (Perez & Soete, 1988). For local firms, however, reinventing mature technology is of critical importance when MNEs are unwilling to sell mature technology, or when the cost of buying mature technology is too high. In these circumstances, it is desirable for local firms to invest in reinventing mature technologies. Conditions that are conducive to the reinvention of mature technologies are (1) much information about mature technologies is widely available: for example, it is possible to collect pertinent information by reading academic publications and industrial publications, examining patent data and studying products that embody these technologies and (2) application of new tools, such as computer-integrated manufacturing systems (CIMS), makes the process of reinventing mature technologies less time consuming and less costly.

Empirical studies also show that reinvention of mature technologies can be a very effective way for local firms to improve their technological capabilities and technological resources. One example is the development and application of CIMS in China. This has helped many local firms to improve their product design, development and manufacturing capabilities. Beijing No.1 Machine Tool Plant actually won the CASA/SME 1995 Industry Leadership Award because of its impressive achievements in developing and applying CIMS technology to improve its technological capabilities (Wu, 2001).

Adaptation opportunities are the possibilities for local firms to improve their technological capabilities by adapting mature technologies to the local environment. The literature on technological change and technology transfer has shown that both product technology and process technology are developed to solve particular problems in particular technical, business, social and political contexts (Tushman & Rosenkopf, 1992; Utterback, 1994). Technology also has implicit and tacit aspects, rather than being fully specified (Pavitt, 1985). Technology must be adapted and modified to fit new contexts or environments.

The need for adaptation creates opportunities for local firms to improve their technological capabilities. There are two possible situations. When there are barriers to MNEs adapting technology, local firms have an opportunity to develop better technologies and gain competitive advantage. Even when there are no barriers for MNEs, the process of adaptation takes time and this could give local firms more time to improve their technological capabilities.

In addition to internal development of emerging technologies and mature technologies, the third opportunity for local firms to improve their technological capabilities is to transfer technology. One important means of technology transfer is to buy production machinery from suppliers. According to Pavitt (1985), sources and patterns of technological progress vary among sectors. In supplier-dominated sectors and scale-intensive or production-intensive sectors, machinery suppliers play crucial roles in technological progress and it is possible for local firms to improve their technological capabilities by buying production machinery from suppliers. Many Japanese firms in the petrochemical and steel industries have been very successful in buying production machinery from suppliers to close the technological gaps between themselves and leading U.S. and European firms, or even to leapfrog these leading companies (Dertouzos, Lester, Solow and the MIT Commission on Industrial Performance, 1989; Hikino, Harada, Tokuhisa & Yoshida, 1998).

Technology transfer can also play an important role in helping local firms to develop emerging technologies, reinvent mature technologies or adapt technology to the local environment. Emerging technology examples are Samsung, which became a major player in CDMA by transferring technology from Qualcomm (Chung & Lee, 1999), and Sony, which developed its famous Trinitron by transferring technology from the US (Lyons, 1976). In mature technologies, technology transfer played a significant role in helping many local firms in Japan and South Korea to improve their technological capabilities (Cusumano, 1985; Lee & Lim, 2001).

COMPETING WITH MNEs BY INNOVATING

From the resource-based perspective, a firm's competitive advantage comes from its superior resources. A firm is therefore advised to choose its strategy according to its resources. We believe that, although a firm should consider not only its technological resources, it makes sense to analyse local firms' technology strategies by focusing on comparing the local firms' technological resources with those of MNEs, when choosing its technology strategy. This is based on two observations. First, existing literature on catch-up has shown that one key difference between local firms and MNEs is that local firms do not have proprietary technologies while MNEs do (Amsden, 2001; Amsden & Chu, 2003; Hymer, 1976; Kim, 1997, 1998; Westphal *et al.*, 1985). Second, some anecdotal evidence shows that it is very hard for local firms in China to develop proprietary technologies – but some firms are extremely successful in developing other competencies, such as marketing capabilities (Gao, 2003; Xu, 2002).

Using the analysis of barriers to appropriability and opportunities for improvement in the previous section, we now turn to the discussion of possibilities and conditions for local firms to compete with MNEs by developing innovation capabilities and proprietary technologies.

We first look at the situation when there are high barriers to appropriability. Under these conditions, it will be difficult for MNEs to use their superior technological resources to gain a competitive advantage over local firms. For example, high barriers to appropriability will make it hard for MNEs to set up manufacturing facilities in the host country. As a result, MNEs will not be able to enjoy the benefits of low labour costs, familiarization with the local business and social environment, and adaptation of their technology and products to the local market. Even if MNEs were able to export their products to the host country, they would have to pay high tariffs and their products are not customized to meet the local market demands.

In these circumstances, local firms can compete against MNEs by developing strong manufacturing capabilities, even when it is hard for local firms to develop their own proprietary technologies because there are few opportunities for improvement. This is because local firms will enjoy all the benefits of low labour costs, familiarity with the business and social environment, and adaptation of technology and products to the local market. They also do not have to pay extra taxes (tariffs). Accordingly, the development of strong manufacturing capabilities will make it possible for local firms to offer low-cost, mature products with adequate quality to compete against MNEs. This might help to explain how many Japanese and Korean firms have been able to compete with MNEs by developing strong manufacturing capabilities in protected domestic markets (Amsden, 2001; Dertouzos, Lester, Solow and the MIT Commission on Industrial Performance, 1989).

When there are high barriers to appropriability, local firms might be able to compete against MNEs by developing innovation capabilities and proprietary technologies if there are opportunities for improvement. The existence of opportunities for improvement might help local firms to narrow the gap in proprietary technologies between themselves and MNEs. In these circumstances, local firms can offer not only low-cost, mature products but also advanced and new products.

We now look at the situation of low barriers to appropriability. In this situation, it will be easy for MNEs to use their superior technological resources in the host country to compete with local firms. For example, low barriers to appropriability will make it easy for MNEs to set up manufacturing facilities. As a result, MNEs will be able to enjoy the benefits of low labour costs, familiarization with the local business and social environment, and adaptation of their technology and products to the local market. When barriers to appropriability are low, it will be hard for local firms to compete against MNEs by developing strong manufacturing capabilities, because local firms will have no advantages in producing low-cost, mature products. Local firms might be able to survive in some niche markets that MNEs find unattractive. In this situation, local firms might be able to compete against MNEs by developing innovation capabilities and proprietary technologies if there are opportunities for improvement. This is because the existence of opportunities for improvement might help local firms narrow the gap in proprietary technologies between themselves and MNEs.

The discussion above indicates that local firms might be able to compete against MNEs by developing innovation capabilities and proprietary technologies if there are opportunities for improvement. In fact, when there are low barriers to appropriability and many opportunities for improvement, the effective strategy for local firms competing with MNEs is to develop innovation capabilities, and the strategy of developing manufacturing capabilities will not be effective.

THE EMERGENCE OF CHINESE TELECOM EQUIPMENT FIRMS

In the following we illustrate our argument that local firms might be able to compete against MNEs by developing innovation capabilities and proprietary technologies. Our emphasis is on the importance of developing innovation capabilities and proprietary technologies, which has not been well studied in the existing literature. We use three cases from the telecom equipment industry to illustrate our argument. These cases are ZTE Corporation (ZTE), Datang Telecom Technology and Industry Group (Datang) and Huawei Technologies Co. (Huawei). These are three of the local firms that have successfully employed strategies of developing innovation capabilities and proprietary technologies to compete with MNEs.

Data were collected mainly through interviews, although archival data were used as well. Interviews focused on the three firms' technology strategies and their effectiveness, and on the barriers to appropriability and opportunities for improvement in the telecom equipment industry. Interviewees include top management (CEO, senior VP), board members (Chair and Vice Chair of the Board of Directors), middle-level managers (mainly R&D managers), R&D researchers and manufacturing engineers. Interviews ran from about an hour to as long as four hours, but typically lasted 90 to 120 minutes. We took extensive notes during interviews because most of the interviewees preferred not to be taped. We stopped when we believed that we had collected sufficient data, and we had 35 interviews with 15 people. Before analysing the three telecom equipment firms' technology strategies, we first briefly introduce these three firms.

ZTE was founded in 1985. The registered capital was 4.8 million RMB Yuan (about US$1.7m in 1985). In its early days ZTE did not have its own products but acted as OEM for some Hong Kong firms. Today ZTE is China's largest listed telecom equipment manufacturer and one of the 520 key state-owned enterprises supported by the central government. ZTE's development to date can be divided into two stages: before 1996 and since 1996. In the first stage, although ZTE was competing on the low-end market, it invested intensively in R&D to develop its own technologies and products. In the first period, ZTE not only accumulated a lot of experience in technology and product development, but also attracted a group of excellent researchers and engineers. This helped to built up ZTE's initial technological capability basis for its long-term development. The second stage of ZTE's development started from early 1996. After the successful development of ZXJ10 and ZXA10, ZTE believed that it had developed the capability, especially technological capability, to make strategic transitions in its development. In February 1996, it decided to make the following transitions: (1) from making a single product to making multiple products; (2) from focusing on the rural market to focusing on the urban market; (3) from focusing on the domestic market to starting to explore the overseas market.

With continuous technological innovation, ZTE has made impressive progress in implementing the transition planned in 1996. It has expanded its product line from mainly switches and access equipment to mobile network, data communication network, optical transport network, intelligent network and a next generation network. ZTE is also making rapid progress in the world market. Its products have been used in over 40 countries and regions, including Cyprus, Bangladesh, Myanmar, Thailand, the USA, Russia, Egypt, Kenya, Congo, Zambia and Hong Kong.

Datang was officially registered in 1999, although the Chinese Academy of Telecom Science and Technology (CATT), the main body of the group, was founded in 1957, and many of its subsidiaries were set up much earlier. For example, Xian Datang was set up in 1993 and Beijing Gaohong was set up in 1995. The transforming of CATT into Datang Group is based on the central government's policy requesting that R&D at public research organizations closely reflects market demand. Datang Group now has more than 10 000 employees, one central research academy, one engineering design academy, one listed company, eight non-listed subsidiaries, and 11 research institutes. The Group has very strong technological capability. For example, SP30 is one of the the key large-scale public digital switches developed domestically. Time Division Synchronous Code Division Multiple Access (TD-SCDMA) is one of the three international standards for 3G wireless communications.

Huawei was established in 1988 and is now the largest local telecom equipment supplier in China. It has 220 000 employees, with 85 per cent of them holding a bachelor's degree or above. Its sales revenue in 2001 is 25.5 billion RMB Yuan (about US$3.1 billion in 2001). Similar to ZTE, Huawei's development can be divided into two periods: from the founding of the company in 1988 to 1995; and 1996 to now.

When Huawei was first founded in 1988, its main business was to sell switches imported from Hong Kong. This was a very profitable business because the telecom service industry was growing very rapidly at that time, and the demand for telecom equipment including switches was very strong. In this stage, contrary to most firms, Huawei invested almost all its profits in R&D to develop its own technology and products. In 1992, Huawei developed its own small-scale switch and in 1995 developed its own large-scale switch.

Huawei's development entered the second period from 1996 with the successful development of the C&C08 large-scale digital switch in 1995. As a manager commented, 'Huawei entered a new era with the successful development of the C&C08 machine.' In the second period Huawei has not only expanded and strengthened its switch business but also made rapid progress in optical transmission, wireless and data communication. Huawei has transformed itself from a telecom equipment supplier to a telecom solution provider.

The Strategy of Developing Innovation Capabilities and Proprietary Technologies

In the past two decades, the public telecom service industry in China has been expanding rapidly, and this rapid development has created a huge market for various kinds of telecom equipment. Facing this huge market opportunity, a few local firms, including ZTE, Datang and Huawei, chose to develop innovation capabilities and proprietary technologies rather than relying mainly on buying technology and developing manufacturing capabilities as the majority of the local firms did. It turned out that the local firms that chose to develop innovation capabilities developed proprietary technologies and emerged as the leading firms in the industry. These firms not only outperform the local firms choosing to buy technology and develop manufacturing capabilities but also are able to compete with MNEs at the high-end market.

To implement the strategy of developing innovation capabilities and proprietary technologies, ZTE, Datang and Huawei have taken numerous actions. These include taking the risks to invest heavily in R&D, close collaboration with the knowledge frontier, focused R&D, and hiring and keeping talented people.

First action: taking the risks to invest in R&D

We first look at Huawei. According to a manager at Huawei, Ren Zhengfei, CEO of the firm, points out that 'there are risks in innovation, but you should not stop doing innovation because of risks. If you do not take risks, you would have to follow the leaders, and have to be a second class or third class firm. You would not be able to compete with MNEs, and would not be able to survive'.

One example showing Huawei's willingness to take risks in developing proprietary technology was the development of HJD48, a PBX switch. To develop this product, Huawei invested almost all its profits from selling imported products from Hong Kong in the first few years of its operation. Another example was the development of JK1000, a public switch, in 1993 and 1994. Huawei invested dozens of millions of RMB Yuan in the development of the product. Unfortunately, the development work failed. Huawei accepted this failure and continued its investment in developing proprietary technology.

A third example is the development of IC design capability. In the 1980s and early 1990s, the government tried hard to develop the IC industry but made little progress. Although there were a lot of difficulties, Huawei started to invest in IC design in the mid-1990s. In 1997, Huawei became one of the National IC Design Centers. Today Huawei is able to design the key ICs for its products.

Similarly, ZTE is also willing to take the risks to invest in R&D. In commenting on why ZTE has developed a lot of products based on internally developed key technologies, a manager at ZTE said, 'ZTE and a lot of firms in the telecom equipment industry made a lot of money in the early days of their business. The difference between ZTE and other firms at that time was that ZTE was willing to take the risks and invest a lot of money in R&D continuously.'

The development of specific products also shows that ZTE is willing to take risks in R&D. For example, a manager pointed out that ZTE decided to develop ZXJ10 in 1993, although they had 'no experience in developing large scale digital switches' at that time. The development of Softswitch also involves a lot of risks. A manager pointed out,

> Softswitch is a software identity, and is much more complex than ZXJ10 . . . there are a lot of difficulties in developing this new product . . . this is a new product, so it is not tested in the market. Although there is agreement on the basic framework of the product among the industry leaders, there is no agreement on the details such as the definition of calling model, API, criteria for performance, OSS, etc.

The leading local telecom equipment firms have invested heavily in R&D, and believe that this is a key reason for their successful development of advanced technology and products. According to an R&D manager at ZTE, 'one of the key reasons that we are able to develop ZXJ10 and ZXSS10 is that we have spent a lot of money in R&D to develop the products'. In fact, from 1997 to 2001, ZTE's investment in R&D was 16 to 22 per cent of its sales revenue.

Similarly, Huawei claims that it invests more than 10 per cent of its sales revenue in R&D, and believes that it is continuous high investment that makes it possible for it to develop key technology and competitive products. Huawei also has a lot of practices in place to make sure that its investment in R&D is at a high level. For example, 'profit maximization' is not Huawei's goal, improving the firm's core competency is. Another practice is that it is a rule in Huawei for the R&D department to use up the allocated R&D budget. According to a manager, 'this forces the managers and employees in the R&D department to innovate'.

Second action: developing close collaboration with the knowledge frontier

In the second half of 1987, ZTE decided to develop a PBX digital switch. The problem was, they only had three to four people whose major was in switch technology or computer science and technology. To strengthen their R&D capability, ZTE collaborated with the Beijing Institute of Post and Telecommunications. In November 1989, ZX500 was successfully developed. ZTE claimed that this was the first domestically developed digital switch.

ZTE institutionalized the practice of collaborating with the knowledge frontier mainly by setting up research institutes close to the knowledge frontier. For example, it has set up research institutes in Nanjing, Beijing, Shanghai, Xi'an and Chongqing. Close collaboration with the knowledge frontier has been very effective. For example, ZXJ10 was developed by the Nanjing Institute set up in September 1993. This product was one of the key large-scale digital switches developed domestically. This product also helped to lay the foundation for ZTE's transforming from a low-end product maker to a major telecom equipment firm.

Similarly, although Datang Group itself was formed from a research organization, it is very active in collaborating with other parts of the knowledge frontier. For example, Datang Group is the first domestic firm that has been able to develop backbone routers by collaborating with the University of National Defense (UND), which is very strong in computer science and engineering.

Third action: concentrating R&D in limited areas

Huawei has highly focused R&D and sticks to 'the principle of intensity of pressure' and focuses on limited areas – software technology, common purpose technology and IC design. Huawei will use all its resources in the development of one product or technology at one time to make sure that it will be successful in this product or technology. According to Ren Zhengfei, CEO of the firm, 'Huawei is not strong enough to catch up in all areas. We have to focus on some key technologies. We put all the eggs in one basket – to invest all the resources in a specific technology and product – hoping that high investment intensity will lead to breakthroughs in the targeted area.'

The 'principle of intensity of pressure' has proved to be very effective. For example, following this principle, Huawei invested almost all its profit in the early days of its operation and developed HJD48, a PBX switch, in 1992. Following the same principle, Huawei developed the C&C08, a large-scale office switch, in 1995. Today Huawei has leading technologies and products in SDH, access network, intelligent network, signal network, internet access server, and advanced technology and products in DWDM, C&C08iNet,

router and mobile telecommunication. Huawei is one of a few companies in the world that are able to supply next-generation exchange systems.

In order to stick to the 'principle of intensity of pressure', Huawei chooses to give up a lot of opportunities. For example, in the early 1990s, the real-estate business and investing in stock market were highly profitable, but Huawei chose not to enter these businesses. In fact, Huawei makes it a rule not to invest in telecom service business in the 'Huawei Constitution' formulated in 1995.

Similarly, focused R&D is one feature of ZTE's technology strategy. This helps ZTE to compensate for its disadvantage of being a small firm. According to Hou Weigui, CEO of ZTE, 'ZTE is a very young company. There is still a big gap between its overall strength and that of the giant MNEs.' This is a critical issue ZTE and other Chinese telecom equipment firms have to carefully deal with. It's not practical for ZTE to be a technology leader in all areas. On the other hand, we can be the leader in some areas. We can make breakthroughs in some areas.'

For example, although the telecom equipment industry is changing rapidly and there are a lot of potential opportunities, ZTE chooses to focus its R&D on areas that are of strategic importance. Software is one area. About 70 per cent of ZTE's R&D people are working on software.

One example is Softswitch. Although this is an emerging technology, ZTE is early in predicting the technology trend and believes this is a technology of strategic importance. According to a manager of the firm, 'we attached a lot of importance to this product, hoping that it is a breakthrough for us to be a first class telecom equipment provider'. ZTE has spent 200 million RMB Yuan (about $24 million) in R&D in this technology, and has made impressive progress. Now ZTE is one of the industry leaders in the world in this area.

Another example is IC design. According to Hou Weigui, CEO of the firm, 'ZTE is developing system-based products. Software and IC are the key technologies in our firm, especially software . . . We have six to seven hundred people doing IC design. I think IC design is as important as IC manufacturing. For example, Qualcomm has very strong capability in IC design, but it does not manufacture IC by itself.' Now ZTE is able to design ICs for its major products.

Fourth action: hiring and keeping a lot of talented people

Huawei believes that human capital is much more important than financial capital. Hiring and keeping talented people is critical for Huawei's survival and development.

Since 1996, Huawei has been hiring a great number of students. In many cases, it goes to a university and hires almost all the students with a degree in telecommunication. For example, in 2000, Chongqing University of Post and Telecommunication had just over 40 graduating students with a degree in telecommunication. Huawei hired 39 of them. The University of the Southeast had rather more than 30 graduating students with a degree in radio technology, and Huawei hired 25 of them.

Huawei not only hires a lot of talented people but also has a lot of policies and practices to keep them. The first policy is capability-based fast promotion. Huawei is very 'generous' in promoting its employees if they show high potential to contribute to the firm. Because of this policy, many very young people were promoted quickly to very high positions, especially in the early days of the company. For example, Li Yinan graduated and joined Huawei in 1994. He was promoted to engineer on the second day, chief engineer in two weeks, vice-president of Central Research in six months, and president of Central Research in one year. The next year he became the youngest vice-president of the company – he was only 23 years old at that time.

The second policy is to create an attractive work environment for the employees. For example, Huawei is very generous in buying academic and technical books, journals and other materials published in English. Huawei gives its R&D people the autonomy to order equipment and tools. Huawei offers its R&D people effective tools such as high-end notebooks. It also creates a lot of opportunities for its R&D people to go to other countries to learn. Compared with their experiences in universities and the conditions in most domestic firms, the R&D people at Huawei are very happy to have such a good work environment, and are motivated to work hard.

The third policy is to offer high salary and stock options. To attract and keep talented people, Hauwei is willing to pay a very high salary. For example, in 1995, a newly graduated student with a Master's degree was able to make 5000 Yuan/month (about US$600). This is comparable to income in firms with foreign investment, which is much higher than that of most local firms. Salary increase in Huawei is also very fast. Huawei also gives shares to its employees, making them shareholders of the firm.

The salary policy also gives better treatment to R&D people (and the marketing people). Their salary is higher than that of testing people and production people, whose salary is higher than that of people in functions such as administration and secretariat. Similarly, ZTE believes that the key to a high-tech firm's development is to hire and keep talented people. According to Hou Weigui, CEO of the firm, 'It is important to have an appropriate property right arrangement for ZTE, but that is not sufficient. The heart and sole of a high-tech company is technological innovation . . . In the past ZTE has been growing very fast and one of the most important factors is that ZTE has attracted a lot of excellent people.'

ZTE has taken a lot of actions to hire and keep talented people. For example, it has a lot of research institutes in Beijing, Shanghai, Nanjing and other cities. A key reason for this is that ZTE's headquarters is at Shenzhen, and many people, especially those who have lived in cities such as Beijing, Shanghai or Nanjing for a long time, are not willing to move to work and live in Shenzhen. To solve this problem, ZTE decided to set up research institutes in these cities. As a result, it is now possible for ZTE to hire the talented people who do not want to go to Shenzhen; as for the people hired, they are able to work in ZTE but do not have to leave the cities they like most.

Explaining the Effectiveness of the Innovating Strategy

The effectiveness of the technology strategy employed by ZTE, Datang and Huawei offers evidence to support our argument that local firms might be able to compete against MNEs by developing innovation capabilities if there are barriers to appropriability and opportunities for improvement, and if the local firms are effective in taking advantage of these barriers and opportunities.

First, the effectiveness of the strategy of developing proprietary technologies relates to the fact that that there are opportunities for improvement in the telecom equipment industry. For instance, there exist 'reinvention opportunities' for the domestic firms to develop digital switches. By the mid-1980s, digital switch technologies had become relatively mature, and a lot of related technical information is widely available. In fact, it was in this situation that the HJD04 switch, the first locally developed large-scale public digital switch, was successfully developed at the end of 1991 by Great Dragon. This machine was the fastest digital switch in the world at that time. As a result of the successful development of the HJD04 switch, the price of imported switches and switches from Joint ventures (JVs) dropped dramatically.

There are also opportunities for improvement in emerging technologies. One example is the third-generation wireless technologies. Although China was backward in 1G and 2G wireless communications, the emergence of the 3G wireless communications since the 1990s has offered a golden opportunity for China to catch up. The result is that Datang developed TD-SCDMA, which is now one of the three key international standards for 3G wireless communications. In fact, compared with the other two standards, TD-SCDMA has outstanding technological and economic advantages such as high spectrum efficiency, high capacity, high suitability for asymmetrical data services, and low cost.

Softswitch is another example. This technology is still emerging. Seeing this as an opportunity to improve its technological capability significantly,

ZTE started its R&D in this technology in 1998. Through heavy financial investment and human resources investment, ZTE has become one of the industry leaders in this technology.

Second, the effectiveness of the strategy of developing proprietary technologies also relates to the fact that there exist barriers to appropriability in the telecom equipment industry. The first barrier is regulatory, and the most obvious example is the government's direct helping of the domestic telecom equipment firms to capture market share after they were able to produce high-end products. For example, in 1993, after the domestically produced large-scale digital switches passed the inspection of the government, the Ministry of Post and Telecommunications asked the service providers to use domestically produced products when their performance was equivalent to that of products from MNEs. More important, in 1996 and 1999, the Ministry organized two coordination meetings to encourage the use of domestically produced large-scale digital switches. The results were that the domestic telecom service providers placed orders of nine million lines and 18 million lines, respectively. In fact, these two meetings were the turning points for the domestic telecom equipment firms. They began to replace MNEs as the major suppliers of large-scale digital switches in the fixed line area. According to Hou Weigui, CEO of ZTE, 'the government knows how to support domestic firms. The two coordination meetings were important. They helped us get firmly established in the fixed market. Now the government is helping with the mobile sector, hoping that domestic firms will capture 20 per cent market share in two to three years, and 50 per cent by 2003'.

The Ministry of Post and Telecommunications also organized a meeting to promote domestically made GSM equipment after the domestic telecom equipment firms developed the capability to supply the equipment. The service providers decided to buy mobile switches of nine million lines.

The second barrier is commitment. Although the MNEs are very active in exporting products to the local market and have joint ventures with local firms, a common problem is that these MNEs are not active in localization, especially through adding more R&D functions to their subsidiaries or JVs. As one manager of a MNE's subsidiary commented, 'we are not allowed by the headquarter to do R&D. We are not able to respond quickly to the local market'. Relatively lower engagement in the local market of MNEs in the telecom equipment industry has created a lot of opportunities for domestic firms to innovate and grow.

The story of Huawei's development of its access server is an example. A telecom service provider bought equipment from an MNE but the equipment was designed without considering the special requirements of the Chinese market. The telecom service provider asked Huawei if they could make equipment that could fulfil their special requirements. Huawei responded quickly and added a new product category.

ZTE's success in CDMA system is another example. Compared with the much bigger MNEs, ZTE is not only small but also a latecomer in CDMA equipment. On the other hand, ZTE is good at taking into consideration the customers' detailed requirement in its product design. ZTE has been very successful in China Unicom's CDMA system bidding.

Third, the effectiveness of the strategy of developing proprietary technologies internally also relates to the fact that the three local firms are very effective in taking advantage of the opportunities for improvement and barriers to appropriability. For example, ZTE was very effective in seizing opportunities in mature technology and developed its ZXJ10, the basis for ZTE to change from a small low-end market player to a leading firm in the telecom equipment industry. ZTE was also very effective in seizing opportunities in emerging technologies to develop its Softswitch technologies. More specifically, in the case of ZXJ10, according to an expert of the company, the key reasons for the successful development of the product include the following:

- the company attached great importance to the development of the product and invested heavily;
- a dedicated research institute was set up with many talented people;
- getting help from research institutes and universities, especially the Nanjing University of Post and Telecommunication, to develop the key technologies.

In the case of ZTE's Softswitch technology, the key reasons for the success include the following:

- the company correctly predicted the evolution of the technology and the market, and started to do R&D very early (in 1998);
- the company invested heavily in the R&D of the product;
- the product is mainly a software identity, although very complex.

Local firms have also been very effective in taking advantage of the barriers to appropriability. Huawei is a good example. For example, although MNEs have good services in central cities, they are weak in other areas. Huawei seized this opportunity and sent out a lot of service people to make sure that it had an advantage in competing with MNES in these areas. For example, in Helongjiang province, one large MNE once had only three to four people, mainly selling their products, while Huawei had more than 200 people. The result is, once its products go wrong, Huawei is able to have people to solve the problem quickly. For example, during the spring festival of 2000, a local network broke down, and it was not clear which machine had problems because there were a lot of different models of switches operating in the network. Huawei quickly sent out its technicians and helped fix the problem, although it turned out the problem was not with Huawei's machine. In contrast, the firm whose product went wrong was late in responding to the problem.

CONCLUSION

The existing literature suggests that local firms should choose to compete against MNEs by developing strong manufacturing capabilities. In this chapter we analysed the possibility and the conditions for local firms to compete with MNEs by developing innovation capabilities and proprietary technologies, and argued that local firms might be able to compete against MNEs by developing these if there are opportunities for improvement. The practice of ZTE, Datang and Huawei offered support for our argument.

Our study offers new insights into the analysis of local firms' technology strategies. First, the bulk of the existing literature has been conducted on the assumption that MNEs have knowledge-based assets, unique resources and proprietary technologies, while local firms do not. Our study indicates that local firms should apply the concept of barriers to appropriability to analyse the difficulties MNEs face when trying to compete in a host country's market, and apply the concept of opportunities for improvement to explore the opportunities for local firms to improve their technological resources.

Second, a large portion of the existing literature has been conducted mainly in countries such as South Korea and Japan where, because of protection of the local markets, competition from MNEs was limited, especially in the early stages of local firms' development. Our study indicates that, in the era of globalization, when barriers of appropriability are likely to be low, it will be hard for local firms to compete effectively against MNEs by developing strong manufacturing capabilities. When there are opportunities for improvement, it might be possible for local firms to compete against MNEs by developing innovation capabilities and proprietary technologies.

Third, our study also relates to the impact of competition and ownership on technological capability development. The literature on economic transition argues that one of the key problems of the traditional planned economy is that there is no competition between firms, and the result is that these firms do not feel the pressure to engage in technological progress (Shi, 1998).

Increased competition is necessary for the firms to develop new technology and new products. The literature on economic transition also suggests that public ownership under the traditional socialist economy system has difficulty in inducing innovative activities in firms. Ownership change, especially privatization, is necessary for the firms to pursue technological progress (Shi, 1998; Shleifer and Vishny, 1994, 1998; Steinfeld, 1998).

Our study indicates that controlled competition in the form of limited protection of the local firms is helpful for local firms to develop innovation capabilities and proprietary technologies. Without the government's organizing the two coordination meetings to encourage the use of domestically produced large-scale digital switches, the local firms' innovation will be limited.

Our study also indicates that public ownership could play a positive role in technological innovation. For example, Datang, the leading domestic firm in wireless technology, is state-controlled. ZTE has a mixed ownership, with both state ownership and private ownership. Only Huawei is a private firm. In fact, although Huawei is a private firm, state support in the form of bank loans is important for its innovation. Furthermore, according to Ren Zhengfei, CEO of Huawei, 'profit maximization never is the goal'. The goal is 'to be a world class telecom equipment supplier'.

Our study also has limitations and further research is needed. First, we analysed three 'successful' firms' technology strategies to illustrate our argument that local firms might be able to compete against MNEs by developing innovation capabilities and proprietary technologies if there are opportunities for improvement. Since there are no local firms which are as successful as the firms we studied by not developing innovation capabilities, our conclusion is likely to have the 'success' bias. In addition, by studying more cases, future research could further test our argument.

Second, this chapter focused mainly on the analysis of two kinds of environmental factors, barriers to appropriability and opportunities for improvement. Firms (local or MNE) might develop different technology strategies based on different strategic postures toward the two environmental factors. Some firms might be insensitive to these factors, while others might be not only aware of these factors but also proactively creating and using these factors (Chen, 1996; Grimm and Smith, 1997). Future research could analyse these interactive competitive behaviours between local firms and MNEs.

Third, recognizing the crucial importance of technological resources for local firms, we focused on the relationship between technological resources and technology strategies. Future research could examine further the impact of other kinds of resources on the selection of technology strategies.

Fourth, we introduced two concepts, barriers to appropriability and opportunities for improvement, to study local firms' technology strategies. Future research could explore whether it is possible to extend these two concepts to the analysis of local firms' corporate strategies.

Finally, our study is based on the study of local Chinese firms. Future study could explore whether it is possible to apply our findings and conclusions to firms in other developing countries.

REFERENCES

Amsden, A. (2001), *The Rise of 'the Rest': Challenges to the West from Late-industrializing Economies*, Oxford: Oxford University Press.

Amsden, A. and W.-W. Chu (2003), *Beyond Late Development: Upgrading Policies in Taiwan*, Cambridge, MA: The MIT Press.

Barney, J.B. (1986), 'Strategic factor markets: expectations, luck, and business strategy', *Management Science*, **32** (10), 1231–41.

Bartlett, C.A. and S. Ghoshal (1989), *Managing Across Borders: The Transnational Solution*, Boston, MA: Harvard Business School Press.

Buckley, P.J. and M. Casson (1976), *The Future of the Multinational Enterprise*, London: Macmillan Press.

Burgelman, R.A. (2002), *Strategy is Destiny*, New York: The Free Press.

Carroll, P. (1993), *Big Blue: The Unmaking of IBM*, New York: Crown Publishers, Inc.

Chandler, A.D. (1962), *Strategy and Structure: Chapters in the History of the American Industrial Enterprise*, Cambridge, MA: The MIT Press.

Chen, M.-J. (1996), 'Competitor analysis and inter-firm rivalry: toward a theoretical integration', *Academy of Management Review*, **21** (1), 100–134.

Christensen, C.M. and J. Bower (1996), 'Customer power, technology investment, and the failure of leading firms', *Strategic Management Journal*, **17** (3), 197–218.

Christensen, C.M. and R.S. Rosenbloom (1995), 'Explaining the attacker's advantage: technological paradigms, organizational dynamics, and the value network', *Research Policy*, **24** (2), 233–57.

Chung, K. and K. Lee (1999), 'Mid-entry technology strategy: the Korea experience with CDMA', *R&D Management*, **29** (4), 352–63.

Clark, K.B. (1985), 'The interaction of design hierarchies and market concepts in technological evolution', *Research Policy*, **14** (5), 235–51.

Cohen, W.M. and D.A. Levinthal (1990), 'Absorptive capacity: a new perspective on learning and innovation', *Administrative Science Quarterly*, **35** (1), 128–52.

Cusumano, M.A. (1985), *The Japanese Automobile Industry: Technology and Management at Nissan and Toyota*, Cambridge, MA: Harvard University Press.

Dertouzos, M.L., R.K. Lester, R.M. Solow and the MIT Commission on Industrial Performance (1989), *Made in America: Regaining the Productive Edge*, Cambridge, MA: The MIT Press.

Dierickx, I. and K. Cool (1989), 'Asset stock accumulation and the sustainability of competitive advantage', *Management Science*, **35** (2) 1504–11.

Doz, Y. (1986), *Strategic Management in Multinational Companies*, Oxford: Pergamon Press.

Dunning, J.H. (1992), *Multinational Enterprises and the Global Economy*, Reading, MA: Addison-Wesley Publishing Company.
Foster, R.N. (1986), *Innovation: The Attacker's Advantage*, New York: Summit Books, Simon and Schuster.
Freeland, R.F. (1996), 'The myth of the M-form? Governance, consent, and organizational change', *American Journal of Sociology*, **102** (2), 483–526.
Gao, X. (2003), 'Technological capabilities catching up: follow the normal way or deviate', PhD, dissertation, MIT Sloan School of Management.
Gao, X., P. Zhang and X. Liu (forthcoming), 'Competing with MNEs: developing manufacturing capabilities or innovation capabilities', *Journal of Technology Transfer*.
Grimm, C.M. and K.G. Smith (1997), *Strategy as Action: Industry Rivalry and Coordination*, Cincinnati, OH: South-Western College Publishing.
Hamel, G., Y. Doz and C.K. Prahalad (1989), 'Collaborate with your competitors – and win', *Harvard Business Review*, **67** (1), 133–9.
Henderson, R.M. (1993), 'Underinvestment and incompetence as responses to radical innovation: evidence from the photolithographic alignment equipment industry', *Rand Journal of Economics*, **24** (2), 248–70.
Henderson, R.M. and K. Clark (1990), 'Architectural innovation: the reconfiguration of existing product technologies and the failure of established firms', *Administrative Science Quarterly*, **35** (1), 9–30.
Hikino, T., T. Harada, Y. Tokuhisa and J. Yoshida (1998), 'The Japanese Puzzle', in Ashish Arora, Ralph Landau and Nathan Rosenberg (eds), *Chemicals and Long-term Economic Growth: Insights from the Chemical Industry*, New York: John Wiley & Sons, Inc., pp. 103–36.
Hobday, M. (1990), *Telecommunications in Developing Countries: The Challenge from Brazil*, London: Routledge.
Hymer, S. (1976), *The International Operations of National Firms: A Study of Direct Foreign Investment*, Cambridge, MA: The MIT Press.
Katz, R and T.J. Allen (1982), 'Investigating the not invented here (NIH) syndrome: a look at the performance, tenure, and communication patterns of 50 R&D project groups', *R&D Management*, **12** (1), 7–19.
Kim, L. (1997), *Imitation to Innovation: The Dynamics of Korea's Technological Learning*, Boston, MA: Harvard Business School Press.
Kim, L. (1998), 'Crisis construction and organizational learning: capability building in catching-up at Hyundai motor', *Organization Science*, **9** (4), 506–21.
Lall, S. (1982), *Developing Countries as Exporters of Technology: A first look at the Indian experience*, London: Macmillan Press.
Lee, K. and C. Lim (1998), 'Technological regimes, catching-up and leapfrogging: findings from Korea industries', *Research Policy*, **30** (3), 459–83.
Lester, R.K. (1998), *The Productive Edge: A New Strategy for Economic Growth*, New York and London: Norton and Company.
Lou, Y. (2000), 'Dynamic capabilities in international expansion', *Journal of World Business*, **35** (4), 355–78.
Lu, Q. (2000), *China's Leap into the Information Age*, Oxford: Oxford University Press.
Lu, Q. and W. Lazonick (2001), 'The organization of innovation in a transitional economy: business and government in Chinese electronic publishing', *Research Policy*, **30** (1), 55–77.
Lyons, N. (1976), *The Sony Vision*, New York: The Crown Publishers, Inc.

Magazine, I.C. and M. Patinkin (1989), 'General Electric: the heart of an American product', in Arthur A. Thompson, William E. Fulmer and A.J. Strickland (eds), *Readings in Strategic Management*, Boston, MA: Irwin, pp. 189–221.

Mowery, D. and R. Nelson (eds) (1999), *Sources of Industrial Leadership: Studies of Seven Industries*, Cambridge: Cambridge University Press.

Pavitt, K. (1985), 'Technology transfer among the industrially advanced countries: an overview', in Nathan Rosenberg and Claudio Frischtak (eds), *International Technology Transfer: Concepts, Measures, and Comparisons*, New York: Praeger, pp. 3–23.

Penrose, E.T. (1959), *The Theory of the Growth in the Firm*, Oxford: Oxford University Press.

Perez, C. and L. Soete (1988), 'Catching up in technology and windows of opportunity', in Giovanni Dosi, Christopher Freeman, Richard Nelson, Gerald Silverberg and Luc Soete (eds), *Technical Change and Economic Theory*, London and New York: Pinter Publishers, pp. 458–79.

Peteraf, M.A. (1993), 'The cornerstones of competitive advantage: a resource based view', *Strategic Management Journal*, **14** (3), 179–91.

Prahalad C.K. and G. Hamel (1990), 'The core competence of the corporation', *Harvard Business Review*, **68** (3), 79–91.

Shen, X. (1999), *The Chinese Road to High Technology: A Study of Telecommunications Switching Technology in the Economic Transition*, London: Macmillan Press.

Shi, Y. (1998), *Chinese Firms and Technology in the Reform Era*, London: Routledge.

Shleifer, A. and R.W. Vishny (1994), 'Politicians and Firms', *Quarterly Journal of Economics*, (109), 995–1025.

Shleifer, A. and R.W. Vishny (1998), *The Grabbing hand: Government Pathologies and Their Cures*, Cambridge, MA: Harvard University Press.

Steinfeld, E. (1998), *Forging Reform in China: the Fate of State-owned industry*, Cambridge, UK: Cambridge University Press.

Studer-Noguez, I. (2002), *Ford and the Global Strategies of Multinationals: The North American Auto Industry*, London and New York: Routledge.

Teece, D.J., G. Pisano and A. Shuen (1997), 'Dynamic capabilities and strategic management', *Strategic Management Journal*, **18** (7), 509–33.

Tushman, M.L. and L. Rosenkopf (1992), 'Organizational determinants of technological change', *Research in Organizational Behavior*, **14**, 311–47.

Utterback, J.M. (1994), *Mastering the Dynamics of Innovation*, Boston, MA: Harvard Business School Press.

Wang, X. (1999), *Experiences in Developing the Huaguang Electronic Publishing System* (in Chinese), Beijing: Beijing University Press.

Wernerfelt, B. (1984), 'A Resource-based view of the firm', *Strategic Management Journal*, **5** (2), 171–80.

Westphal, L.E., L. Kim and C.J. Dahlman (1985), 'Reflection on the Republic of Korea's acquisition of technological capabilities', in Nathan Rosenberg and Claudio Frischtak (eds), *International Technology Transfer: Concepts, Measures, and Comparisons*, New York: Praeger, pp. 167–221.

Womack, J.P., D.T. Jones and D. Roos (1991), *The Machine That Changed the World*, New York: Harper Perennial.

Wu, C. (2001), 'Theoretical basis for a modern manufacturing system', (in Chinese), unpublished manuscript, Tsinghua University, Beijing.

Xu, L. (2002), 'Resources and capabilities in high-tech entrepreneurship develop-

ment: a study on Chinese high-tech startups', PhD dissertation, MIT Sloan School of Management.
Zaheer, S. and E. Mosakowski (1997), 'The dynamics of the liability of foreignness: a global study of survival in financial services', *Strategic Management Journal*, **18** (6), 439–64.
Zhang, Q. (2000), *PTIC: Made by China* (in Chinese), Beijing: China Yanshi Press.

9. The effect of organizational reward and control on firm innovation and financial performance in Chinese technology firms

Yuan Li, Dan Li, Haiyang Li and Yongbin Zhao

INTRODUCTION

Recent literature has witnessed increasing attention to the growth of high technology firms in emerging markets such as China, India and Russia (Bruton & Rubanik, 2002; Li & Atuahene-Gima, 2001; *The Economist*, 2004). The development of technology firms represents a driving force for economic transformation in these emerging markets. Technology firms not only help develop new wealth, but also create new employment opportunities. For example, in Zhongguancun Science Park in Beijing, China, there were more than 9500 firms with 400 000 employees and industrial outputs of US$22.8 billion up to 2002. This output contributed two-thirds of the industrial growth of Beijing (Beijing Bureau of Statistics, 2002).

The importance of technology firms for wealth and job creation has led researchers to search for factors that predict these firms' survival and success. Several scholars have demonstrated that firm innovation in the form of new products and processes has become increasingly important as a way for technology firms to achieve competitive advantages (Eisenhardt & Schoonhoven, 1990; Li & Atuahene-Gima, 2001). However, firm innovation does not occur in a vacuum. To ensure success in the development of new products or new processes, it is important for technology firms to manage effectively (for example, with reward and control) their innovating employees to improve firm performance.

Previous studies on the organizational reward–innovation relationship have reported mixed results. Some researchers have shown that material rewards (such as bonuses and pay increases) can encourage innovation (for example, Baer, 1997; Eisenberger & Cameron, 1996) while others have disclosed evidence to the contrary. For example, Amabile and colleagues (Amabile, 1983; Amabile, Hennessey & Grossman, 1986; Hennessey &

Amabile, 1998) found that material rewards are detrimental to innovation. The mixed findings may stem from the fact that prior research has defined innovation very broadly and failed to distinguish between different types of innovations (that is, incremental innovations versus radical innovations). Thus, without specifying the nature of innovation type, the linkage between reward mechanisms and innovation cannot be theoretically developed.

Organizational control literature describes control as a process of monitoring and evaluating behaviours and outcomes (Jaworski, 1988; Ouchi, 1980). Drawing upon Jensen's (1993) argument that the major source of failure of the market for corporate control regarding innovation is internal control systems, Hitt, Hoskisson, Johnson and Moesel (1996) examined the effect of internal managerial control systems (that is, strategic controls and financial controls) on firm innovation. While insightful, their study is limited to the context of merger and acquisition and they focused on control systems for business-level or divisional managers. So far, no study has examined organizational controls of innovating employees and how these controls may affect the development of different types of innovations (for example, incremental and radical innovations), particularly in an emerging market context.

In this study, we attempt to advance the literature in several ways. First, we draw upon the innovation literature by making a distinction between two types of firm innovations: incremental innovation and radical innovation (Dewar & Dutton, 1986; Hill & Rothaermel, 2003). We add to the literature by simultaneously examining the effects of both organizational rewards and controls on these two types of firm innovations. Second, the literature has suggested that innovation can help firms catch up with the opportunities in uncertain environments to build/maintain their competitive advantage, and sequentially benefit their long-term performance. Although some scholars (for example, Foster, 1986; Hill & Rothaermel, 2003; Tripsas & Gavetti, 2000) have studied the relationships between innovation and firm performance, it is still unclear whether radical innovation and incremental innovation have the same effects on firm performance.

Third, our focus on technology firms in China's emerging market also represents a contribution. Previous studies on the relationship between organizational reward, control and innovation have mainly been done in the developed markets which are characterized by individualism (that is, people look after their own interests rather than the interest of in-groups) and low uncertainty avoidance (that is, people tolerate ambiguity and uncertainty). Different from the Western developed countries, China is a collectivist and high uncertainty avoidance society (Hofstede, 1997). In the West, managers have the power to control some behaviours but employees retain control over others. However, the Chinese managers may control every sphere of life of their employees (Atuahene-Gima & Li, 2002). Thus how firms reward and control employees has more far-reaching implications for innovation in the Chinese context than in the Western countries. Therefore our focus on Chinese technology firms can enrich our understanding of how the effects of organizational reward and control mechanisms on innovation and further firm performance can be different in a different national culture context.

The rest of the chapter is organized as follows. We first review the relevant literature on organizational reward, organizational control, innovation and firm performance. We then develop our research hypotheses. Figure 9.1 illustrates the theoretical model proposed in this chapter. In the empirical sections, hypotheses are tested using a sample of 194 technology firms in China. The chapter ends with a discussion of the results and implications for both researchers and practitioners.

THEORY AND HYPOTHESES

Theoretical Background

Firm innovations have been generally classified into two types: *radical* and *incremental innovations* (Dewar & Dutton, 1986; Hill & Rothaermel, 2003; Sheremata, 2004). Radical innovations are based on new design concepts that break paradigms, whereas incremental innovations are based on minor improvements or adjustments in the current product or technology. Radical innovations are based on the knowledge that is largely different from current knowledge, or comes from a mix of current knowledge and entirely new knowledge. In other words, radical innovations involve fundamental changes or development in current product or process. In contrast, incremental innovations are sustaining of the existing status quo (Hart & Christensen, 2002; Tushman & Rosenkopf, 1996). They are a development of current product and process based on current knowledge and they involve minor improvements or simple adjustments in existing product technology (Hill & Rothaermel, 2003). The major difference between radical and incremental innovations is the degree of novel technological process content embodied in the innovations. A radical innovation represents a clear departure from existing products or process. It incorporates a large degree of new knowledge and involves a high level of risk and resource consumption (Dewar & Dutton, 1986).

Designed to promote employees' efforts in innovation, organizational rewards can vitalize the latent ability of employees by arousing their passion

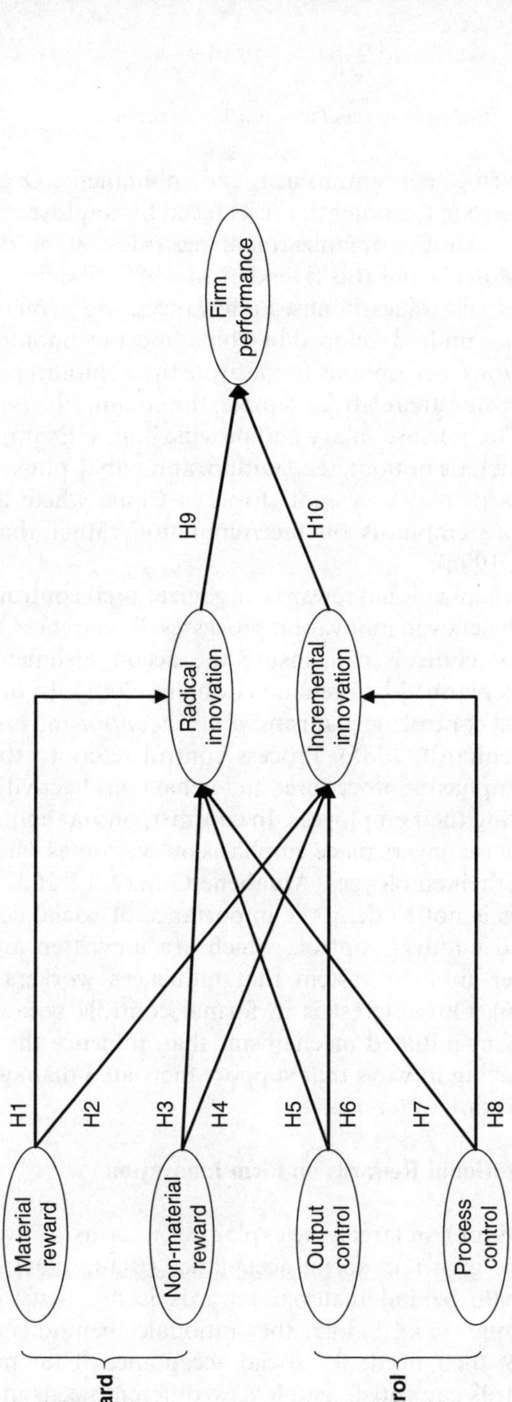

Figure 9.1 Theoretical framework

for work and inspiring their enthusiasm and spontaneity. Organizational rewards can compensate the potential risks faced by employees in developing innovations. Generally, organizational rewards can be divided into *material* and *non-material* rewards (Henderson, 1989). Examples of material rewards include basic wages, bonuses, allowances, and so on. Because the capital market is still underdeveloped in China, most technology firms can not use stocks and/or stock options to motivate their employees. Therefore, the reward mechanisms are relatively simple; the commonly used organizational rewards are to increase salary and provide bonus. Examples of non-material rewards include honour, recognition, and public praise, and so on. Non-material rewards play an essential role in China where the national culture places more emphasis on face/reputation rather than financial possession (Chow, 1992).

In addition to organizational rewards, organizational controls are necessary to enhance efficiency in innovation processes. Researchers have argued that organizational controls can ensure the accomplishment of firms' innovative goals as planned beforehand (Robbins, 2001). In this study, we focus on two formal control mechanisms: *process control* and *output control* (Ouchi, 1980; Eisenhardt, 1985). Process control refers to the extent to which managers emphasize procedures and behavioural activities in monitoring and evaluating their employees. In contrast, output control refers to the extent to which managers place emphasis on outcomes when monitoring and evaluating their employees (Atuahene-Gima & Li, 2002). Our focus on formal controls is not to deny the importance of social controls such as self-control and cultural controls which are unwritten and typically represent a worker-initiated system that influences workers' behaviour (Jaworski, 1988:26). Our interest is in formal controls because they are written, management-initiated mechanisms that influence the probability of employees behaving in ways that support the stated managerial objectives such as innovation outcomes.

Effects of Organizational Rewards on Firm Innovation

Individuals' behaviour can largely be explained in terms of two dominant interests: economic gain and social acceptance (Blau, 1964; Henderson, 1989). The rationale behind material rewards is to satisfy individuals' needs for economic gain, while the rationale behind non-material rewards is to meet their needs for social acceptance. Thus material and non-material rewards can satisfy employees' different needs in developing innovations.

We argue that material reward is negatively related to radical innovations, but positively related to incremental innovations. Radical innovations involve a high level of risks and uncertainties. It usually takes more time and more resources to develop radical innovations but the outcomes are uncertain. Amabile (1983) suggests that, in situations involving a large amount of uncertainties, the overemphasis on external and objective performance reduces the intrinsic incentive and curiosity needed for innovation work. A series of studies by Amabile and her colleagues (Amabile, 1983; Amabile *et al.*, 1986; Hennessey & Amabile, 1998) indicate that extrinsic rewards – concrete tangible rewards such as bonuses, pay increases and awards – are detrimental to innovation. This is particularly true in China where employees in technology firms typically receive a higher salary than employees in other industries. Therefore, for employees who have already enjoyed high salaries, material rewards may be less effective in motivating them to engage in radical innovations involving high risks. Further, in a country emphasizing honour and reputation, employees' confidence and feeling of achievement may be harmed by organizations which only provide material rewards. In such situations, employees may try to avoid radical innovations to reduce their responsibilities in case the radical innovation fails.

In contrast, incremental innovations involve minor adjustment and development of current products and processes, which have a low degree of risk and high likelihood of success. Thus it is relatively fast and easy to observe and evaluate the outcomes of incremental innovations. When material reward is utilized and objective goals are set, employees consider not only the possible economic gain from developing innovation but also the potential loss if their innovations fail. Therefore the employment of material rewards can prompt employees to focus on developing innovations that are easy to accomplish, most of the time, incremental innovations.

H1: In Chinese technology firms, material reward is negatively associated with radical innovation.

H2: In Chinese technology firms, material reward is positively associated with incremental innovation.

Organizational rewards, to be effective, must be consistent with commonly held cultural values about work (Chow, 1992). Non-material reward plays an essential role in China where employees consider face/reputation more important than financial possession. When employees receive non-material rewards such as public praise, they feel that they are acknowledged by their organizations and their face/reputation is upheld. As a consequence, employees are more willing to conduct innovations including both radical and incremental innovations Non-material reward is particularly important for radical innovations, which are more venturesome. The development of radical innovations usually takes longer than that of incremental innovations. Also the outcome of employees' innovation efforts is less predictable for radical innovations than for incremental innovations. Innovating employees need time to appraise and select multiple goals, requiring firms to provide flexibility to develop their work activity (Mumford, 2000). Additionally, for employees developing radical innovations, personal growth and self-realization are more valuable than receiving material reward. Non-material rewards can work better than material rewards in motivating employees to develop radical innovations.

H3: In Chinese technology firms, non-material reward is positively associated with radical innovation.

H4: In Chinese technology firms, non-material reward is positively associated with incremental innovation.

Organizational Controls and Firm Innovations

Technology firms use different organizational controls to ensure that innovations will be effectively developed and commercialized. Here we use agency theory to explain the relationships between organizational controls and firm innovations. Agency theory assumes that both the agent (for example, innovating employee) and the principal (for example, technology firm) are rational and self-interested and the agent is both effort- and risk-averse (Bloom & Milkovitch, 1998; Eisenhardt, 1989). This creates a moral hazard, in which the agent tends to maximize compensation without exerting the effort required to maximize the principal's goals.

Output control represents a 'hands-off' approach to managing innovating employees, in that they are given a great deal of autonomy and independence to perform their duties and are compensated for the output they achieve (Atuahene-Gima & Li, 2002: 65). Thus output control shifts substantial innovation risks to the innovating employees because their outputs may be affected by environmental and company factors beyond their control. As Whitener *et al.* (1998: 515) argue, to the extent that the employee is compensated on the basis of outcomes beyond his or her control, the performance risk to him or her is higher. When developing radical innovations, employees face considerable uncertainties about future outcomes. Under this condition, both the firm and the employees tend to minimize their own risks. Thus a purely output-based reward structure is likely to be counterproductive for employees developing radical innovations because it places an excessive amount of risk on the employees. Further, when an organization emphasizes output control, the appraising standards are usually short-term oriented (Hayes & Abernathy, 1980). Previous studies have shown that the organization whose incentive is based on short-term financial outcomes always has comparatively little investment in R&D (Hoskisson, Hitt & Hill, 1991; Hoskisson, Hitt, Johnson & Moesel, 1993). These results suggest that output control can be detrimental to radical innovations.

In contrast, employees are more likely to conduct incremental innovation when the output control is utilized. Incremental innovation involves less risk and takes less time and thus employees face less uncertainty about future outcomes in developing this type of innovation. In this situation, employees can take advantage of the autonomy and independence given by the firm to perform their innovation activities without worrying about the potential risks.

H5: In Chinese technology firms, output control is negatively associated with radical innovation.

H6: In Chinese technology firms, output control is positively associated with incremental innovation.

Process control reflects the extent to which a firm emphasizes procedures and behavioural activities in monitoring, evaluating and rewarding employees. Process control ensures that employees receive rewards as long as process requirements are met, irrespective of the output achieved (Atuahene-Gima & Li, 2002). It therefore reduces the pressure and risks in developing radical innovations because it is the firm, rather the innovating employee, who assumes *much* of the performance risk. Consequently, although process control may limit autonomy and independence of the employees, it sends a positive signal of the firm's concern, care and support to the employees who are developing radical innovations.

By using process control, employees at different organizational hierarchical levels can exchange information more efficiently (Hoskisson, Hitt & Ireland, 1994; Hitt *et al.*, 1996). Top managers who stress process control always take part in and direct subordinates' activities, which is helpful to constructing a system to share responsibility for innovations between managers and employees. Because employees can feel the support from the managers, and believe their performance will be appraised according to the efforts devoted rather than short-term outcomes, they will be more willing to take risks and responsibilities involved in radical innovations (Hitt *et al.*, 1996). Similarly, Scott (1995) argues that directing process and approach, rather than specifying a single desired outcome, better manages radical innovations.

However, introducing process control into an incremental innovation process may increase unnecessary costs and discourage the development of innovations. As we argued previously, a potential disadvantage of process control is that it could be perceived by employees as limiting their self-control and autonomy. Particularly in developing incremental innovations which has characteristic of low risk and small steps on adjustment of existing technologies or products, too much monitoring from the managers can make the employees feel that their managers do not trust their capabilities of innovation and that they have limited autonomy. Under this condition, process control can be negatively related to incremental innovation.

H7: In Chinese technology firms, process control is positively associated with the radical innovation.

H8: In Chinese technology firms, process control is negatively associated with the incremental innovation.

Innovation and Firm Financial Performance

Innovation is important to value creation. Prior studies have shown that firm innovation has enormous effects on firms' survival and success (Damonpour, 1991; Dougherty & Hardy, 1996; Nohria & Gulati, 1996). In technology-intensive industries, firms face quickly changing market and technology environments where new products are developed at any moment. Technology firms have to innovate constantly to maintain long-term competitive advantages. Any firm which cannot innovate or whose innovations are not successful in time would be eliminated from the market. However, not all kinds of innovations are helpful for firm performance, particularly for long-term financial performance.

In general, radical innovations are critical for firm performance because, once a radical innovation succeeds, it will represent a high obstacle for rivals. Thus radical innovations can help firms achieve long-term competitive advantage. However, radical innovations are expensive. They are characterized by tremendous resource requirements and a high level of risks. Investing a tremendous amount of financial and human resources in radical innovations may drag down a firm's performance in the short run. Yet radical innovation can bring firms long-term competitive advantage and continually increase firm performance in the long run. In contrast, incremental innovation can bring firm performance advantage in the short run, but cannot produce sustained competitive advantage for the firm in the long run. As a matter of fact, the emphasis on incremental innovations can be harmful to a firm's financial performance if the firm cannot catch up with its competitors by adopting radical innovations.

The above argument is particularly valid in the Chinese context. In a recent study, Hitt, Li and Worthington (2005) argue that, because of the potential of the Chinese market, both local firms and foreign entrants are experiencing significant competition in this market. Thus both local firms and foreign entrants must acquire the necessary knowledge as rapidly as possible to remain competitive. This requires an exploratory learning process which involves experimentation with new alternatives and developing radical innovations. Thus, we propose the following:

H9: In Chinese technology firms, radical innovations are positively associated with firm financial performance in the long run.

H10: In Chinese technology firms, incremental innovations are negatively associated with firm financial performance in the long run.

METHODOLOGY

Data Collection and Sample

We randomly selected 300 firms from the list of technology firms provided by the Committee of Economy and Commerce from eight provinces – Shaanxi, Henan, Shanghai, Guangdong, Liaoning, Sichuan, Shandong and Shanxi. We contacted chief executive officers of these firms to introduce the study and encourage their participation. A total of 280 firms agreed to participate in this project. We collected data through on-site interviews with a questionnaire. We assured all respondents of confidentiality to encourage their candid responses. Each interview took about 60 to 90 minutes. Our data collection efforts yielded 194 completed questionnaires for a participation rate of 64.7 per cent (=194/300). A brief description of sample characteristics is summarized in Table 9.1.

Measures

Material reward

We used three items to measure material reward. The respondents were asked to indicate the extent to which they agree that the following reward approaches work for them in conducting innovation activities: (1) increasing individual material fortune, (2) increasing opportunity to gain economic

Table 9.1 Sample characteristics (N = 194)

Sample characteristics	Percentage
Industries	
Material industry	5.2
Chemical industry	9.3
Electronic industry	40.2
Engineering industry	25.2
Medicine	12.9
Other industry	7.2
Employee numbers	
50 or less	18.5
51–200	37.1
201–500	19.1
501–1000	9.3
1000 or more	16.0

interest, and (3) guaranteeing the future living of home members. Responses were based on a scale ranging from 1 (strongly disagree) to 7 (strongly agree). The Cronbach alpha for this measure was 0.85.

Non-material reward

Non-material reward was also measured by using three items. The respondents were asked to indicate the extent to which the following reward approaches work for them in conducting innovation activities: (1) acquiring social acceptance, praise and honour, (2) obtaining individual opportunity by accepting the challenge of the innovation, and (3) obtaining individual self-development. Responses were based on a scale ranging from 1 (strongly disagree) to 7 (strongly agree). The Cronbach alpha for this measure was 0.78.

Process control

We developed the measures of process control by modifying the items from Xu and Wang (1997). We asked the respondents to indicate the extent to which they agree that top managers in their organization control firm innovation activities by: (1) permitting the employees' mistake taking place in the innovation process, (2) developing highly trusting relationship between managers and employees, and (3) monitoring the extent to which employees follow established procedures for innovations. Responses were based on a 7-point scale ranging from 1 (strongly disagree) to 7 (strongly agree). The Cronbach alpha for this measurement was 0.88.

Output control

The measures of output control are mainly taken from the research of Hitt *et al.* (1996). We asked the respondents to indicate the extent to which they agree that top managers in their organization control innovation activities by: (1) exerting high requirement on ROI (return on investment) of the innovation, (2) ensuring cash currency being abundant through the innovation, and (3) requiring the increase of the net assets in the innovation process. Responses were made on a 7-point scale ranging from 1 (strongly disagree) to 7 (strongly agree). The Cronbach alpha was 0.74.

Radical innovation

We measured radical innovation by using four items that were modified from Ettlie *et al.* (1984) and Dewar & Dutton (1986). The questions are, to what extent do you agree with the following statements: (1) your firm develops new products which have completely new functions in the market, (2) your firm conducts product innovation by frequently introducing new technologies, (3) your firm conducts process innovations by introducing completely new technologies in the industry, and (4) your firm conducts process innovations based on breakthrough concepts. Responses were based on a 7-point scale ranging from 1 (strongly disagree) to 7 (strongly agree). The Cronbach alpha was 0.77.

Incremental innovation

We measured incremental innovation by using three items modified from Ettlie *et al.* (1984) and Dewar & Dutton (1986). The questions are, to what extent do you agree with the following statements: (1) your firm develops new products with some new style and services in the market, (2) your firm conducts process innovation by frequently modifying and improving the firm's existing technologies, and (3) your firm conducts innovations by improving current products and processes based on current knowledge. Responses were based on a 7-point scale ranging from 1 (strongly disagree) to 7 (strongly agree). The Cronbach alpha was 0.66.

Firm performance

We measured firm's financial performance by using multiple items. Reliance upon multiple performance measures is important, as no one indicator reasonably captures firm financial performance (Bourgeois, 1980; Chakravarthy, 1986; Weiner & Mahoney, 1981). We asked the respondents to indicate the extent to which they agree that their firms have achieved the following outputs in the past five years (1997–2001): (1) increasing rate of sale revenue; (2) increasing rate of profit; (3) increasing rate of ROI (return on investment); and (4) increasing rate of market share. Responses were based on a 7-point scale ranging from 1 (strongly disagree) to 7 (strongly agree). The Cronbach alpha was 0.85.

RESULTS

We tested the hypotheses using structural equation modelling (SEM) techniques as implemented in AMOS 4.0. Our analysis follows the two-stage procedure recommended by Anderson and Gerbing (1988): (1) estimating the model's reliability and validity, which can ensure that the variables in the following analysis are reliable and valid and (2) testing the theoretical model.

Table 9.2 reports the reliability of all items for our measurements. Generally, the item's loading should not be less than 0.4. In Table 9.2, all loadings are greater than 0.4. Although the Cronbach alpha of incremen-

Table 9.2 Results of confirmatory factor analysis

Factors	Items	Loading	alpha
Material reward	1. increasing individual material fortune	0.796	
	2. increasing opportunity to gain economic interest	0.862	0.85
	3. guaranteeing future living of home members	0.866	
Non-material reward	4. acquiring social acceptance, praise and honour	0.796	
	5. obtaining individual opportunity by accepting the challenge of the innovation	0.830	0.78
	6. obtaining individual self-development	0.871	
Process control	7. permitting the employees' mistake taking place in the innovation process	0.933	
	8. developing highly trusting relationship between managers and employees	0.935	0.88
	9. monitoring the extent to which employees follow established procedures for innovations	0.840	
Output control	10. exerting high requirement on ROI of the innovation	0.590	
	11. ensuring cash currency being abundant through the innovation	0.928	0.74
	12. requiring the increase of the net assets in the innovation process	0.912	

Table 9.2 (continued)

Factors	Items	Loading	alpha
Radical innovation	13. develops new products which have completely new functions in market	0.741	0.77
	14. conducts product innovations by frequently introducing new technologies	0.791	
	15. conducts process innovations by introducing completely new technologies in the industry	0.764	
	16. conducts process innovations based on breakthrough concepts	0.781	
Incremental innovation	17. develops new products with some new style and services in the market	0.679	0.66
	18. conducts process innovation by frequently modifying and improving the firm's existing technologies	0.837	
	19. conducts innovations by improving current products and processes based on current knowledge	0.790	
Firm performance	20. increasing rate of sales revenue	0.898	0.85
	21. increasing rate of profit	0.926	
	22. increasing rate of ROI	0.876	
	23. increasing rate of market share	0.661	

tal innovation is 0.66 (lower than 0.70), all item loadings are more than 0.68. Thus the reliability of incremental innovation should not be a concern (Li & Atuahene-Gima, 2001). Table 9.3 presents the descriptive statistics of all factors generated from our measurement model, including the means, standard deviations and correlations.

Table 9.4 summarizes the structural fitness statistics. The χ^2 of our structural equation model is 177.412. The goodness-of-fit index (GFI) is 0.928, which is larger than the commonly accepted 0.9. The adjusted goodness-of-fit index (AGFI) is 0.885; the incremental fit index (IFI) 0.998; the normed fit index (NFI) 0.933; the Tucker–Lewis index (TLI) 0.997; the comparative fit index (CFI) 0.998; and the root mean square error of approximation (RMSEA) 0.013, lower than the commonly accepted level of 0.05. These show that the fit statistics of SEM are acceptable. Table 9.5 summarizes the results of our hypothesis testing.

Table 9.3 Descriptive statistics

Variables	Mean	S. D.	1	2	3	4	5	6	7
1. Material reward	4.63	1.04	1.00						
2. Non-material reward	4.93	1.01	0.61**	1.00					
3. Output control	4.94	1.14	0.42**	0.49**	1.00				
4. Process control	5.34	1.16	0.47**	0.58**	0.73**	1.00			
5. Radical innovation	3.81	1.53	0.12	0.18*	0.09	0.15*	1.00		
6. Incremental innovation	4.30	1.37	0.11	0.22**	0.17*	0.21**	0.64**	1.00	
7. Firm performance	3.51	1.92	−0.89	−0.07	−0.05	−0.04	0.17*	0.11	1.00

Note: N = 194; significance level: * $p < 0.05$; ** $p < 0.01$.

Table 9.4 Structural fitness test

Indicator	Desired value	Value
χ^2	—	177.412
P	>0.05	0.373
GFI	Near or greater than 0.9	0.928
AGFI	Near or greater than 0.9	0.886
RMSEA	<0.035	0.013
Incremental fit index (IFI)	Near or greater than 0.9	0.998
Normed fit index (NFI)	Near or greater than 0.9	0.933
Tucker–Lewis index (TLI)	Near or greater than 0.9	0.997
Comparative fit index (CFI)	Near or greater than 0.9	0.998

Table 9.5 Summary of hypothesis testing

Hypothesis	Description of path	Standardized path coefficient	P-value	Conclusion
H1	Material reward → radical innovation	−0.549*	0.021	supported
H2	Material reward → incremental innovation	−0.279	0.149	not supported
H3	Non-material reward → radical innovation	0.597*	0.021	supported
H4	Non-material reward → incremental innovation	0.605*	0.022	supported
H5	Output control → radical innovation	−0.368*	0.032	supported
H6	Output control → incremental innovation	0.483*	0.026	supported
H7	Process control → radical control	0.383*	0.042	supported
H8	Process control → incremental innovation	−0.565*	0.047	supported
H9	Radical innovation → firm performance	0.541**	0.002	supported
H10	Incremental innovation → firm performance	−0.356*	0.042	supported

Note: Significance level: * $p < 0.05$; ** $p < 0.01$.

Organizational Rewards and Firm Innovation

The first set of hypotheses (Hypotheses 1–4) predicts the effects of organizational rewards on firm innovations. Hypothesis 1 argues that the material reward is negatively associated with radical innovation in Chinese technology firms. This hypothesis received strong empirical support (-0.549, $p < 0.05$). Hypothesis 2 claims that the material reward is positively associated with incremental innovation in Chinese technology firms. This hypothesis was not supported ($p > 0.10$). Hypothesis 3 predicts that non-material reward is positively associated with radical innovation in Chinese technology firms. This hypothesis was strongly supported (0.597, $p < 0.05$). Hypothesis 4, which predicts a positive relationship between non-material reward and incremental innovation, was also supported (0.605, $p < 0.05$). That is, for Chinese technology firms, the use of non-material reward can enhance both radical and incremental innovations.

Organizational Controls and Firm Innovations

The second set of hypotheses (Hypotheses 5–8) predicts the effects of organizational controls on firm innovations. Hypothesis 5 posits that output control is negatively associated with radical innovation in Chinese technology firms. This hypothesis received strong support (-0.368, $p < 0.05$). Hypothesis 6, predicting a positive relationship between output control and incremental innovation, was also supported (0.483, $p < 0.05$). Hypothesis 7 claims that process control is positively related to radical innovation. This hypothesis was supported (0.383, $p < 0.05$). Hypothesis 8, which predicts a negative relationship between process control and incremental innovation, was also supported (-0.565, $p < 0.05$). These results suggest that, in Chinese technology firms, the focus on process control can enhance radical innovation, while the emphasis on output control can advance incremental innovation.

Firm Innovations and Long-term Performance

Hypotheses 9 and 10 predict the effect of firm innovations on long-term financial performance. Hypothesis 9 predicts that radical innovation is positively related to a firm's long-term performance; this hypothesis was strongly supported (0.541, $p < 0.01$). Hypothesis 10, arguing for a negative relationship between incremental innovation and a firm's long-term performance, was also supported (-0.356, $p < 0.05$). The results show that it is radical innovation, not incremental innovation which can help Chinese technology firms to achieve better long-term financial performance.

DISCUSSION AND CONCLUSION

With a sample of technology firms in China, in this study we examined the effects of organizational reward and control on different types of firm innovations and further firm performance. There are several interesting findings. First, our results indicate that, in Chinese technology firms, material reward has a negative relationship to radical innovation; on the contrary, non-material reward promotes both radical and incremental innovations. Such findings are consistent with the results reported by Amabile and her colleagues (Amabile, 1983; Amabile *et al.*, 1986; Hennessey & Amabile, 1998), but contrary to the conclusions of Baer (1997) and Eisenberger & Cameron (1996). It seems that, in technology-intensive industries which involve high levels of risks and uncertainties for innovation, non-material rewards are more likely to motivate employees to take risks and engage in firm innovations (including both incremental and radical innovations). As we noted earlier, Chinese culture places a significant emphasis on face/reputation rather than financial possession. Thus, compared with financial possession, social acceptance and self-development offered by non-material rewards can make innovating employees in Chinese technology firms more fulfilled.

Consistent with our hypotheses, the second group of findings shows that process control in technology firms can encourage radical innovation, but discourage incremental innovation, while output control may encourage incremental innovation but discourage radical innovation. These results show that process control and output control play differential roles in a firm's innovation process. It seems that well designed process control mechanism can inspire employees to work on the significant product or technology innovations which need a long time to enter the market and earn returns (Goold & Campbell, 1987; Hoskisson *et al.*, 1991). By implementing process control, top managers share responsibilities for innovation, and employees can expect their efforts to be valued. For technology firms in the Chinese transition economy, developing radical innovations is more difficult and risky than it is in developed economies. Thus process control becomes more critical for radical innovations in this context. However, output control can lead to incremental innovations. Since output control shifts substantial innovation risks to the innovating employees, employees are more likely to conduct incremental innovations when the output control is utilized.

The third group of findings is that, in Chinese technology firms, radical innovations can enhance firms' long-term financial performance, but incremental innovations may damage their long-term performance. These results show that technology firms' highly different radical innovations can help create high barriers for existing competitors and potential new entrants. This helps firms to maintain their competitive advantage and then advance long-term performance. Technology firms can survive through incremental innovations in the short run, but, in the long run, technology firms which depend on incremental innovations will be driven out of the market in the rapidly changing technological environment. Our results demonstrate that, for technology firms, not all innovations are critical for firms' long-term performance. While radical innovations can advance a firm's long-term performance, incremental innovations may decrease the firm's long-term performance.

Our study contributes to the literature in several regards. First, by making a distinction between radical and incremental innovations, we examined how material rewards and non-material rewards may have different effects on incremental and radical innovations. As noted earlier, while some researchers have shown that material rewards can encourage innovation, others have disclosed opposite evidence. Our study advanced this literature by suggesting that material reward may be detrimental to radical innovation but may have no relationship to incremental innovation. Thus our findings highlight that, without specifying the nature of innovation type, the linkage between reward mechanisms and innovation cannot be theoretically developed.

Second, our study also contributes to the organizational control literature. Although prior research has argued that control mechanisms may have some relationships to innovation (Hitt *et al.*, 1996), the ways in which different control mechanisms affect various types of firm innovation are not clear. The study presented here disaggregates control mechanisms and firm innovations and provides unique empirical findings because they represent, we believe, the first evidence in the management literature on the linkages between organizational controls and firm innovations. Our results add to this line of research by showing that process control and output control play different roles in effecting radical innovation and incremental innovation.

Third, by differentiating incremental and radical innovation, we were also able to offer an in-depth analysis of the way different types of firm innovations affect firm performance, particularly long-term financial firm performance. Our results show that, in Chinese technology firms, radical innovations are positively related to firm long-term performance while incremental innovations are negatively related to firm long-term performance. These findings send a strong message to managers in technology firms that, to maintain their sustainable competitive advantages, it is important for their firms to focus on radical innovations rather than incremental innovations.

Finally, we believe that this is the first empirical study which has systematically examined the relationship between organizational rewards/control mechanisms and firm innovations in the context of technology firms in

China's emerging market. Note that our theoretical model is built upon theories developed from the Western market economies (such as the US). By applying this model to China, we found evidence to support the model. However, rather than saying that our study has extended the Western literature to the Chinese context, we believe that research of this kind would enrich our understanding of the nature of rewards/control–innovation linkage across different institutional contexts.

Limitations and Future Research

Despite the implications and contributions to the relevant literature, this study has several limitations that open up future research avenues. First, as discussed earlier, our results can be biased by the sample, in which more than 40 per cent of the sample firms are from the electronics industry. Also, to improve the generalizability of our findings, future research should build on these results and examine the relationship between organizational rewards and controls and firm innovations in low-technology industries. It would be interesting to examine how the relationships we demonstrated here may differ between high-technology industries and low-technology industries.

Second, the use of self-report data may pose such potential problems as the limited recall of the respondents, biased perceptions of past realities, and common method issues. However, we took several actions during data collection to improve both the reliability and the validity of the retrospective reporting. Our post hoc examination and validation analysis indicated no serious common method problems.

Finally, the cross-sectional data used in the study do not allow for causal interpretations among the variables. Nor do they allow us to examine the 'dynamic' nature of the relationships we examined here. Future studies can benefit from a longitudinal research design. In conclusion, organizational rewards and controls can exert important effects on a firm's innovations and further financial performance. We expect this study to serve as a starting point for further efforts in enriching the understanding of the relationship between organizational rewards and controls, firm innovations, and performance in the context of transition economies.

REFERENCES

Amabile, T.M. (1983), *The Social Psychology of Creativity*, New York: Springer-Verlag.

Amabile, T.M., B.A. Hennessey and B.S. Grossman (1986), 'Social influences on creativity: the effects of contracted-for-reward', *Journal of Personality and Social Psychology*, **50**, 14–23.

Anderson, J.C. and D.W. Gerbing (1988), 'Structural equation modeling in practice: a review and recommended two step approach', *Psychological Bulletin*, **103**, 411–23.

Atuahene-Gima, K. and H. Li (2002), 'When does trust matter? Antecedents and contingent effects of supervisee trust on performance in selling new products in China and the United States', *Journal of Marketing*, **66** (July), 61–81.

Baer, J. (1997), 'Gender differences in the effects of anticipated evaluation on creativity', *Creativity Research Journal*, **10**, 25–32.

Beijing Bureau of Statistics (2002), 'Firm innovations in the Zhongguancun Science Park: an international comparison' (www.stats.gov.cn).

Blau, P.M. (1964), *Exchange and Power in Social Life*, New York: John Wiley.

Bloom, M. and G.T. Milkovitch (1998), 'Relationship among risk, incentive pay, and organizational performance', *Academy of Management Journal*, **41** (3), 283–97.

Bourgeois, L.J. (1980), 'Performance and consensus', *Strategic Management Journal*, **1**, 227–48.

Bruton, G.D. and Y. Rubanik (2002), 'Resources of the firm, Russian high-technology startups, and firm growth', *Journal of Business Venturing*, **17**, 553–76.

Chakravarthy, B.S. (1986), 'Measuring strategic performance', *Strategic Management Journal*, **7**, 437–58.

Chow, I.H. (1992), 'Chinese workers' attitudes towards compensation practices in the People's Republic of China', *Employee Relations*, **14** (3), 41–55.

Damonpour, F. (1991), 'Organizational innovation: a meta-analysis of effects of determinants and moderators', *Academy of Management Journal*, **34** (3), 555–90.

Dewar, R.D. and J.E. Dutton (1986), 'The adoption of radical and incremental innovations: an empirical analysis', *Management Science*, **32** (11), 1422–33.

Dougherty, D. and C. Hardy (1996), 'Sustained product innovation in large, mature organizations: overcoming innovation-to-organization problems', *Academy of Management Journal*, **39**, 1120–54.

The Economist (2004), 'Science and technology: southern comfort, eastern promise; third-world biotechnology', 11 Dec., 83.

Eisenberger, R. and J. Cameron (1996), 'Detrimental effects of reward: reality on myth?', *American Psychologist*, **51**, 1153–66.

Eisenhardt, K.M. (1985), 'Control: organizational and economic approaches', *Management Science*, **31** (2), 134–49.

Eisenhardt, K.M. (1989), 'Agency theory: an assessment and review', *Academy of Management Review*, **14**, 57–74.

Eisenhardt, K.M. and C.B. Schoonhoven (1990), 'Organizational growth: linking founding team, strategy, environment, and growth among U.S. semiconductor ventures, 1978–1988', *Administrative Science Quarterly*, **35** (3), 504–29.

Ettlie, J.E., W.P. Bridges and R.D. O'Keefe (1984), 'Organization strategy and structural differences for radical versus incremental innovation', *Management Science*, **30** (6), 682–95.

Foster, R.N. (1986), *Innovation: the Attacker's Advantage*, New York: Summit Books.

Goold, M. and A. Campbell (1987), *Strategies and Styles*, London: Blackwell.

Hart, S. and C.M. Christensen (2002), 'Driving innovation from the base of the global pyramid', *MIT Sloan Management Review*, **44** (1), 51–6.

Hayes. R.H. and W.J. Aberbathy (1980), 'Managing our way to economic decline', *Harvard Business Review*, **58** (7), 67–77.

Henderson, R. (1989), *Compensation Management: Rewarding performance*, 5th edn, Englewood Cliffs, NJ: Prentice-Hall.

Hennessey, B.A. and T.M. Amabile (1998), 'Reward, intrinsic incentive, and creativity', *American Psychologist*, **53**, 674–5.

Hill, C.W. and F.T. Rothaermel (2003), 'The performance of incumbent firms in the face of radical technological innovation', *Academy of Management Review*, **28** (2), 257–74.

Hitt, M.A., H. Li and W.J. Worthington (2005), 'Emerging markets as learning laboratories: learning behaviors of local firms and foreign entrants in different institutional contexts', *Management and Organization Review*, **1** (3), 353–80.

Hitt, M.A., R.E. Hoskisson, R.A. Johnson and D.D. Moesel (1996), 'The market for corporate control and firm innovation', *Academy of Management Journal*, **39**, 1084–1119.

Hofstede, G. (1997), *Cultures and Organizations: Software of the mind*, New York: McGraw-Hill.

Hoskisson, R.E., M.A. Hitt and C.W.L. Hill (1991), 'Managerial risk taking in diversified firms: an evolutionary perspective', *Organization Science*, **3**, 296–314.

Hoskisson, R.E., M.A. Hitt and R.D. Ireland (1994), 'The effects of acquisitions and restructuring (strategic refocusing) strategies on innovation', in G. von Krogh, A. Sinatra and H. Singh (eds), *Managing Corporate Acquisitions*, London: Macmillan Press, pp. 144–69.

Hoskisson, R.E., M.A. Hitt, R.A. Johnson and D.D. Moesel (1993), 'Construct validity of an objective (entropy) categorical measure of diversification strategy', *Strategic Management Journal*, **14**, 215–35.

Jaworski, B.J. (1988), 'Toward a theory of marketing control: environmental context, control types, and consequences', *Journal of Marketing*, **52**, 23–39.

Jensen, M.C. (1993), 'The modern industrial revolution, exit, and the failure of internal control systems', *Journal of Finance*, **48**, 831–80.

Li, H., K. Atuahene-Gima (2001), 'Product innovation strategy and the performance of new technology ventures in China', *Academy of Management Journal*, **44**, 1123–34.

Mumford, M.D. (2000), 'Managing creative people: strategies and tactics for innovation', *Human Resource Management Review*, **10** (3), 313–51.

Nohria, N. and R. Gulati (1996), 'Is slack good or bad for innovation?', *Academy of Management Journal*, **39** (5), 1245–64.

Ouchi, W.G. (1980), 'Markets, bureaucracies and clans', *Administrative Science Quarterly*, **25**, 129–41.

Robbins, S.P. (2001), *Organizational Behavior*, 9th edn, Englewood Cliffs, NJ: Prentice-Hall.

Scott, R.K. (1995), 'Creative employees: a challenge to managers', *Journal of Creative Behavior*, **29**, 64–71.

Sheremata, W.A. (2004), 'Competing through innovation in network markets: strategies for challengers', *Strategic Management Journal*, **29** (3), 359–77.

Tripsas, M. and G. Gavetti (2000), 'Capability, cognition, and inertia: evidence from digital imaging', *Strategic Management Journal*, **21**, 1147–61.

Tushman, M.L. and L. Rosenkopf (1996), 'Executive succession, strategic reorientation and performance growth: a longitudinal study in the U.S. cement industry', *Management Science*, **42** (7), 939–53.

Weiner, N. and T.A. Mahoney (1981), 'A model of corporate performance as a function of environmental, organizational, and leadership influences', *Academy of Management Journal*, **24**, 453–70.
Whitener, E.M., S. Brodt, A. Korsgaard and J.M. Werner (1998), 'Managers as initiators of trust: an exchange relationship framework for understanding managerial trustworthy behavior', *Academy of Management Review*, **23**, 513–29.
Xu, X. and Y. Wang (1997), 'Ownership structure, corporate governance, and corporate performance: the case of Chinese stock companies', The World Bank Economic Development Institute Office of the Director.

PART III

Entry of foreign ventures and globalization of local ventures in China

10. Location, political risk and foreign ventures in the PRC

Jonathan Brookfield and Jun Li

INTRODUCTION

The operations of Western multinationals in the People's Republic of China (PRC) are concrete manifestations of a kind of corporate entrepreneurship. With nearly all such ventures having been enabled by relatively recent political changes in the PRC, most represent new ventures. Moreover, given the economy's market potential and attractiveness as a base for production, the business activities of Western multinationals in the PRC have the potential to significantly reorient the strategic perspective and competitive position of such companies. If corporate entrepreneurship encompasses two types of phenomenon including 'the birth of new businesses within existing organizations' and 'the transformation of organizations through renewal of the key ideas on which they are built' (Guth and Ginsberg, 1990) and corporate entrepreneurship may be defined as 'the process whereby an individual or groups of individuals, in association with an existing organization, create a new organization, or instigate renewal or innovation within that organization' (Sharma and Chrisman, 1999:18), then the activities of Western multinationals in the PRC seem to fit comfortably within established notions of corporate entrepreneurship.

As Dess, Ireland, Zahra, Floyd, Janney and Lane (2003:363) point out, however, 'companies can no longer simply export their domestic business practices to foreign markets and expect to reap the full benefits of internationalization'. One important class of change for companies as they internationalize is changes in the political environment. Differences in political environment can give rise to differences in political risk, which often lead to differences in firm political strategies, and such differences may be especially broad when considering, say, the home environment of US companies relative to the political environment surrounding their investments in the PRC. According to a recent PriceWaterhouse-Coopers (2004) survey, conducting a more thorough risk analysis before investing topped the list of replies companies offered to the question of what such companies would do differently if asked to invest in China again.

Political risk comes in many different forms, and firms have developed a number of different strategies to cope with it. This study begins by considering some different notions of political risk and then looks at the experience of several foreign ventures in the PRC. The chapter focuses on the companies' political activities and the ways in which they have dealt with issues of political risk, especially bureaucratic risk, as they have deepened their involvement in the local economy. As an important interface between an organization and its environment, political strategies – in their in-betweenness – also represent a kind of entrepreneurial behaviour. As a result, this study holds out the promise of overlapping with the concerns of corporate and international entrepreneurship in a number of different ways.

In this study, bureaucratic risk is seen as a conceptual sub-set of political risk which stems from the many ways governments and government officials can make things difficult for firms in terms of rules, regulations and fees. Broadly speaking, bureaucratic risks may be thought of as government activities which, at the extreme, can lead to a kind of creeping expropriation (Binkowski, 1980).

As a political strategy, location has received some attention in the United States at both the macro level (for example, the decision of Japanese firms to invest and set up operations in the US to avoid potential trade friction) and also the micro level (through studies of the decisions foreign firms have made to establish factories in specific locations as a result of favourable negotiations with state and local officials). By focusing on location decisions in the PRC, a country with an institutional structure quite different from that of the United States, we believe this chapter makes a distinct contribution to research regarding location and political strategy.

In what follows, this chapter considers the political activities and, in particular, the political risk mitigation strategies of six US companies with operations in the PRC. This chapter first provides a discussion of political risk and political strategy. We then give a basic overview of the six companies and their operations in the PRC, after which we discuss some of the political activities and political risk mitigation strategies these companies have adopted. We then focus on evidence regarding the use of location as a political strategy. A short discussion and conclusion follows. Available evidence seems to suggest that location can be used as a political strategy.

POLITICAL RISK

As Fitzpatrick (1983) has noted, definitions of political risk run from the very broad – 'political risks are all "non-business" risks' (Truitt, 1974:12) – to the extremely narrow; for example, political risk concerns 'the loss of control over ownership or loss of benefits of enterprise by government action' (Commission on Foreign Investments, 1962).

While a number of useful definitions of political risk exist, including Howell (1998:3) and Kobrin (1982:29), for the purposes of this study, we see risk as 'a condition of uncertainty, the outcome of which could have a negative effect on an entity or organization'. Political risks may therefore be seen as 'conditions of uncertainty arising from political events which could have a negative effect on an entity or organization'. Political risk assessment may therefore be seen as an analysis of political events with an eye towards the potential negative effects that a political event could have on a company or investment and, in this, our approach is similar to that of Griffin and Pustay (2002:72), which is to say, we see political risk stemming from changes in the political environment that may adversely affect a firm. We think the above definition helps overcome the problem of an overly narrow conception of political risk and also helps focus attention, not just on political events, but on their potential impact on a specific investment.

Given the definition, it may be helpful to note explicitly several distinctions embedded in the above language. First, it should be clear that we have taken a broad view of the drivers of political risk. While some have a somewhat more focused view of political risk (for example, Weston and Sorge, 1972:60), seeing it as stemming specifically from government action, we think political risk may arise, not just from government action, but from political events more broadly. Second, we do not think political instability need necessarily lead to political risk, at least not from a corporate point of view. As many have commented with regard to post-war Italian politics, while some places may see a great deal of political instability, that instability need not be a problem and so need not lead to political risk. In short, changes in the political environment that have no impact on a particular firm or investment are not sources of risk and therefore should not be seen as political risks (keeping in mind that the uncertainty that may result from certain political activities can, at times, have a dramatic effect on asset values and a major indirect impact on a firm, even if the direct effects are modest). Third, in this chapter, we see risk as being associated with the potentially negative things that could happen to a business or investment. In this respect, we think our approach dovetails nicely with common parlance, but we also recognize that there exists another view of the phenomenon, based on a kind of volatility approach to the concept (c.f. Haendel,

West & Meadow, 1975:xi; Miller, 1992:311). Fourth, we make a distinction between things that may affect the value or profitability of a firm and things that may negatively affect a firm. We see the second approach as being broader and more inclusive and therefore less likely to lead to a failure to consider, for example, the consequences of political risk for employees.

Finally, we note the essential role that uncertainty must play in any notion of political risk. This is important for three reasons. First, things that are expected can, at times, be easily controlled for, and so may turn out to have little impact on a firm. Second, even if they cannot be controlled for, there may be some political activities which negatively affect a firm but should not necessarily be considered political risks. Taxes, for example, though obviously having a negative effect on firm profitability, should probably not be classed as a political risk (though unexpected changes in tax policy may well qualify). Third, this distinction is important because it is also a kind of claim that political risk is, and in fact has always been, more about uncertainty than about risk. If we accept a Knightian perspective that risk can be analysed probabilistically, but true uncertainty is very difficult to know what to do with, then it would seem that political risk – oftentimes dealing with highly complex future events built upon contingency and chance – has often had, at its heart, a kind of inescapable uncertainty.

Another way that may be helpful to think about the above definition of political risk may be to consider an alternative. If political risk is 'variation in corporate performance which can not be predicted beforehand and stems from activities where the wills of two or more people are harnessed to a particular task' (De Jouvenel, 1957; Miller, 1992:311), three things are immediately evident. First, risk is a firm-level outcome variable. Second, the above definition contains a volatility notion of risk which, though commonly used in the finance community, is somewhat at odds with the way the term is used in basic conversation (in general, the risk connotes possible negative consequences, and so a windfall profit would not typically be considered a risk, though it would so qualify under the definition above). Third, it would appear that the existence of risk depends on a lack of foreknowledge. Finally, under such a definition of risk, many of the 'risks' discussed by the international business community become 'sources of risk' (or 'uncertainties', to use Miller's nomenclature).

As Henisz and Story (2003) note, use of the above definition of politics does avoid 'the trap of state-centrism', but only at the cost of a dramatic enlargement of the concept. If economics can be thought of as 'the study of how societies use scarce resources to produce valuable commodities and distribute them among different groups' (Samuelson and Nordhaus, 1989:5) and little economic activity today is of the Robinson Crusoe sort, use of the above definition of politics would seem to make it impossible to separate politics from economics (something which may not be altogether a bad thing, but is not the way academia has recently thought about such topics).

Ultimately, definitions of political risk seem to depend on the issue at hand, and there appear to be two basic directions. First, there is an area studies approach that lends itself to a 'country risk' perspective. This approach tends to define political risk quite broadly – including a wide range of political, economic and social issues – but simultaneously confines the locus of inquiry to a particular country or region. There is also a 'politics first' approach. In this case, analyses tend to be somewhat narrowly confined to issues of government behaviour, though the governments in question may at times include host governments and home governments, as well as other third party governments. Overall, we tend to agree with Kobrin (1979) that it might make sense to drop the term 'political risk' from use, as thinking in terms of a broader notion of 'investment risk' would allow for a more integrated consideration of business risk as per Miller (1992), but we acknowledge that such a change seems unlikely.

In an effort to help connect a definition of political risk to the real world, this chapter provides a basic typology of political events (Table 10.1). The events themselves are divided conceptually into host government-related events, home government-related events, third party government events, and non-government-related activities. Within the home government category, events related to political instability are grouped together, as are those related to government policy. While prices and other macroeconomic variables typically are not solely the result of government activity, government actions and policies often do have a strong influence on them, and so they are also included here. In a similar vein, while corruption, by definition, is not a government policy, it is typically located within a government's bureaucracy and so is also included. While it is certainly true that not all government-related activity is necessarily detrimental to a firm (import restrictions, for example, typically benefit incumbent foreign firms), the categories listed in the table have been chosen to conform to the definition of political risk used in the chapter.

A basic listing of the risks potentially associated with each political activity is also included. Root (1972) considers events in terms of ownership, operational and transfer risks. As Shubik (1983) notes, the most important thing is to establish what is at stake, and he classifies company resources into real assets, financial assets and personnel. Using the above, we look at events in terms of employee risk, balance sheet risks, profit risks and opportunity costs. Hopefully, the modified terminology is less radical than it may appear at first glance. By our way of thinking, balance sheet risks involve situations where real assets may be at risk of being lost or destroyed. Violent political events are often associated with such risks, but

Table 10.1 A classification of political events and their potential risks

Host government issues	Employee risk	Balance sheet risks	Profit risks	Opportunity costs
Political instability				
War	X	X	X	X
Civil war	X	X	X	X
Revolution	X	X	X	X
Leadership change		X	X	X
Coup d'etat		X	X	X
Government policies				
Expropriation		X		
Local ownership requirements		X		
Pressure for joint ventures		X		
Arbitrary arrests	X			
Taxation			X	
Tariffs			X	
Contractual breaches			X	
New regulations			X	
Local content requirements			X	
Import/export restrictions				X
Payment delays			X	
Input cost regulations			X	
Lack of copyright enforcement			X	
Hiring quotas			X	
Countertrade requirements			X	
Export requirements			X	
Distribution restrictions			X	
Support of competition			X	
Currency inconvertibility				X
Currency restrictions				X
Profit repatriation restrictions				X
Market entry restrictions				X
Bureaucratic issues				
Unexpected fees			X	
Uneven enforcement of regulations			X	
Delays in issuing licensing & permits			X	X

Table 10.1 (continued)

Host government issues	Employee risk	Balance sheet risks	Profit risks	Opportunity costs
Macroeconomic concerns				
Recession			X	
Inflation			X	
Currency devaluation		X	X	
Other				
Bribery			X	
Corruption			X	
Anti-foreign court rulings		X	X	
Bureaucratic delays			X	X
Home government issues				
Pressure for divestment		X		
Export restrictions				X
Technology transfer controls				X
Blockades				X
Embargoes				X
Third-party government issues				
War	X	X	X	X
Blockades			X	X
Economic sanctions			X	
Non-government issues				
Kidnapping	X		X	
Assault, murder	X			
Sabotage	X	X	X	
Terrorism	X	X	X	
Riots	X	X	X	
Strikes			X	
Demonstrations			X	
Boycotts			X	

a 'peaceful' nationalization of company assets would also qualify. Profit risks are associated with events that are likely to increase costs or reduce incoming cash flows, but do not directly impair a company's assets (of course, by reducing profits, such actions will almost certainly reduce the company's value and, if sufficiently extreme, such activities can destroy the viability of a venture). Opportunity costs are associated with activities that do not necessarily immediately affect an existing venture's profitability but, by reducing the company's overall flexibility, make it harder for the firm to follow up promising opportunities. A good example of activities that can lead to such opportunity cost risks are profit repatriation restrictions (although, from a certain perspective, seeing them in conjunction with an active black market makes them appear a bit like a government-imposed tax that is collected by private sector actors). Employee risk concerns potential dangers to the wellbeing of company employees.

At its broadest, political risk can look an awful lot like a catch-all for all non-business risks. Of course, such a broad conceptualization begs the question of what exactly constitutes business risk. While equating business risk with Miller's (1992) notions of industry and firm uncertainty seemingly takes care of much of this void in definitional strategy, some challenges remain. For example, it is not immediately clear how natural catastrophes would fit into such a schema. Clearly a broad concept, political risk may not, in fact, be infinitely elastic.

FROM POLITICAL RISK TO POLITICAL STRATEGY

Ultimately, whether a particular political event holds any risk for an investment depends on the nature of that investment. From a corporate perspective, as can be seen from Table 10.2, a wide variety of strategies exist for firms seeking to mitigate political risks. We see companies as being able to choose from five basic strategies, which is to say, we think firms may look to control, forestall, co-opt, cope with or avoid potential political risks.

Conceptually, one aspect of political risk that is particularly challenging is recognizing how quickly the activities and skills surrounding political risk mitigation become promising resources for affecting a proactive political strategy to shape public policy. In other words, it is remarkable how quickly a good defence turns into a strong offence and how quickly the desire to avoid state sanction can turn into an operation to curry state favour and obtain public subsidy.

Unfortunately, there is currently something of a mismatch between the literature's consideration of political risk and political strategy. Since political risk mitigation strategies tend to focus on avoiding the negative consequences of political activities, they can be blind to opportunities present in the political arena. On the other hand, since most discussions of political strategy have tended to focus on actions to influence policy makers and public policy, the importance of non-governmental political events tends to be ignored.

In an effort to construct a comprehensive taxonomy of political strategies, Hillman and Hitt (1999) introduced three dimensions to help make distinctions among political strategies. Specifically, they introduced the notion of relational versus transactional approaches, individual versus

Table 10.2 Strategies for political risk mitigation

Control:	purchase related activities; bring in the gunboats
Forestall:	spin (have good PR); lobby; threaten
Co-opt:	establish long-term contracts; find partners; establish joint ownership, with local business, with other foreign firms, with local government, with employees, list locally Alliances; form coalitions with other businesses, find local allies (e.g. suppliers, employees, customers); give to charities, buy off troublemakers, be a good corporate citizen; purchase locally, employ host country citizens
Cope:	Insure; hedge; displace the risks onto someone else; keep a low profile, be invisible, imitate – be indistinguishable Modify structure of organization and assets; ensure local assets depend on other assets for value; incorporate redundancy into system design; improve security measures for people and property
Avoid:	decline to invest; refuse to put money to work in the location Diversify; have new money avoid the location; reduce resource commitments; reduce financial commitment, find ways for currently committed cash to flee, lease, rather than own, capital equipment, shorten payback period Use other people's money; list on local stock exchange, borrow locally, have local entity borrow anywhere it can Reduce headcount, evacuate personnel, exit

collective levels of participation, and informational, constituency-building and financial strategies. While their framework covers a great deal, it is ultimately focused really only on public policy-shaping strategies, to the neglect of passive and anticipatory strategies (Weidenbaum, 1980) – something they themselves recognize. In this chapter, we use the term 'political strategy' in a broad sense potentially to include passive and anticipatory strategies as well as proactive ones. Furthermore, we believe that the political actions taken by actors outside of government can represent both opportunities and challenges for companies and so ought to be taken into account as much as possible when constructing a political strategy.

LOCATION AS A POLITICAL STRATEGY

One interesting anticipatory political strategy is the possible use of location to mitigate political risks. While considerations of location, especially at a micro level, are almost non-existent in most studies of political risk and political strategy, international business research suggests location may be quite important (see, for example, Dunning, 1998). At the macro level, by establishing production operations abroad, Japanese car makers have used location to avoid protectionism and exclusion from key markets. Moreover, for a company like Toyota, in the United States, such an approach has evolved into a sophisticated dispersion of operations across a number of Congressional districts, which has – among many things – had the effect of increasing the company's leverage with politicians in Washington. It has also led the company from a goal of ensuring market access to success in obtaining a number of government perks like the designation of certain Toyota factories in the United States as site-specific foreign trade sub-zones. While blind imitation of such a strategy is a non-starter in a one-party, authoritarian state like the People's Republic of China, use of location as a political strategy may still be possible. In particular, at the micro level, the location of manufacturing operations in state-level industrial parks seems to hold out promise for insulating a company from a substantial amount of local bureaucratic risk, and so may merit further investigation. According to a study by PriceWaterhouseCoopers (2004), 75 per cent of firms surveyed agreed that China's legal and political environment represented a key challenge to investing in China, and advice on China's legal framework, statutory requirements and tax compliance topped the list of assistance desired by companies in the survey.

INDUSTRIAL PARKS IN THE PRC

Since opening its economy to the outside world, the People's Republic of China has established a number of different types of industrial parks and development zones to encourage economic growth (China Knowledge Holding, 2004). In particular, the PRC has seen the development of Special Economic Zones (SEZs), Economic and Technological Development Zones (ETDZs), High-technology Industrial Development Zones (HIDZs), Free Trade Zones (FTZs) and Export Processing Zones (EPZs). In addition to these zones, the PRC also has border and economic cooperation zones, software development parks, university science parks and resort development zones, though these tend to be more specialized types of areas.[1]

In the early 1980s, four SEZs were established in Shenzhen, Xiamen, Zhuhai and Shantou. In these regions, special policies were put into place and preferential tax policies were enacted in an attempt to attract foreign investment, especially investment from overseas Chinese. Between 1984 and 1988, 14 Economic and Technological Development Zones ETDZs were established in 12 coastal cities. Unlike the SEZs, the ETDZs were specific areas of land, typically located in the suburbs of major cities, within which special policies were adopted and activities were overseen by a special administrative committee. ETDZs may thus be seen as the PRC's first industrial parks.

With the establishment of the Pudong New District in Shanghai in 1990, a second wave of ETZDs began to emerge and, between 1992 and 1993, 18 state-level ETDZs were established. In 2000, a third wave of ETDZ expansion began, with the central government giving the nod to an additional 11 ETDZs in the PRC's central and western regions. In addition to the regular ETZDs, there exist at least two special zones: the Ningbo DaXie Development Zone, founded in 1993 and administered by China International Trust and Investment Corporation (CITIC), and the Suzhou Industrial Park, a joint venture between Singapore and the PRC, established in 1994. Moreover, encouraged by the success of the state-level ETDZs, a large number of provincial, municipal and even county-level ETDZs have been established across the PRC (in 2004, the number of such local ETDZs was estimated to be over 4000).

In the late 1980s, the Ministry of Science and Technology began a new programme to utilize the resources of universities and research institutes to develop and commercialize new high technology products by working with business enterprises in High-tech Industrial Development Zones (HIDZs). In 1988, the first HIDZ was established in Zhongguancun, in northwest Beijing, and in the period between 1991 and 1992 over 50 more were established. An important characteristic of these zones is the requirement of each zone to include a university-based research centre, an innovation centre and partnerships with commercial enterprises. Moreover, it is expected that the firms which are allowed in such zones should be certified as high technology firms (Li, Schoonhoven and Zhang, 2004).

Free Trade Zones (FTZs) were first established at the beginning of the 1990s in order to facilitate export processing, trade, logistics and bonded warehousing. Companies in FTZs are eligible for tax refunds on exports, exemptions from import duties and concessions on the PRC's value added tax (VAT). Currently, this is the only kind of zone in which foreign trading companies can be registered, and today there are roughly 15 state-level FTZs in 12 coastal cities. More recently, Export Processing Zones (EPZs) have also been established. Created to facilitate export processing, the first was opened in Kunshan in 2000, and currently there are at least 25 EPZs in the PRC. Generally speaking, one often finds many FTZs and EPZs to be part of ETDZs (though physically separated from them by barriers monitored by customs officials).

All of the various industrial parks, free trade areas and economic development zones offer a number of advantages, including the prospects of a relatively developed infrastructure, efficient administration and attractive business terms. To that end, industrial parks tend to be subject to a number of preferential policies. In particular, companies operating in such areas are often partially or (for a period of time) fully exempt from corporate income taxes and real estate taxes, exempt from customs duties (in the free trade zones), exempt from equipment licensing rules, and are eligible for VAT refunds for exported products. Generally speaking, one important difference between the various industrial parks lies in the level of governmental sponsorship, as state-level industrial parks typically have better infrastructure, better management and better prospects for continuity in official policies than their local counterparts. As noted by the EIU (2005), national zones have greater bureaucratic autonomy, ranging from enhanced customs processes to faster and more transparent investment approval procedures. For example, as per McGregor (2004), in reference to the Tianjin Economic Development Area, Jiang Weiming, president of Novozymes in the PRC, says Teda is attractive because 'it is results driven and service minded, and takes practical approaches when it comes to administration of the park. There are a lot of areas with rather grey political agreements, which have no substance.'

CASE STUDIES

This chapter looks at the experience of six American multinationals in the People's Republic of China (see Table 10.3). All are large, publicly listed companies with 2003 revenues ranging from $8 billion to $45 billion (USD). Two are in the telecom equipment industry; two are in pharmaceuticals and two are consumer products companies. All generate a substantial portion of the sales outside of the US (around 40 per cent in most cases, but closer to 50 per cent in the case of P&G and more than 70 per cent in the case of Colgate). While all the companies have long histories (either independently or, in the case of Lucent, as part of AT&T), in all cases, their entry into the PRC is relatively recent, paralleling the PRC's decision to open its economy to the world. The relatively recent entry notwithstanding, the companies in question have already invested hundreds of millions to billions of dollars in the PRC.

Political Activities and Political Strategies

In looking across the wide sweep of activities of American multinationals in the PRC, we see examples of companies taking both transactional and relational approaches to political issues in the country. On the one hand,

Table 10.3 Basic company information (value in billions USD)

	Motorola	Lucent	Pfizer	Merck	P&G	Colgate
Worldwide sales (2003)	$27.06	$8.47	$45.19	$22.49	$43.38	$9.90
Non US sales (2003)	46.8%	39.5%	40.6%	40.8%	49.6%	72.3%
Sales in PRC (2003)	$3.87	$0.94	$0.16	NA	1.00	$0.24
PRC sales as % non US sales	30.6%	27.9%	0.9%	NA	4.6%	3.4%
Year company established	1928	1875 (1996)	1849	1891	1837	1806
Year firm enters PRC	1987	1984	1989	1992	1988	1992
Year 1st mfg. plant est. in PRC	1992	1990	1989	1994	1988	1992
Total amount invested in PRC	$3.40	$2.90	$0.5 +	NA	$0.30	$0.11
Location of PRC HQ	Beijing	Beijing	Shanghai	Shanghai	Guangzhou	Shenzhen

Sources: company annual reports, company websites.

we see great banquets and dinners, all of which speak to a personal, relational approach to political action. On the other hand, we also see groups of foreign companies working together to strengthen intellectual property enforcement in the People's Republic of China: tactical alliances, which though perhaps requiring years to see progress, still speak of an issue-oriented approach to political action. Similarly, we can see evidence of political action at the executive, corporate and collective level.

In terms of the six companies examined in this chapter, two political initiatives seem most prominent, specifically: work to be a good corporate citizen (that is, build a positive image) and make efforts to strengthen the local enforcement of intellectual property rights. As far as building a positive image goes, all the companies studied have engaged in such activities. In some cases, this has been by making donations to local causes. For example, Motorola has donated over 11 million (RMB) for university scholarships. P&G has donated over 40 million (RMB) for the public welfare. Lucent spent over $2 million to help build a lab for environmental sciences, and, in 2003, Pfizer donated over $450 000 to fight SARS. Moreover, support for certain initiatives has been particularly popular. Motorola, Lucent and P&G have all contributed to Project Hope, an initiative to help poor students get an education. Similarly, both Motorola and P&G helped support Beijing's bid for the 2008 Summer Olympics.

One particularly interesting twist on the efforts of these companies to build a positive local image may be found in some of their lobbying efforts back in the United States. Specifically, Lucent, Pfizer and Motorola have all, in the past, been active lobbying the US government to grant the PRC most-favoured-nation (MFN) status.

As far as the strengthening of local intellectual property rights enforcement goes, P&G, Colgate-Palmolive, Merck and Pfizer have, as members of the Quality Brands Protection Committee (QBPC) – an association of foreign invested enterprises in the PRC established to leverage their collective voice against counterfeiting – encouraged the PRC's central government to use stronger measures. Similarly, Pfizer has on its own initiative lobbied Beijing on the issue, as well as urged US officials to take a tougher stand on intellectual property rights (IPR) more generally.

The Location of Manufacturing Operations in the PRC

American companies have invested a great deal in the PRC, and the six companies examined in this chapter have been at the forefront of such investment. According to available information, the six companies have established a total of 33 manufacturing ventures in the PRC (see Table 10.4). Of those ventures, six are wholly owned subsidiaries. The rest are joint

Table 10.4 Manufacturing operations in the PRC

	City	Province	Type of venture	In industrial park?	Park sponsor	Year established
Motorola						
	Tianjin		WOS	Yes	State	1992
	Shanghai		JV	Yes	State	1995
	Leshan	Sichuan	JV	Yes	State	1995
	Hangzhou	Zhejiang	JV	No		1996
	Hangzhou	Zhejiang	JV	Yes	State	1996
	Beijing		JV	Yes	State	1998
	Beijing		JV	Yes	State	1998
	Guangzhou	Guangdong	JV	Yes	State	1998
	Shanghai		JV	Yes	State	1999
	Tianjin		WOS	Yes	Municipality	2000
Lucent						
	Shanghai		JV	Yes	State	1990
	Shanghai		WOS	Yes	State	1993
	Beijing		JV	Yes	State	1993
	Qingdao	Shandong	JV	Yes	State	1993
	Tianjin		WOS	Yes	State	1995
	Qingdao	Shandong	WOS	Yes	State	1998
	Shanghai		JV	Yes	State	1999
Pfizer						
	Suzhou	Jiangsu	JV	Yes	State	1985
	Dalian	Liaoning	JV	Yes	State	1989
	Wuxi	Jiangsu	JV	n.a.	n.a.	n.a.
Merck						
	Hangzhou	Zhejiang	JV	Yes	State	1994
P&G						
	Guangzhou	Guangdong	JV	Yes	State	1988
	Guangzhou	Guangdong	JV	Yes	State	1990
	Guangzhou	Guangdong	JV	Yes	State	1992
	Beijing		JV	Yes	Municipality	1993
	Chengdu	Sichuan	JV	Yes	State	1993
	Tianjin		JV	No		1993
	Guangzhou	Guangdong	JV	Yes	State	1994
	Guangzhou	Guangdong	JV	Yes	State	1995
	Tianjin		JV	Yes	Municipality	1996
Colgate						
	Guangzhou	Guangdong	JV	Yes	State	1992
	Guangzhou	Guangdong	WOS	Yes	State	1995
	Yangzhou	Jiangsu	JV	No		2000

Note: n.a: not available.

Sources: company websites, China Association of Development Zones.

ventures. Motorola has invested in Beijing, Shanghai and Tianjin as well as Zhejiang, Sichuan and Guangdong provinces. Lucent has invested in Beijing, Shanghai and Tianjin along with Shandong province. Pfizer has manufacturing in Liaoning and Jiangsu provinces. Merck is in Zhejiang province. Procter & Gamble has manufacturing operations in Beijing and Tianjin as well as Guangdong and Sichuan provinces, and Colgate has manufacturing in Guangdong and Jiangsu provinces. Despite the wide variety of geographic locations, most of the companies' manufacturing operations share a similarity – they tend to be located in industrial parks, specifically industrial parks established by the central government. In 29 of the 32 cases where information is available (roughly 90 per cent of the cases), manufacturing operations are located in an industrial park, and in 26 out of 29 of those cases (again, about 90 per cent of the cases), the operations are located in industrial parks sponsored by the central government. Given that there are only about 100–200 state-level industrial parks, but recent estimates peg the number of local industrial parks at more than 4000, the propensity of foreign invested manufacturing operations to be located in state-level industrial parks is far higher than what one would expect just by chance.

For the companies investigated in this study at least, there definitely appears to be a strong tendency for manufacturing operations to be located in state-level industrial parks. While there are a number of potential reasons for this relationship, given the difficulties foreign companies have faced in terms of unexpected fees, licences and taxes in the PRC, and given what has been characterized as relatively efficient industrial park administration, the results are consistent with the use of location as a political risk mitigation strategy.

DISCUSSION AND CONCLUSION

American firms have adopted a number of different political strategies for coping with conditions in the PRC, and some of their most visible political activity has included work to deepen their local roots by building a positive image as a good corporate citizen as well as efforts, generally collective efforts, to push for stronger intellectual property rights enforcement. One somewhat surprising aspect of the political behaviour of American multinationals in the PRC has been the degree to which their political activities have seemed to hinge on lobbying the US government, whether it be over a desire for US trade officials to take a stronger line on IPR enforcement or an attempt to help the PRC secure most-favoured-nation status for its trade with the United States.

Another interesting aspect of the firm behaviour examined in this chapter has been the degree to which strategies for coping with political risk have become intertwined and indistinguishable from political strategies that seek to benefit from government actions and favour. As far as locational choices are concerned, on the one hand, location in an industrial park seems to insulate firms to some degree from the hazards of bureaucratic risk, and so location appears to be a kind of political risk mitigation strategy. On the other hand, given government investments in industrial park infrastructure and other government inducements for locating in such a region, it is difficult to separate choices made to cope with the negative effect of government action from decisions made in order to reap the benefit of government favour.

Speaking generally, location and geographical proximity have long been considered important to the development of social capital, and there is no reason to think that location should be any less important to the development of *guanxi*. While further research is needed to determine the relative importance of different industrial park characteristics for firm location decisions, the preference of US firms for locating their manufacturing operations in industrial parks sponsored by the central government is consistent with a notion that location may be used as a political strategy. Moreover, while the exact strategies pursued in the People's Republic of China appear somewhat different from what one might expect in, say, a representative democracy, the notion of location as a political strategy seems robust to institutional variation.

As Baron (1995) notes, it is important for a firm to include market and non-market considerations in the process of developing firm strategies. While location is usually seen in terms of its economic effects, its political implications also need to be considered. The same may be said of joint ventures, which have both obvious economic consequences and important implications for political risk in terms of their potential to provide some defence against political predation as well as their ability to reduce a company's overall asset exposure (Liu and Bjornson, 1998). From an integrated perspective, it should be possible for choices regarding entry mode and firm location to work together to reduce a company's overall political risk, an issue that may be particularly important for many firms in high-tech industries, where capital requirements are often substantial and so exposure to political risk may be particularly great.

Taken as a whole, this chapter has discussed a number of issues related to political risk, political strategy and location. In particular, it has both provided a broad framework for thinking about political risk and uncovered support for location as a political strategy, an anticipatory corporate political behaviour that has been relatively neglected up to now.

NOTE

1. Most of this section is based on information from China Knowledge Holding (2004), but also makes use of information from Li, Schoonhoven and Zhang (2004).

REFERENCES

Baron, D.P. (1995), 'Integrated strategy: market and non-market components', *California Management Review*, Winter, **37** (2), 47–65.

Binkowski, E. (1980), 'Tax consequences of creeping expropriation', *International Tax Journal*, December, **7** (2), 117–26.

China Knowledge Holding (2004), 'China's industrial parks promise attractive business terms'; retrieved 01–06–05 from http://www.chinaknowledge.com/viewnews.asp?ID=27.

De Jouvenel, B. (1957), 'Sovereignty: an inquiry into the political good', University of Chicago Press.

Dess, G.G., D.R. Ireland, S.A. Zahra, S.W. Floyd, J.J. Janney and P.J. Lane (2003), 'Emerging issues in corporate entrepreneurship', *Journal of Management*, **29** (3), 351–78.

Dunning, J.H. (1998), 'Location and the multinational enterprise: a neglected factor?', *Journal of International Business Studies*, **29** (1), 45–66.

Economist Intelligence Unit (EIU) (2005), 'China industry: Beijing's BDA remains the place to be in the capital', 26 January.

Fitzpatrick, M. (1983), 'The definition and assessment of political risk in international business: a review of the literature', *Academy of Management Review*, **8** (2), 249–54.

Griffin, R. and M. Pustay (2002), *International Business: A managerial perspective*, Upper Saddle River, NJ: Prentice-Hall.

Guth, W. and A. Ginsberg (1990), 'Guest editors' introduction: corporate entrepreneurship', *Strategic Management Journal*, 11, 5–15.

Haendel, D.H., G.T. West, with R.G. Meadow (1975), *Overseas Investment and Political Risk*, Philadelphia, PA: Foreign Policy Research Institute.

Henisz, W.J. and J. Story (2003), 'Corporate risk assessment and business strategy: a prime task for senior management', in P.K. Cornelius and B. Kogut (eds), *Corporate Governance and Capital Flows in a Global Economy*, New York: Oxford University Press, pp. 217–38.

Hillman, A. and M. Hitt (1999), 'Corporate political strategy formulation: a model of approach, participation, and strategy decisions', *Academy of Management Review*, **24**, 825–42.

Howell, L.D. (1998), 'Country and political risk analysis: foundations for foreign direct investment', in L. D. Howell (ed.), *The Handbook of Country and Political Risk Analysis*, East Syracuse, NY: PRS Group.

International Bank for Reconstruction and Development & International Chamber of Commerce: Commission on Foreign Investments (1962), 'International investment assurance: analysis of replies received', Document No. 111/109, 12.11.1962-mtp., International Bank for Reconstruction and Development.

Kobrin, S.J. (1979), 'Political risk: a review and reconsideration', *Journal of International Business Studies*, **10** (1), 67–80.

Kobrin, S.J. (1982), *Managing Political Risk Assessment: Strategic response to environmental change*. Berkeley, CA: University of California Press.

Li, H., K. Schoonhoven and Y. Zhang (2004), 'Factors affecting the growth of China's high technology development zones: evidence from 1991–2000', working paper, Texas A&M University.

Liu, Y. and B. Bjornson (1998), 'Managing exposure of direct foreign investment to political risk: the case of food businesses in China', *International Food and Agribusiness Management Review*, **1** (3), 359–72.

McGregor, R. (2004), 'The regions: fierce competition for jobs', *Financial Times*, 6 December.

Miller, K.D. (1992), 'A framework for integrated risk management in international business', *Journal of International Business Studies*, **23** (2), 311–31.

PriceWaterhouseCoopers (PWC) (2004), 'Doing business in China: a survey of issues encountered by foreign companies doing business or investing in China', March.

Root, F. (1972), 'Analyzing political risks in international business', in A. Kapoor and P. Grub (eds), *Multinational Enterprise in Transition: Selected readings and essays*, Detroit, MI: Darwin Press, pp. 354–65.

Samuelson, P. and Nordhaus, W. (1989), *Economics*, 13th edn, New York: McGraw-Hill.

Sharma, P. and J.J. Chrisman (1999), 'Toward a reconciliation of the definitional issues in the field of corporate entrepreneurship', *Entrepreneurship Theory and Practice*, **23** (3), 11–27.

Shubik, M. (1983), 'Political risk: analysis, process, and purpose', in R.J. Herring (ed.), *Managing International Risk: Essays commissioned in honor of the centenary of the Wharton school*, University of Pennsylvania/New York: Cambridge University Press, pp. 109–38.

Truitt, J.F. (1974), 'Expropriation of private investment', Division of Research, Graduate School of Business, Indiana University.

Weidenbaum, M. (1980), 'Public policy: no longer a spectator sport for business', *Journal of Business Strategy*, **3**, 46–53.

Weston, F.V. and B.W. Sorge (1972), *International Managerial Finance*, Homewood, IL: Richard D. Irwin.

11. International new ventures' learning in culturally distant markets: the case of China

Shaker A. Zahra

Both the internationalization process (Johanson & Vahlne, 1977, 1990) and 'born global' (Oviatt & McDougall, 1994) theories emphasize learning as a key factor in motivating companies to go abroad. The process theory (Johanson & Vahlne, 1977) posits that experiential learning is a key factor that determines how firms enter new foreign markets. This theory depicts internationalization as a series of sequential decisions in selecting the scope and scale of a firm's foreign operations. It also highlights the importance of experiential learning in determining a firm's ability to choose progressively complex modes of foreign market entry (Anderson, 1993; Zahra & George, 2002). These modes usually start with exporting, proceed to licensing and then evolve into acquisitions and greenfield investments. According to the process theory, organizational learning allows firms to move from one mode to the next. Moreover, firms learn from and through their international operations.

The 'born global' theory suggests that new ventures that go abroad at their inception can also learn a great deal from and in their foreign markets (McDougall & Oviatt, 2000). This process of learning enhances what these companies are doing and how they use what they have learned to assemble the resources required for further international expansion. Autio, Sapienza and Almeida (2000) propose that new ventures have major advantages in learning over their well established rivals. Born global ventures have organic structures that allow them to assimilate what they learn quickly and proceed to exploit it by creating a source of competitive advantage. The fact that owner managers are at the helm of these 'born global' ventures enables them to transform the knowledge theory gained from international markets into a competitive advantage. Consistent with this view, Zahra, Ireland and Hitt (2000) have empirically shown that new ventures learn a great deal about technological development in international markets. Still, a recent review of research on the 'born global' phenomenon reports that researchers have not delineated *how* new ventures learn in their international markets (Zahra, 2005). This is especially true in culturally distant markets, those markets that differ significantly from their own home markets. These distant markets usually have different structures and decision rules from new ventures' domestic markets, complicating these firms' ability to understand and assimilate the knowledge that resides within them. If there are 'learning advantages of newness' (Autio *et al.*, 2000), these advantages might be hard to attain and sustain in a culturally distant market. This begs the questions: how do born global new ventures learn in culturally distant markets? How does this learning influence new venture performance (NVP)? These issues are the focus of this chapter.

OBJECTIVE AND CONTRIBUTIONS

The objectives of this chapter are twofold. First, it examines how new ventures learn in culturally distant markets, distinguishing between experiential and vicarious organizational learning. Experiential and vicarious learning are two of the approaches most widely used by organizations (Miller, 1996). Experiential learning means that the firm learns by doing. Vicarious learning means that the firm learns by observing and assessing the experiences and actions of other companies and groups. This study seeks to establish the extent to which born global new ventures use these two approaches. Second, the chapter examines how born global firms vary in their experiential and vicarious learning, focusing on new ventures from France, Sweden and the US that have entered the Chinese market.

This chapter makes three contributions to the literature. The first is separating vicarious from experiential approaches of organizational learning that new ventures might use to understand and penetrate the Chinese market. The literature is replete with classifications of organizational learning strategies but these classifications have not been subjected to systematic empirical testing, a gap this study seeks to fill. It is also important to appreciate the potential complementarity and trade-offs that might exist between experiential and vicarious organizational learning.

The second contribution is linking vicarious and experiential learning to NVP, an issue that has not received much attention in past research. While some researchers have recognized the importance of organizational learning for innovativeness and financial performance, little systematic attention has been given to documenting these affects empirically, especially financial performance. If learning improves NVP, then it is important to establish this link empirically. This link could prove to be especially crucial as new venture managers seek to capitalize on their own unique resources and internal processes. Further, new ventures that go abroad early in their life cycle are apt to experience serious shortages in their resources and therefore have to reach profitability quickly in order to maintain their solvency. If learning improves NVP, then it is important to harvest this link.

The third contribution of the chapter is exploring the moderating effects of country of origin on the relationships between new ventures' learning approaches and their performance. Organizational learning is grounded in national cultures that determine how companies and their managers determine which types of knowledge they value and how they exploit this knowledge. National cultures also shape the way managers collect and interpret data about their markets and competition. Given that new ventures often lack the internal systems to analyse the competition and market trends, the effect of national cultures on their learning and the uses of this learning could be profound. Evidence is lacking about this effect and its implications for born globals' performance.

The next section of the chapter develops the study's theoretical background, emphasizing the knowledge-based theory of the firm. This section culminates in the study's hypotheses. Once this section is concluded, an empirical study is presented. This is followed by the study's results and key findings, highlighting the effect of organizational learning in the Chinese market on NVP. The final section of the chapter discusses the implications of the study's findings for managerial practice and future research.

THEORY AND HYPOTHESES

The knowledge-based theory (KBT) suggests that companies can develop superior and enduring competitive advantages based on their knowledge (Grant, 1996; Leonard-Barton, 1995; Zander & Kogut, 1995). Most of this knowledge is embedded in the firm's employees and other aspects of intellectual capital such as the organization, its processes and systems. This knowledge serves as the foundation of the routines that companies can use to develop dynamic capabilities that they could exploit in pursuing their goals (Teece, Pisano & Shuen, 1997).

Both the stocks and flows of knowledge are important for value creation. 'Stocks' refer to the various types of knowledge that the firm has and how it shapes the strategy. The 'flows' of knowledge are especially important for creating and sustaining the firm's competitive advantage (Dierickx & Cool, 1989). Knowledge flows accumulate from alliances, observing competitors and learning by doing in the firm's different operations. These flows keep the firm's knowledge base current, allowing it to develop new strategies, products, systems and business models that foster its foreign expansion and growth (Grant, 1996).

International expansion could increase the flows of knowledge into born global new ventures' operations. It gives these ventures an important arena in which they can observe new markets, customers, competitors and systems of innovation that can inspire learning (Zahra, 2005). Born global new ventures often encounter new competitive landscapes as they internationalize their operations. They are likely to gain exposure to other countries from their own home market seeking to enter the same markets they have targeted. Born global ventures also deal with very different regulatory and legal environments from their own home markets, a factor that could encourage them to learn different and new skills. Given their limited experiences, born global ventures need to learn new skills in order to survive in international markets, seize opportunities in these markets and achieve superior profitability.

One of the key motivators for companies and entrepreneurs to venture outside their domestic markets is to learn and gain new knowledge (Zahra & George, 2002). According to the KBT, new knowledge can help companies improve their operations by combining incoming knowledge with their own to develop new products and services (Kogut & Zander, 1992). Combinative knowledge is a key source of competitive advantage in today's global markets (Kogut & Zander, 1992). In addition to capitalizing on opportunities in foreign markets, new ventures can learn a great deal about technological development and evolution in foreign markets (Zahra *et al.*, 2000). This allows new ventures to broaden their knowledge base and conceive of new and perhaps radically new products. Given the youth of born globals, incoming knowledge flows do not have to compete with existing rigid knowledge structures and routines, as would be the case in established multinationals (Autio *et al.*, 2000; Sapienza, Autio, George & Zahra, 2005). Consequently, born global ventures could move quickly to transform these knowledge flows into useful and marketable applications, which could give them a competitive advantage.

Given the potential learning advantages of newness of born global ventures (Autio *et al.*, 2000), we do not know how these companies learn, especially in culturally distant international markets (Zahra, 2005). These markets are very different in their structures and decision models from those usually found in new ventures' home industries. Markets and the rules that govern them develop over time, as would be the case in the Chinese market. Many of these rules are culturally grounded, subtle and learned experientially. Outsiders to a culture might not even be aware of these rules or err in interpreting their cultural foundations (Hofstede, 1980, 1991). These rules could challenge those born global new ventures seeking to build their operations internationally. Failing to learn about foreign markets can blind entrepreneurs to potential opportunities and could cause the demise of their ventures. Born global ventures might also fail to understand the subtleties associated with decision making and consumer behaviour and err in defining their competitors.

Born global ventures may also have limited absorptive capacity, which would constrain their ability to understand and decipher knowledge flows from international operations. Very few ventures have the resources or the time needed to process and assimilate this incoming knowledge or to incorporate it in their operations. These firms, however, have to learn quickly and capitalize on those opportunities that exist in their foreign markets (Sapienza, De Clercq & Sandberg, 2005). Some of these markets are subject to considerable dynamism because of changing technologies and entry by foreign competitors. This is the case today with many of the Chinese markets, especially in high technology industries where considerable local innovation is already taking place because of the influx of foreign competition.

Several factors have encouraged US and other Western companies to enter the Chinese market. The most important is the potential phenomenal growth of the Chinese economy, which can give companies a great opportunity to achieve growth (Li & Atuahene-Gima, 2001). Few markets have the potential to grow as fast as the Chinese market. The country's vast population and diverse cultural heritage also create opportunities for effective market segmentation. This gives born global ventures an opportunity to enter that market and compete as niche players and expand over time. Success in the Chinese market can also become a focal point in penetrating South East Asian markets, another important arena for economic and technological development. The Chinese government has actively promoted high technology industries and has taken several measures to transform China into an economic and technological powerhouse. Therefore the Chinese government has created the infrastructure necessary for modern communication and transporting goods and services (Lau, Tse & Zhou, 2002). This has made it easier for some born global new ventures to enter the Chinese market. Foreign companies can also collaborate with domestic and local companies in offering products and services that are unique to the Chinese market and then adapt them to the rest of the Southeast Asian market.

Competing in the Chinese market could be frustrating for US and other Western companies because of the language and other cultural differences. Outdated ideological frames of reference also continue to dominate thinking about competition and the role of the government in the national economy. There is also a persistent concern about the future trajectory of

China's growth and development policies. As a result, success of US and other Western companies might hinge on learning about the Chinese culture, the economic system that exists in China, and the structure of local markets and their growth potential. These forces make learning an important requirement for a new venture's success in the Chinese market (Sohn, 1994).

Learning in the Chinese Market

Companies learn in a variety of ways (Huber, 1991) and different typologies of organizational learning exist in the literature (for example, Miller, 1996). In the context of new ventures, it is convenient to separate vicarious from experiential learning. *Vicarious learning* centres on observing others and imitating their behaviour. It occurs by analysing others' behaviour or actually collaborating with them and joining their networks (Coviello & Munro, 1995). These analyses enable new ventures to understand how other companies think about the structure of the market and the dynamics of competition and then imitate their behaviour. This approach clearly reduces the time and cost necessary to learn about the market and competition. This is especially important because new ventures usually do not have the resources or time to probe the market and thoroughly analyse their competition. Still, by using this approach, born global ventures might err in making causal attributions about what other companies or competitors think or why they behave the way they do. Thus born global news ventures could learn the wrong things or fail to understand the factors that define or shape the behaviours of other companies in the Chinese market.

Experiential learning, on the other hand, focuses on learning by doing where new ventures learn by trial and error and then decipher and articulate what they have learned by institutionalizing it in their operations (Sohn, 1994). Experiential learning occurs also through improvisation where born global venture managers develop creative responses to unexpected circumstances and then probe the utility of these strategies. If proved to be viable, these strategies are then formalized and institutionalized into the firm's operations.

Effects of National Origin

A firm's national origin influences where and how it competes (Bartlett & Ghoshal, 2000; Calori, Johnson & Sarnin, 1992; Zahra & George, 2002). Research highlights the importance of psychic distance in how a firm selects the countries into which it ventures (Johanson & Vahlne, 1977, 1990). 'Psychic distance' refers to the proximity of the culture of the host country (or target market) to the firm's national cultures. Some research suggests that, as the firm gains experience in dealing with other cultures, it progressively ventures into culturally distant markets (Thomas & Mueller, 2000). With new ventures entering the Chinese market at their inception, it is logical to believe that their national cultures will shape the way they will view this market. Like other companies, born global ventures are likely to have mental models that mirror the assumptions embodied in their own national cultures about the Chinese market. These models are likely to influence the way new ventures relate to customers and competition (Zahra, Korri & Yu, 2005), influencing their choices of the approaches they use to learn about the Chinese market.

Some cultures (such as the Swedish) are relation-based, favouring first-hand interactions with others before developing trusting or even economically beneficial associations. These cultures value first-hand and personal contacts and companies that come from countries with these cultures are likely to emphasize interactive and experiential rather than vicarious approaches. Experiential learning generates rich information that promotes a deeper understanding of the market and customers. This could be especially beneficial in the Chinese market, where a 'high context' culture exists. In this culture, it is important to rely on direct and personal means to understand the subtle rules of decision making and how people think (Hofstede, 1991; Zhao & Aram, 1995). Other national cultures (such as the US) might favour the use of multiple approaches to assess the trustworthiness of strangers and new situations. These cultures have a utilitarian (rather than long and enduring personal) focus and usually stress the importance of speedy decision making. Transient, rather than enduring, relationships are the norm. Companies from these countries are likely to employ several experiential and vicarious learning approaches as they penetrate the Chinese market. New ventures established in the US are more likely to typify this group.

The above observations suggest that born global ventures from different national cultural backgrounds will emphasize different ways to learn about the Chinese market, with the US and Sweden representing the two extremes of a continuum (and France falling in the middle). France has had a long history of dealing with several Asian cultures. French companies and managers emphasize personal contacts in their dealings with others, which would favour the use of experiential learning methods. Still, like other Western companies, they value analysis and rational decision making. These qualities promote vicarious learning. This discussion leads to the following two hypotheses:

H1: Swedish ventures will use experiential learning methods in the Chinese market, more than French and US ventures, respectively.

H2: US new ventures will use vicarious learning methods in the Chinese market, more than Swedish and French ventures, respectively.

Organizational Learning and NVP

A major weakness of past research is its failure to document the effect of experiential and vicarious organizational learning on NVP. If learning matters, then its effect on NVP should be explored, especially in view of the fact that many new ventures fail as they internationalize their operations. Organizational learning allows new ventures to understand their competitive landscape and proceed to build their positions in the Chinese market. It also makes it possible for these firms to target the right customers with the right media and message. New ventures also gain insights into how their Chinese customers make and plan new product development and introduction decisions. Organizational learning is conducive to product and process innovations that allow the firm to gain market acceptance and acquire market shares that improve the profitability of their Chinese operations. Learning also facilitates building new skills and competencies that allow the organization to target new customers and segments in the ever-expanding Chinese market. Over time, these variables could improve NVP, as Zahra *et al.* (2000) found earlier. This suggests the following hypotheses:

H3: Experiential learning is positively associated with NVP.

H4: Vicarious learning is positively associated with NVP.

Moderating Effect of National Cultures

A logical conclusion of the preceding four hypotheses is the central importance of national culture as a moderator of the link between the organizational learning approaches and born global firms' NVP in the Chinese market. This conclusion is grounded in the notion of fit, where the firm chooses an appropriate learning approach that matches its market needs and mental models; these models are shaped by national cultures. National cultures also influence the way firms distil the lessons they learn from their markets (Zahra & George, 2002). Some cultures value close interactions and building enduring relationships (Hofstede, 1991). In those cultures that value rich and first-hand information, entrepreneurs are likely to consider what they see and observe, reflect on the strategic relevance of what they know, and proceed to exploit that information. These entrepreneurs are likely to pay less attention to the information gleaned from vicarious sources, believing there could be misattributions. Vicarious learning is often handicapped by 'causal ambiguity' in the 'high context' Chinese culture. In that culture, multiple layers of assumptions and beliefs are likely to coexist with facts, making it difficult to separate assertions from realities (Hofstede, 1991). Even customers' and competitors' verbal clues do not always reveal their true intent, magnifying the causal ambiguity that might exist regarding others' strategic moves. A final complicating factor regarding causal ambiguity is the fact that the Chinese market is undergoing a major economic and ideological transformation and the rules of the competitive game keep changing, encouraging companies and entrepreneurs to 'hedge' their bets, making reading other companies' strategic intent challenging. These observations lead to the following two hypotheses:

H5: Experiential learning in the Chinese market is conducive to higher NVP for Swedish than for French and US ventures.

H6: Vicarious learning in the Chinese market is conducive to higher NVP for US than for French and Swedish ventures.

METHOD

Sample and data

To test the above hypotheses, I used data collected from new ventures that have competed in high technology industries in France, Sweden and the US. It should be acknowledged that the US portion of the database comes from a larger study (see Zahra *et al.*, 2000). Data collection combined a two-wave mail survey, archival sources and phone and e-mail contacts with companies and trade associations to reduce source bias and obtain valid data. Given the potential difficulties identifying international new ventures, four criteria were used to identify a representative sample. First, to increase the findings' generalizability, we selected new ventures from multiple high technology industries. These industries are important for global competitiveness, job creation and innovativeness.

Second, company age was examined. The literature suggested that firms six years old and younger are considered to be new ventures in the US (Brush & Vanderwerf, 1992). Others have used different cut-off points such as 12 years (Covin, Slevin & Covin, 1990). This age cut-off is widely used in European research, therefore I used the six-year cut-off age in the US, but in France and Sweden, I surveyed companies 12 years old and younger.

Third, I examined company ownership. I included both independent (private) and corporate new ventures because both compete directly in several markets. Privately owned ventures are established and managed by individual entrepreneurs. Corporate ventures are created by companies, subsidiaries or business units of established companies. However, I excluded joint ventures and units that have been acquired or spun-off.

Fourth, I compiled a list of companies with international operations. An international new venture has been defined as one that obtained at least 10 per cent of its sales from foreign markets (McDougall, 1989), a criterion used in research on established companies (for example, Hitt, Hoskisson & Kim, 1997; Tallman & Li, 1996). However, this figure is high for young ventures that are early in their internationalization. The 10 per cent figure might also restrict the range of international sales, making it difficult to determine the effect of varying degrees of internationalization on technological learning. To capture a broader spectrum of international business operations among new ventures, I used a minimum of 5 per cent of a firm's sales being generated from foreign operations. In the three countries, different secondary sources were used to develop mailing lists. These sources included computerized searches (including visits to companies' Internet home pages), trade and business publications, and contacts with regional and national trade associations. I also reviewed trade articles, scanned announcements of new products, and read ads for new products to identify potential companies.

Data collection in the US

As reported elsewhere (Zahra *et al.*, 2000), data collection in the US targeted new ventures in 12 high technology industries that included biotechnology, computer software, factory automation, telecommunication, environmental technologies, medical and surgical equipment, pharmaceuticals, speciality chemicals, aerospace, test measurements, advanced material and semiconductors. This process yielded 1417 companies. Before mailing the final questionnaires, however, research assistants contacted 73 companies in which we used questions to verify their active involvement in international markets. Of these, 29 indicated that they did not have significant international operations and were eliminated, reducing the target population to 1388. Of the questionnaires mailed, 34 were returned as undeliverable, reducing the sample to 1354. The two mailings produced 321 completed questionnaires (a 23.8 per cent response rate), which compares favourably with those reported in past studies (for example, Chandler & Hanks, 1993; McDougall, 1989). Responding firms averaged 3.4 years of age (sd=1.1), 71.9 employees (sd=59.3) and 17.3 per cent of their sales (sd=11.5) from international operations. Respondents included CEOs and presidents (68 per cent), vice-presidents for international operations (19 per cent) and senior-vice presidents (13 per cent). These respondents were among the most informed about their companies' international operations and learning.

Table 11.1 Overview of the samples and companies studied

Variables	Countries of origin		
	US	France	Sweden
Number of companies in the sample	321	171	403
Response rate (%)	24	23	59
Average age in years	3.4	3.7	4.3
Average number of employees	79	18	13
Number of high-tech industries in the sample	12	11	13
Year of internationalization (mode)	3	2	1
Number of companies identifying China as their only foreign market	37	29	56

Zahra *et al.* (2000) also reported various tests of response bias. These included examining the differences among respondents to the first and second mailings. The t-test showed no significant differences between the two groups based on age, size or sales growth. The X^2 test, moreover, showed no significant association between respondent status (first versus second mailing) and the firm's location, industry type or ownership. The same analyses (using the X^2 and t-tests) compared responding and non-responding companies. These tests were also insignificant, suggesting that responding firms represented their populations. Readers interested in additional tests conduced to validate the data and measures may consult Zahra *et al.* (2000). Table 11.1 reports that, on average, companies were 3.4 years old, had 79 employees, competed in 12 high technology industries and had internationalized their operations in their third year of existence. Thirty-seven companies reported having operations only in China.

Data collection in France

A similar data collection approach to that used in the US was followed in France, targeting the same high technology industry groups. The exception was the use of a regional sample because of cost constraints. Following the same four-step process described above, 775 firms were identified and questionnaires were sent to them. A second mailing was sent a month later to the same group of companies. These two mailings generated 171 valid replies, for a response rate of 23 per cent (32 companies could not be reached for various reasons and were excluded). Follow-up phone calls to 34 managers clarified and validated responses. Tests for response bias, as described under the US sample, revealed that there was no response bias (p=0.23). As Table 11.1 reports, companies averaged 3.7 years, employed

18 people (on average) and competed in 11 high technology industries. Companies were quick to internationalize their operations, going abroad in their second year of existence. Of the 171 responding companies, 29 indicated that they had operations only in China.

Data collection in Sweden

I used a data collection approach similar to that used in the US in gathering data in Sweden, targeting the same high technology industry groups located in the southern part of the country. Following the same four-step process described for the US, I identified 725 firms and then sent questionnaires to them. A second mailing was sent a month later to companies that had not responded. These two mailings produced 403 valid replies, for a response rate of 59 per cent (note that 42 companies could not be reached for various reasons). Follow-up phone calls to 29 managers validated the responses. Tests for response bias, as described earlier under the US sample, revealed that there was no response bias ($p=0.19$). As Table 11.1 suggests, on average, responding companies were 4.3 years old, had 13 employees, competed in 13 high technology industries and most indicated that they internationalized their operations in their first year of existence. Of the 403 companies, 56 had operations only in China.

Measures

Several measures were developed to capture organizational learning, NVP, national origin and several control variables, as follows.

Organizational learning

Responses to surveys were used to construct the measures for organizational learning. Respondents were asked to 'indicate the extent to which their firm emphasized the following techniques in learning about its markets, customers and competition'. A five-point response rate was used (1 = Not at all, 5 = A great deal of emphasis). Responses to the items capturing each dimension were averaged and an overall index was used in the analyses reported later in the chapter.

Vicarious learning had two dimensions: competitive analysis and learning from relationships. The 'competitive analysis' index had four items ($\alpha = 0.70$): tracking the actions of the competition, conducting formal competitive analysis of local companies, conducting formal competitive analysis of foreign companies, in the market, and analysing competitors' products and brands. The index (four items; $\alpha = 0.66$) for 'relationship-based learning' included the following items: joining alliances with domestic companies, joining alliances with foreign companies in the markets, discussing problems encountered in the market with alliance partners, and discussing problems encountered in the market with local friends and associates.

Experiential learning had two dimensions: learning by doing and improvisation. Items for the learning by doing measure (three items, $\alpha = 0.64$), were learning through trial and error, experimentation and analysis of decisions made and results achieved. Items for improvisation (three items; $\alpha = 0.67$) were trying new things that have never been tried before, improvising as the situation would dictate and allowing employees to come up with their own solutions, as needed, in view of market conditions.

NVP

Two widely used measures captured NVP: return on assets (ROA) and growth in sales. Both were measured two years after the survey data collection. ROA served as a key indicator of a new venture's profitability. Sales growth was a major indicator of the firm's success in penetrating its markets and achieving growth. Data for both measures came from multiple sources, including the mail surveys and company reports and publications.

Moderator variable: country of origin

As will be reported shortly, some of the analyses aggregated the three data sets to test specific hypotheses. In these analyses, dummy codes were entered to control for country of origin effects, with France serving as the reference group.

Control variables

Analyses controlled for five factors that could influence the results. These variables were (1) venture age, measured in years; (2) venture size, measured by full-time employees; (3) industry type (using dummy codes, as appropriate); (4) ownership, where independent new ventures were coded 1 and corporate-sponsored ventures were coded 0; and (5) past average industry performance, using the three-year average ROA for the firm's home market. Data for the first three variables came from the surveys but were validated, where possible, using secondary data. Data for industry-wide ROA were extracted from national secondary sources, including trade associations.

ANALYSIS AND RESULTS

Table 11.1 reports the key characteristics of the study's three samples. Swedish firms were older than US and French ventures ($p < 0.05$). US ventures were larger than French and Swedish firms ($p < 0.01$).

Table 11.2 Differences in organizational learning approaches by national origin

National origin	Organizational learning approaches			
	Competitive analysis	Relationship-based	Learning by doing	Improvisation
US (U)	2.6	3.1	2.9	4.1
France (F)	3.1	3.6	3.2	3.0
Sweden (S)	2.5	4.1	4.2	2.5
P<	0.01	0.01	0.01	0.001
Scheffe comparison	F> S, U	S>F, U	S>F, U	U>F, S F>S

Differences in Organizational Learning Approaches

To test for differences across the three samples in organizational learning approaches, a two-step approach was used. Initially, discriminant analysis determined whether significant differences existed in the various learning methods used by US, French and Swedish companies. The results were significant ($p < 0.001$), indicating that new ventures from the three countries varied in their learning approaches in the Chinese market. Next, I used multiple analysis of covariance (MANCOVA) to pinpoint the differences in these learning approaches across the three countries. The results appear in Table 11.2.

Table 11.2 shows that born global new ventures differed significantly in their use of the four organizational learning approaches (all at $p < 0.01$). To determine the source of these differences, I used the Scheffe test of comparisons among means. Table 11.2 shows that French companies emphasized the use of formal competitive analysis more than US and Swedish new ventures (at $p < 0.05$ or better). Swedish companies used the relationship-based *and* 'learning by doing' approaches more than French and US ventures (at $p < 0.05$ or better). Finally, US companies emphasized improvisational learning more than the French and Swedish companies ($p < 0.05$). French companies also had a higher score than Swedish companies in improvisation-based learning ($p < 0.05$).

The NVP Implications of Organizational Learning

To test the study's final set of hypotheses, I pooled data from the three data sets and ran regression analyses, as reported in Table 11.3. Besides the five

Table 11.3 Effect of organizational learning approaches on new venture performance[1]

Variables	ROA	Sales growth
Intercept	0.11	0.10
Past industry ROA	0.39**	0.31**
Company age	0.09	0.05
Company size	0.03	−0.08
Independent ownership (=1)	0.23*	0.10
US	0.14	0.11
Sweden	−0.09	−0.03
Competitive analysis	−0.04	0.01
Relationship-based	0.19*	0.27*
Learning by doing	0.28**	0.16@
Improvising	0.09	0.23*
Adjusted R^2	0.19	0.14
F	8.43***	5.19**

Note: @ P<0.10, * p<0.05, ** p<0.01, *** p<0.001.

control variables discussed earlier, the analyses contained dummy codes for country of origin, with the French companies being the reference group.

As Table 11.3 suggests, the results for ROA were significant ($p<0.001$), explaining 19 per cent of the variance. Of the control variables, two were positive and significant: past industry ROA ($p<0.01$) and independent ownership ($p<0.05$). Of the organizational learning approaches examined, the coefficients for relationship-based learning ($p<0.05$) and learning by doing ($p<0.01$) were positive and significant. The results for sales growth were also statistically significant ($p<0.01$), explaining 14 per cent of the variance. Past ROA was the only significant control variable ($p<0.01$); it also had a positive coefficient. Relationship-based and improvisation-based learning were positively and significantly associated with sales growth ($p<0.05$) and learning by doing was marginally and positively significant ($p<0.10$).

Next, follow-up tests helped to better understand the implications of organizational learning for NVP. Because the French sample had only 29 companies, I used partial correlations to serve as the primary analytical tool. The results reported in Table 11.4 show that competitive analysis is positively and significantly associated with ROA in the French ($p<0.05$) and US ($p<0.01$) samples. Relationship-based learning is positively associated with ROA among French ($p<0.05$) and Swedish ($p<0.01$) new ventures. Learning by doing is positively associated with ROA in the US and Swedish samples, both at $p<0.05$. Improvisation is positively associated

Table 11.4 The association between organizational learning approaches in the Chinese markets and ROA[2]

National origin	Organizational learning approaches			
	Competitive analysis	Relationship-based	Learning by doing	Improvising
US	0.23*	0.05	0.24*	−0.09
France	0.31**	0.26*	−0.04	0.45**
Sweden	−0.11	0.45**	0.27*	−0.12

Note: * P<0.05, ** P<0.01.

Table 11.5 The association between organizational learning approaches in the Chinese markets and sales growth[3]

Sample	Organizational learning approaches			
	Competitive analysis	Relationship-based	Learning by doing	Improvising
US	−0.13	0.20	−0.21	−0.02
France	−0.17	0.26*	0.31*	0.09
Sweden	0.11	0.32*	0.17	0.29*

Note: * P<0.05, ** P<0.01.

with ROA only in the French sample ($p<0.01$). As predicted, these results show that there were systematic variations in the associations observed between different organizational learning approaches and ROA across the new ventures studied, depending on their national origin.

Looking at Table 11.5, the correlation between relationship-based organizational learning and sales growth was significant ($p<0.05$) in the French and Swedish companies. Learning by doing was positively and significantly associated with sales growth in the French sample ($p<0.05$). Improvisation was positively and significantly associated with sales growth only in the Swedish sample ($p<0.01$).

DISCUSSION

The effect of internationalization on organizational learning has been the subject of considerable interest, with little empirical documentation

(Zahra, 2005). Though researchers have argued that born global new ventures have unique advantages of their own (Autio *et al.*, 2000), it is not clear how these firms learn in the first place. Noting the difficulties surrounding new ventures' learning in culturally distant markets, this study has focused on the experiences of US, French and Swedish firms in the Chinese market. These firms come from home markets with very different market structures and cultural values from those that prevail in the Chinese market, which would make it difficult for them to learn about the way customers and competitors behave in these markets.

Consistent with the spirit of hypotheses 1 and 2, the results show significant differences among born global new ventures, based on their national origin, in their use of the experiential and vicarious learning approaches in the Chinese market (Table 11.2). As argued, national cultures could profoundly determine the mental models that entrepreneurs carry with them and influence how they relate to customers and competition in the Chinese market. A key contribution of the results (Table 11.2) is showing where new ventures from different countries differ in how they experientially and vicariously learn within the same market, an issue that has received little attention in past research.

The results also show that experiential and vicarious organizational learning matters in terms of NVP, both profitability and sales growth – even after controlling for several key variables. These results support hypotheses 3 and 4, though they are not universally significant. Looking at Table 11.3, it becomes clear that relationship-based learning (a key dimension of vicarious learning) is important for ROA and sales growth. Learning by doing is also significant and positive. Finally, improvisation-based learning is conducive to higher sales growth. Overall, the results partially support hypotheses 3 and 4 and show that born global new ventures could benefit greatly from applying different organizational learning approaches. These results clarify the importance of learning for NVP, especially in foreign markets. These results are consistent with the propositions of the knowledge-based theory of the firm.

Another contribution of the results is adding specificity to the moderating role of national cultures in explaining the relationships between the experiential and vicarious learning approaches and NVP (that is, ROA and sales growth). Tables 11.4 and 11.5 present relevant data, partially supporting hypotheses 5 and 6. Even though the moderating effects are not as uniform or pronounced as predicted, the data provide preliminary evidence that born global new ventures from different countries benefit from different organizational learning approaches in their performance (ROA and sales growth). These results reinforce the propositions of the knowledge-based theory of the firm, that knowledge (gained from learning) could improve NPV.

Limitations

The results should be interpreted with care, given the study's short time frame, which could limit the validity of findings. Longer time frames would be especially helpful in further documenting and clarifying the effects of learning on performance. In addition, there is a need for a broader representation of a more culturally diverse set of new ventures from different countries to better appreciate the role of national cultures. Cultural diversity would make it feasible to test more accurately the various hypotheses presented earlier in the chapter.

Practical Implications

The study highlights the importance of organizational learning for successful NVP. These results have two implications for new venture managers. First, they need to think carefully about the approach they will follow in learning about the Chinese market and how it can help them gain a deeper appreciation of market dynamics. Second, there is a need to articulate what new ventures learn in their foreign markets (for example, China) and capture this learning and use it in building a competitive advantage. Learning per se may not confer an advantage on a new venture; it matters more what entrepreneurs do to exploit this learning as they map out their firm's strategic choices in various markets. Any 'learning advantages of newness' are earned through the effective articulation of what managers have learned, why it matters, and how best to use it.

A key lesson from the study is the subtle but profound effect of the firm's home national culture. It shapes what born globals do abroad, even when these ventures are new and inexperienced. Entrepreneurs sometimes understate the value of their national cultures but there are penalties for this, as new ventures might miss the opportunity to learn and exploit this learning in their operations. Frequently, in their quest for a competitive advantage in their foreign markets, entrepreneurs think more about the target market and might inadvertently overlook the biases they have inherited from their own cultures.

Theoretical Implications and Future Research Directions

The results also suggest the need for more large-scale empirical studies on the organizational learning approaches that born global new ventures might use in their international markets. The literature cited earlier highlights the potential diversity of these approaches, but missing from this literature is a thoughtful and authoritative analysis. In addition, as noted, a broader sample with more diverse national cultures is needed, better to appreciate the potential differences that exist in organizational learning methods. Future research would benefit also from examining ways in which companies and their managers transform their experiential and vicarious learning into a source of competitive advantage, especially in culturally distant foreign markets. This transformation is a key source of superior NVP, as suggested by both the KBT and process theory of internationalization. It is important also to examine the effect of organizational learning on new ventures' survival (Sapienza, De Clercq and Sandberg, 2005). Studying these issues will undoubtedly enrich the literature but will also demand more longitudinal research designs that make it possible to infer cause–effect relationships.

CONCLUSION

The dynamism and growth potential of the Chinese market has attracted companies of different national origin to enter and lay claim to their niches. This chapter has focused on French, Swedish and US high technology new ventures that, at their inception, have entered the Chinese market. The study shows that born global new ventures use different approaches to learn in that market. The results also reinforce and highlight the major influence of national cultures in determining the organizational learning approaches new ventures use in the Chinese markets and the gains they achieve in their profitability and growth.

NOTES

1. New venture performance is measured two years after data collection for the predictor variables. Table does not include results for 12 dummy variables for industry type.
2. ROA is measured two years after the data for 'organizational learning' were collected. Numbers in the table are partial correlations.
3. Sales growth is measured two years after the data for 'organizational learning' were collected. Numbers in the table are partial correlations.

REFERENCES

Anderson O. (1993), 'On the internationalization process of firms: a critical analysis', *Journal of International Business Studies*, **24** (2), 209–31.

Autio, E., H.J. Sapienza and J.G. Almeida (2000), 'Effects of age at entry, knowledge intensity, and imitability on international growth', *Academy of Management Journal*, **43** (5), 909–1014.

Bartlett, C.A. and S. Ghoshal (2000), *Transnational Management*, 3rd edn, New York: McGraw-Hill.
Brush, C.G. and P.A. Vanderwerf (1992), 'A comparison of methods and sources for obtaining estimates of new venture performance', *Journal of Business Venturing*, **7**, 157–70.
Calori, R., G. Johnson and P. Sarnin (1992), 'French and British top managers' understanding of the structure and the dynamics of their industries: a cognitive analysis and comparison', *British Journal of Management*, **3** (2), 61–78.
Chandler, G.N. and S.H. Hanks (1993), 'Measuring the performance of emerging businesses: a validation study', *Journal of Business Venturing*, **8**, 391–408.
Coviello N.E. and H.J. Munro (1995), 'Growing the entrepreneurial firm: networking for international market development', *European Journal of Marketing*, **29** (7), 49–61.
Covin, J.G., D.P. Slevin and T.J. Covin (1990), 'Content and performance of growth-seeking strategies: a comparison of small firms in high- and low-technology industries', *Journal of Business Venturing*, **5**, 391–412.
Dierickx, I. and K. Cool (1989), 'Asset stock accumulation and the sustainability of competitive advantage', *Management Science*, **35**, 1504–11.
Grant, R.M. (1996), 'Toward a knowledge-based theory of the firm', *Strategic Management Journal*, **17**, 109.
Hitt, M.A., R.E. Hoskisson and H. Kim (1997), 'International diversification: effects on innovation and firm performance in product-diversified firms', *Academy of Management Journal*, **40** (4), 767–98.
Hofstede, G. (1980), *Culture's Consequences*, Beverly Hills, CA: Sage Publications.
Hofstede, G. (1991), *Cultures and Organizations: Software of the Mind*, London: McGraw-Hill.
Huber, G.P. (1991), 'Organizational learning: an examination of the contributing processes and the literatures', *Organization Science*, **2**, 88–115.
Johanson, J. and J-E. Vahlne (1977), 'The internationalization process of the firm – a model of knowledge development and increasing foreign commitments', *Journal of International Business Studies*, **8** (1), 23–32.
Johanson, J. and J-E. Vahlne (1990), 'The mechanism of internationalization', *International Marketing Review*, **7** (4), 11–24.
Kogut, B. and U. Zander (1992), 'Knowledge of the firm, combinative capabilities, and the replication of technology', *Organization Science*, **3** (3), 383–97.
Lau, C., D.K. Tse and N. Zhou (2002), 'Institutional forces and organizational culture in China: Effects on change schemas, firm commitment and job satisfaction', *Journal of International Business Studies*, **33** (3), 533–50.
Leonard-Barton, D. (1995), *Wellsprings of Knowledge*, Boston, MA: Harvard Business School Press.
Li, H. and K. Atuahene-Gima (2001), 'Product innovation strategy and the performance of new technology ventures in China', *Academy of Management Journal*, **44** (6), 1123–34.
McDougall, P.P. (1989), 'International versus domestic entrepreneurship: new venture strategic behavior and industry structure', *Journal of Business Venturing*, **4** (6), 387–400.
McDougall, P.P. and B.M. Oviatt (2000), 'International entrepreneurship: the intersection of two research paths', *Academy of Management Journal*, **43**, 902–8.
Miller, D. (1996), 'A preliminary typology of organizational learning: synthesizing the literature', *Journal of Management*, **22** (3), 485–505.

Oviatt, B.M. and P.P. McDougall (1994), 'Toward a theory of international new ventures', *Journal of International Business Studies*, **25** (1), 45–64.

Sapienza, H.J., D. De Clercq and W. Sandberg (2005), 'Antecedents of international and domestic learning effort', *Journal of Business Venturing*, **20** (4), 437–57.

Sapienza, H., E. Autio, G. George and S.A. Zahra (2005), 'The effect of early internationalization on firm profitability and growth', *Academy of Management Review*, in press.

Sohn, J.D. (1994), 'Social knowledge as a control system: a proposition and evidence from the Japanese FDI behavior', *Journal of International Business Studies*, **25** (2), 295–324.

Tallman, S. and J. Li (1996), 'Effects of international diversity and product diversity on the performance of multinational firms', *Academy of Management Journal*, **39**, 179–96.

Teece, D.J., G. Pisano and A. Shuen (1997), 'Dynamic capabilities and strategic management', *Strategic Management Journal*, **18** (7), 509–33.

Thomas, A.S. and S.L. Mueller (2000), 'A case for comparative entrepreneurship: assessing the relevance of culture', *Journal of International Business Studies*, **31** (2), 287–301.

Zahra, S.A. (2005), 'Theory on international new ventures: a decade of research', *Journal of International Business Studies*, **36** (1), 20–29.

Zahra, S.A. and G. George (2002), 'International Entrepreneurship: The Current Status of the Field and Future Research Agenda', in M. Hitt, D. Ireland, D. Sexton and M. Camp (eds), *Strategic Entrepreneurship: Creating an Integrated Mindset*, Oxford: Blackwell Publishers, pp. 255–88.

Zahra, S.A., D.R. Ireland and M.A. Hitt (2000), 'International expansion by new venture firms: international diversity, mode of market entry, technological learning and performance', *Academy of Management Journal*, **43** (5), 925–50.

Zahra, S.A., J.S. Korri and J. Yu (2005), 'Cognition and international entrepreneurships: implications for research on international opportunity recognition and exploitation', *International Business Review*, **14**, 129–46.

Zander, U. and B. Kogut (1995), 'Knowledge and the speed of the transfer and imitation of organizational capabilities: an empirical test', *Organization Science*, **6**, 76–92.

Zhao, L. and J.D. Aram (1995), 'Networking and growth of young technology-intensive ventures in China', *Journal of Business Venturing*, **10** (5), 349–70.

12. Theory and empirical evidence on R&D globalization in Chinese firms

Jin Chen and R. Michael Holmes Jr.

A trend toward research and development (R&D) globalization began nearly two decades ago and continues to the present day (Gerybadze & Reger, 1999; Li & Holmes, 2005). R&D globalization refers to the movement of R&D facilities outside of a firm's home country. A firm's global R&D facilities may be critical to the firm's overall strategy and may enhance the firm's ability to compete in both the short and the long term (Ronstadt, 1977; Håkanson & Nobel, 1993; Nobel & Birkinshaw, 1998). Traditionally, the main participants in R&D globalization have been large multinational enterprises (MNEs) from developed markets (Nonaka & Takeuchi, 1995). Recently, however, emerging markets such as China have surfaced as important players in the global economy (Li & Zhong, 2003) and firms from these markets are becoming increasingly willing to expand R&D operations into foreign markets (Fischer & von Zedtwitz, 2004). Nonetheless, while an impressive literature has emerged to explain R&D globalization by MNEs from developed markets, our understanding of R&D globalization in emerging market firms is somewhat limited.

We attempt to address this issue by investigating the global R&D strategies of a small sample of firms from an emerging market, China. We examine both the forms of global R&D these firms pursue and the coordination mechanisms utilized by the firms to manage their overseas facilities. Forms of global R&D refer to the motivations, organizational structures and entry modes associated with a firm's global R&D strategy. Coordination mechanisms refer to control procedures, organizational systems and processes, and communication patterns used to manage the firm's overseas facilities. Our theoretical explanation is primarily drawn from the resource-based view of the firm and organizational learning theory.

A firm's global R&D strategies are influenced by its resources and capabilities (Li & Holmes, 2005). In the present work, we define resources as tangible and intangible assets to which the firm has access (Barney, 1991) and capabilities as the organizational routines used to deploy these resources (cf. Winter, 2003). We argue that a firm's resources and capabilities influence both the forms of global R&D pursued by the firm and the coordination mechanisms it utilizes. Given the resource and capability gaps that often exist between emerging market firms and MNEs from developed markets (Svetlicic & Rojec, 1994; Meyer, 2004), we believe that it is important to examine the global R&D strategies of emerging market firms. Comparing the strategies used by emerging market firms to the strategies used by MNEs from developed markets may enhance our understanding of the benefits and challenges of global R&D.

We begin by offering an explanation of global R&D strategies using the resource-based view of the firm and organizational learning theory. Given the limited research on R&D globalization by emerging market firms, this section necessarily draws from the literature on R&D globalization by MNEs from developed markets. We identify three prominent forms of global R&D and the coordination mechanisms these forms require. The forms differ in both the resources they make available and in the capabilities required to manage them. We then briefly examine the ways in which the resources and capabilities of emerging market firms may influence their global R&D strategies. Our study and its results are then discussed and compared to the literature on the global R&D strategies of MNEs from developed markets. We conclude with a discussion of the study's implications and limitations and suggest directions for future research.

THEORETICAL DEVELOPMENT

Firms are conceptualized as collections of resources (Penrose, 1959; Wernerfeldt, 1984) and capabilities (Teece, Pisano & Shuen, 1997; Winter, 2003). Competitive advantage often lies in the possession of valuable, rare and inimitable firm-specific resources (Barney, 1991) and in the idiosyncratic capabilities to leverage such resources (Miller, 2003). Firms generally use their international expansion efforts either to access resources available in foreign markets (for example, Shan & Song, 1997; Zahra, Ireland & Hitt, 2000) or to exploit the firm's current resources and capabilities in new markets (for example, Dunning, 1977; 2001). Firms expanding for the first reason have the opportunity to build their firm-specific resources in areas in which they are currently weak (Rugman & Verbeke, 2001). Accessed resources can be combined with those already possessed to create idiosyncratic stocks of resources and capabilities that enhance the firm's competitiveness (Denrell, Fang & Winter, 2003; Sirmon, Hitt & Ireland, 2007). Firms expanding for the second reason have the opportunity to leverage their current resource and capability endowment in pursuit of profit and/or growth opportunities in foreign markets. Such expansion may be more likely among firms that are already endowed with a valuable stock of resources and capabilities.

Despite the benefits of international expansion, it is challenging and can create coordination difficulties that make its benefits difficult to capture (Hitt, Hoskisson & Kim, 1997). However, as firms engage in international expansion, they often develop capabilities that improve their ability to coordinate overseas facilities. This is a form of organizational learning. Learning is often grounded in the firm's unique experiences (Levitt & March, 1988) and involves both the acquisition of new knowledge and the ability to exploit that knowledge in new applications (Cohen & Levinthal, 1990; Zahra & George, 2002). Through experience, firms can develop a set of valuable capabilities to utilize and build upon in future global expansion efforts (Kogut, 1984; Li, 1995). Experience allows the firm to learn about and experiment with new methods of organizing and to develop a knowledge base to anticipate and react to potential problems before they jeopardize the firm's efforts. Accordingly, through experience, firms often become more proficient in coordinating their global facilities, are able to engage in more advanced forms of international expansion, and are more likely to succeed when they do so (Bartlett & Ghoshal, 1989; Lu & Beamish, 2004).

The motivations for pursuing global R&D can be used to identify different forms of global R&D (Kuemmerle, 1997, 1999). Numerous typologies of global R&D forms exist in the literature (for example, Ronstadt, 1977; Håkanson & Nobel, 1993; Chiesa, 1996; Kuemmerle, 1997; Nobel & Birkinshaw, 1998; Gassmann & von Zedtwitz, 1999). While these typologies vary in their level of detail, they generally recognize that global R&D can be motivated by (a) the desire to access resources available in overseas markets, (b) the desire to exploit the firm's current resources and capabilities in a given market, or (c) the desire to support the firms operations in multiple geographic markets. These three forms of global R&D provide different resource opportunities for firms. The first form allows the firm to build its technological resources by gaining exposure to advanced technology available in foreign markets. The second form allows the firm not only to upgrade its technological resources, but also to pursue financial resources by exploiting profit and growth opportunities in a given market. The third form provides resources associated with both of the first two forms, but can also promote innovation and global competitive advantage through extensive knowledge sharing among foreign facilities.

However, coordinating R&D is challenging and coordinating dispersed facilities located in multiple geographic markets is even more challenging. Indeed, although the nature of global R&D often requires the transfer of technology to and/or from overseas facilities, technology transfer is difficult. One reason is that firms generally must invest in isolating mechanisms to guard against expropriation during transfer (Teece, 1981). When technology is transferred to a large number of facilities in multiple countries with potentially heterogeneous intellectual property laws, the firm generally needs to develop more elaborate safeguards to protect its technology. Also technology often has a large tacit component and may therefore require intense personal interaction between employees to facilitate understanding of transferred technology (Teece, 1986). Additional communication may be necessary to integrate the firm's global R&D efforts (Nobel & Birkinshaw, 1998). Conducting R&D across country boundaries, however, can increase transportation and related costs and may create language and cultural barriers that limit the effectiveness of employee interaction and communication. Finally, managing R&D facilities in several countries exposes the firm to multiple and possibly conflicting political demands that serve as additional obstacles to overcome in coordinating global R&D.

Such arguments suggest that a firm's global R&D facilities become more difficult to manage as technology transfer, communication, the number of global R&D facilities or the number of countries in which they are located increase. Thus we argue that, although the three forms of global R&D provide increasingly advanced resources opportunities for firms, these forms also require progressively more resources and increasingly sophisticated capabilities to coordinate. We discuss each form below.

First, firms that seek to access the technological resources available in foreign markets often use their foreign R&D units as 'listening posts' to monitor foreign technology developments (Graham, 1992; Florida, 1997). Facilities established for this purpose are often located in areas of global expertise in a given technology (Kuemmerle, 1999). A primary purpose of such facilities is to improve the firm's knowledge base by monitoring and tapping into emerging technological advances in a given industry, in a given country (Pearce & Singh, 1992; Patel, 1996). The facilities also upgrade the firm's knowledge base by serving as recruiting stations to attract talented scientific personnel (Zejan, 1990). Such facilities may not be heavily engaged in research and development per se and benefit the firm primarily by providing access to external sources of technology.

Facilities established for this purpose are a fast and economical way for firms to gain access to technology that they do not yet have the resources and capabilities to develop. These facilities are often staffed by employees from the home market (Gassmann & von Zedtwitz, 1999), which allows the firm to utilize many of its existing processes in the new facilities and to save on the costs of hiring and training new employees (for example,

Barkema & Vermeulen, 1998). In addition, complex isolating mechanisms to protect the firm's technology may not be necessary, given the limited research and development conducted at the facilities. The R&D that is performed by the employees of such facilities may take place in the laboratory of partners to improve the firm's access to additional technology. This also allows the firm to avoid substantial investment in large numbers of scientific personnel, expensive lab equipment, and other overhead expenses associated with managing a fully functioning global R&D facility. An important line of communication in these facilities is that between the facility and the parent company and critical knowledge generally flows from the foreign facility to the parent (Kuemmerle, 1997). Key decision-making authority often resides in the parent or in centralized R&D facilities located in the home market and managing such facilities can often be accomplished with relatively simple coordination mechanisms.

Second, global R&D facilities can be used to target strategically important markets to establish, maintain, or strengthen competitive positions in such markets. This form of global R&D may be more likely when foreign markets are essential to the firm's overall strategy. While the locus of innovative activity may remain in the home market (Håkanson, 1990), facilities established to service local markets may engage in local product development (Sakakibara & Westney, 1992). Such activity may be required by host governments as a condition of entry (cf. Patel, 1996) and can also indicate commitment to the market (Gelb, 2000). In addition, performing R&D in a market allows the firm to adapt technology developed in centralized R&D facilities to local demands (Chiesa, 1996; Kumar, 1996). Firms can also use such facilities to provide technical services to other subsidiaries in the market (Ronstadt, 1977) and to ease the transfer of technology to local production units (Nobel & Birkinshaw, 1998). Establishing global R&D facilities in a given market can therefore improve the firm's ability to exploit profit and growth opportunities in that market. The resources provided by such facilities can be used to further expand operations in the given market and may also be used to fund initiatives in other markets.

This form of global R&D introduces additional complexity and becomes more difficult to coordinate than the first form. Facilities established to exploit local market opportunities are engaged in more research and development, so additional control must be exercised to prevent knowledge spillovers (cf. Cheng & Bolon, 1993). In addition, although such facilities may remain dependent on R&D facilities in the home market for access to the firm's core technology, utilization of such technology to exploit local market opportunities may require that the foreign facilities share information about the local market with the parent. Because key communication flows both from the parent to the foreign facilities and from the facilities to the parent, coordinating such facilities generally results in the formation of dense ties between the facilities and the parent (for example, Asakawa, 2001). However, concentration of the firm's core technology in the domestic market and discretion over other resources needed to sustain the operations of foreign facilities allow centralized headquarters functions to maintain strict control over the facilities (Asakawa, 2001). Such control may include formalized rules and procedures that are relatively inexpensive to implement and that specify the desired behaviour of dispersed facilities (Nobel & Birkinshaw, 1998).

Third, foreign R&D facilities may be intended to support, not just the market in which the facility is located, but also the firm's global network of operations. Such facilities often have a global mandate (Ronstadt, 1977) and maintain a broader and more creative role in the firm's international operations than other forms of global R&D (Nobel & Birkinshaw, 1998). These facilities may perform basic research, applied research, or both. Basic research involves scientific investigation into cause and effect relationships and, while not associated with a specific commercial objective (Lall, 1979), may be applicable to a wide range of products. Applied research is more commercially driven (Lall, 1979) and may be tailored to specific products and processes. Facilities established to service global markets can be used to build the firm's technological resources and to develop its innovative capabilities. For example, facilities may be charged with developing technology that complements the firm's existing technology (Cantwell & Kosmopoulou, 2001). This helps the firm create new technological resources and capabilities not easily imitated by rival firms (Cantwell, 2001). In addition, the technology developed in such facilities is often designed for use in multiple product and geographic markets and may therefore be a key source of long-term competitiveness (Ronstadt, 1977).

Facilities established for this purpose are often part of a global network of R&D facilities (Håkanson, 1990). The facilities may be highly specialized, relatively autonomous, and may become heavily involved in both the formulation and the implementation of the firm's international strategies (Brockoff, 1998). Coordination difficulties emerge both from the volume of R&D activities that must be integrated and from the number of scientists and engineers that must be managed. Moreover, significant cross-border technology transfer may be necessary to realize the full benefit of the firm's network of facilities. Consequently, control procedures that protect the firm's technology, while preserving the autonomy of foreign facilities, may also be necessary (for example, Cheng & Bolon, 1993). This may demand strong leadership and effective vertical communication to promote goal congruence, encourage cooperation (for example, Moenaert *et al.*, 2000), and provide direction to dispersed facilities. Effective lateral communication may also be required to promote better utilization of the firm's existing technology and to promote joint research and development to produce new technology. The complexity of such networks suggests that designing and maintaining them may require a substantial commitment of resources and well-developed management capabilities.

GLOBAL R&D AND EMERGING MARKET FIRMS

The resource and capability endowments of emerging market firms influence their international expansion efforts (Makino, Lau & Yeh, 2002). Emerging market firms are generally endowed with fewer financial, technological and managerial resources than MNEs from developed markets and often have access to less sophisticated strategic factor markets from which to acquire such resources (Khanna & Palepu, 1997, 1999). Such limitations may produce hesitancy and caution that constrain the amount of international expansion emerging market firms pursue (Bartlett & Ghoshal, 2000; Andreff, 2002). Additionally, emerging market firms and MNEs from developed markets often have different resource needs, given the limited resources of the former (Hitt, Dacin, Levitas, Arregle & Borza, 2000). Further, while many MNEs from developed markets were heavily involved in international expansion decades ago, international expansion is still a relatively recent phenomenon for most emerging market firms. Many emerging market firms are only beginning to acquire and develop the resources and capabilities necessary to compete successfully in overseas markets.

These arguments suggest that emerging market firms' global R&D strategies may differ from those of MNEs from developed markets. For example, the weak resource endowment of many emerging market firms may lead such firms to use their global R&D strategies to acquire resources, rather than to exploit them. That is, while MNEs from developed markets may use their international R&D facilities to exploit market opportunities in foreign markets, emerging market firms may be more likely to use their facilities to gain access to new resources by operating in foreign markets (for example, Rugman & Verbeke, 2001). Specifically, we might expect emerging market firms to emphasize access to foreign technology rather than exploitation of currently owned technology in their global R&D strategies.

Further, many emerging market firms have little experience with global R&D and therefore may not have had the opportunity to build the resources and capabilities to manage large-scale forms of global R&D.

Thus, although we expect that emerging market firms may be in need of many of the benefits an integrated network of innovative and well-dispersed global R&D facilities provides, this form of global R&D may be less common among emerging market firms, given the challenges of coordinating such networks. The weak resources and capabilities of many emerging market firms may also influence their selection of entry modes. Indeed, it is likely that emerging market firms will choose entry modes that require limited resource commitments and that demand less advanced capabilities to coordinate.

Attempts by emerging market firms to build the resources and capabilities global R&D requires may also influence their global R&D strategies. For instance, we might expect such firms to employ a variety of different coordination mechanisms to manage overseas facilities. This allows experimentation with different approaches for coordinating global R&D, as emerging market firms seek to identify an optimal mix of coordination mechanisms. In addition, when firms with limited international experience attempt to expand internationally, their performance often declines before learning improves their ability to address the challenges of operating in foreign markets (Lu & Beamish, 2001; 2004). Applying this logic to the global R&D strategies of emerging market firms, such firms may experience limited success in some areas of global R&D, as the resources and capabilities to manage this activity are developing.

Study Design

Setting

We investigate the global R&D strategies of a small sample of Chinese firms. Chinese firms are an appropriate sample for this study for two reasons. First, China is the world's largest emerging market (Hoskisson, Eden, Lau & Wright, 2000; Li & Zhong, 2003) and is becoming a powerhouse in the global economy. The most recent *Fortune* Global 500 included 15 Chinese firms and, according to the World Bank, China now boasts the world's seventh largest gross domestic product. However, given the country's small per capita income and the fact that large segments of the country's economy still remain under tight government control, China is still classified as an emerging market by most accounts. Second, Chinese firms have recently begun to globalize their R&D strategies and studying such firms allows us to gain insight into their initial global R&D investments. By comparing the strategies of these firms with the literature on MNEs from developed markets, we may enhance our understanding of both global R&D by emerging market firms and global R&D in general.

Sample

The initial sample consisted of 279 manufacturing-based firms headquartered in China. These firms were from the home appliance (100 firms), information technology (47), pharmaceutical (35), mechanical machinery and instrumentation (28), automobile (20), food (18), textile (13), tobacco (11) and petrochemical (7) industries. The questionnaire and interview data used in this study are discussed below.

Questionnaire

In early 2002, questionnaires were distributed to all 279 firms. The questionnaire was addressed to an executive officer (for example, a vice-president) with responsibility over technology in each respective firm. A total of 67 questionnaires were returned, but only 28 included complete data. Two additional firms were removed because they were not yet engaged in global R&D. Of the 26 remaining firms, ten are in the home appliance industry, ten are in the mechanical machinery and instrumentation industry, five are in the information technology industry, and one is in the pharmaceutical industry.

The low response rate likely reflects the relative novelty of global R&D for Chinese firms. That is, it is likely that many of the non-respondent firms have yet to expand technology operations abroad. Also, owing to the highly sensitive nature of global R&D, many firms may consider information on this subject proprietary and some managers may be hesitant to answer questions concerning their firm's global R&D strategies. Given that this is an emergent area of research and that our research design does not require multiple regression or related procedures, we believe that our sample size is adequate to draw preliminary inferences regarding the global R&D strategies of Chinese firms.

The questionnaire contained items related to the global R&D strategies of the sampled Chinese firms. Respondents were asked to answer the items as they pertained to the forms of global R&D and coordination mechanisms used by the respondent's own firm. The items were intended to collect information about the motivations, organizational structures and entry modes associated with the forms of global R&D the firms pursued. The questionnaire also inquired about the control procedures, organizational systems and processes, and communication patterns used to coordinate the firm's global R&D facilities. Each item is scored on a five-point scale. A score of 1 (5) indicates that the respondent rated the given item as least (most) important to his or her firm's global R&D strategy, suggesting that he or she was least (most) satisfied with the issue expressed in the item, or that the issue was minimally (exactly) applicable to his or her firm.

Interviews

Of the 26 firms, nine managers who agreed to participate were also interviewed. These managers were either senior technology officers or chief technology officers in their respective firms. The interviews were structured and lasted between 60 and 90 minutes each. We used these interviews to gain additional insight into the global R&D strategies of the firms in our sample.

Results

We begin by presenting quantitative results collected from our questionnaires. Table 12.1 contains the results pertaining to the forms of global R&D pursued and Table 12.2 contains results pertaining to the coordination mechanisms employed. We then discuss qualitative evidence from some of our interviews. The qualitative section is intended to illustrate each of the aforementioned forms of global R&D and some of the coordination mechanisms utilized by firms in our sample that employ such forms.

Quantitative Results: Forms of Global R&D

Motivations

Table 12.1 suggests that three of the four most popular motivations for engaging in global R&D involve improving the firm's access to technology located in foreign markets. The most popular motivation was to expand the firm's technological potential. The third most popular motivation was absorbing new knowledge by recruiting highly skilled personnel and the fourth most popular motivation was to monitor competitors and foreign technology trends. Responding to competitors and facilitating the implementation of corporate strategies was also a popular motivation. Motivations associated with speeding product introduction into foreign markets, providing technical support to foreign subsidiaries, cooperating on R&D projects with other organizations, and geographical convenience were generally less important to the Chinese firms in our sample than were motivations associated with accessing technology.

Organizational structures

Table 12.1 also indicates that much of the R&D activity of the Chinese firms in our sample remains focused on the home market. Many of the firms in our sample responded that most of their R&D facilities are concentrated in the domestic market and are more strongly oriented towards the domestic market than towards overseas markets. Those global R&D

Table 12.1 Forms of global R&D

Motivations	***Score***
Expanding and strengthening technological potential	3.71
Responding to competition and implementing corporate strategies	3.50
Absorbing new knowledge by recruiting highly skilled personnel	3.00
Monitoring competitors and foreign technology trends	2.86
Introducing products to foreign markets quickly	2.71
Providing technical support for foreign subsidiaries	2.64
Cooperating on R&D projects with local firms and research institutes	2.50
Geographical convenience	2.00
Organizational structures	
R&D facilities are concentrated in the domestic market and focus more on domestic than foreign markets	3.42
Domestic R&D facilities coordinate global R&D and tightly control global R&D facilities	3.17
An integrated R&D network	1.45
Regional R&D centres in technologically advanced areas coordinate global R&D	1.27
Autonomous and globally dispersed R&D facilities, with weak control from the parent	1.09
Entry modes	
Exporting	3.45
Joint venture	3.00
Greenfield operations	2.92
Patent licensing	2.82
OEM arrangements	2.16
M&A	2.09

Note: Each item is scored on a five-point scale.

facilities that do exist are coordinated and tightly controlled by R&D facilities located in the domestic market. Regional, autonomous, and well-integrated dispersed facilities were much less common.

Entry modes

Results in Table 12.1 also suggest that the firms in our sample generally preferred low-cost modes of market entry. For example, exporting and joint ventures are the two most preferred modes of entry, respectively. Greenfield operations and patent licensing are also somewhat preferred. OEM arrangements were not especially common and the most expensive form of entry, mergers and acquisitions (M&A), was the least common entrance strategy.

Table 12.2 Coordination mechanisms

Control procedures	***Score***
Unit formalization	3.20
Detailed planning	3.20
Standardization work roles	3.10
Power over annual R&D budgets	3.00
Extension of parent control over R&D projects through regional technology centres	2.80
Organizational systems and processes	
Promotion of goal congruence among global R&D facilities	3.79
Use of teams to coordinate R&D facilities	3.64
Attempts to develop a culture of information exchange	3.64
Use of multinational teams to facilitate interaction and relationship building among R&D employees	3.36
Use of language and cultural training	2.64
Use of job rotation among global R&D facilities	2.57
Movement of R&D engineers for knowledge transfer among R&D facilities	2.50
Communication patterns	
Communication between managers of the parent and managers of global R&D facilities is emphasized	3.93
Communication between the parent and global R&D facilities depends primarily on information technology	3.73
Communication among global R&D facilities depends primarily on information technology	3.18
Communication between the parent and global R&D facilities is primarily face-to-face	1.90
Communication among global R&D facilities is primarily face-to-face	1.50
Effectiveness of communication between the parent and global R&D facilities	2.80
Effectiveness of communication among global R&D facilities	2.22

Note: Each item is scored on a five-point scale.

Quantitative Results: Coordination Mechanisms

Control procedures

Table 12.2 indicates a variety of control procedures used by the sampled firms. The most common controls involved unit formalization, detailed planning, and standardization of work roles. Control by exercising power over annual R&D budgets and the establishment of regional technology centres to extend the parent's control over R&D projects performed overseas were slightly less common.

Organizational systems and processes
Table 12.2 also reveals a variety of systems and processes used to manage the firms' global R&D facilities. The most popular actions utilized were the promotion of goal congruence among the subsidiaries, the use of teams to coordinate the facilities and to promote personal interaction and relationship building among R&D employees, and attempts to develop a culture of information exchange. Language and cultural training was less common, as was job rotation and the movement of R&D engineers among global R&D facilities.

Communication patterns
Table 12.2 indicates that communication between the parent and managers of the firm's global R&D facilities is heavily emphasized by the firms in our sample. The firms also tended to rely on information technology rather than on face-to-face interaction to facilitate communication. Also, the effectiveness of communication between the parent and the global R&D facilities appears superior to the effectiveness of communication among the global R&D facilities, although neither type of communication appears especially satisfactory.

Qualitative Evidence

We use the global R&D strategies of two firms to illustrate the first form of global R&D discussed above. Haier Group is a Chinese home appliance firm that manufactures products such as freezers, washing machines and televisions. Haier has established global technology alliances with large multinationals such as Philips, Toshiba, and Lucent Technologies and operates six jointly-owned and ten wholly-owned R&D centres throughout North America, Europe and the Asian Pacific. Most of these centres are technology-tracking units designed to monitor the actions of competitors and technological advances in the home appliance industry. The financial and human resources used to operate the facilities are generally supplied by the parent firm.

Holley is a large Chinese conglomerate firm that uses its overseas facilities to access the advanced technology of developed market firms for use in the home market. Holley has a wholly-owned subsidiary in the United States that is designed to identify and acquire foreign high technology firms. We should note that this is not typical of the manner in which many of the other firms in our sample access foreign technology. The company's

American subsidiary recently acquired a CDMA (Code-Division Multiple Access) R&D unit from Philips Electronics that gave the firm the resources and capabilities to engage in systematic chip design for mobile telephones. This acquisition gave Holley not only the unit's tangible assets but also the rights to all intellectual property and in-process R&D at acquired facilities. The firm also gained access to a large number of experienced R&D engineers. This acquisition has improved Holley's ability to service the large and growing mobile telecommunications market in China and represents a potential competitive advantage over rival Chinese firms, a large majority of which do not have the technology to undertake systematic chip design.

The next two firms' global R&D strategies are more consistent with the second form of global R&D. ZTE is a telecommunication equipment manufacturer in China. The firm currently operates three R&D facilities in the United States and one in Korea that are intended to monitor international technology trends and recruit highly skilled technical personnel. However, these facilities also participate in product development efforts to service local markets. These efforts are tightly controlled by a central planning committee and technical centre located in China that must approve all product development projects before they are begun. Although the facilities have some autonomy in the recruitment and hiring of employees, funding for the facilities is provided by the parent company and managers in China exercise full discretion over budget allocation decisions relevant to the facilities.

Wanxiang manufactures quality bearing products and auto parts and has been successful using global R&D to service strategically important foreign markets and to support growth opportunities in such markets. For example, the US is a key market for Wanxiang and the firm owns a large R&D facility in the US. Many of the products used to service the US market originate in this facility and 20 per cent of all R&D engineers Wanxiang employs are located in the US. Wanxiang also recently implemented a 'localization' strategy in its global R&D facilities. To reduce the influence of cultural and psychic distance and to facilitate exploitation of foreign market opportunities (for example, O'Grady and Lane, 1996), many of the R&D engineers in Wanxiang's global R&D facilities are non-Chinese and most (70–90 per cent) are recruited from the local market. R&D projects routinely involve collaboration between R&D engineers with different linguistic and cultural backgrounds. As a consequence, language and cultural training is held regularly. Since Wanxiang introduced these practices, its time-to-market for new products has been reduced by 35 per cent and its foreign sales continue to grow impressively.

The only firm in our sample that operated an integrated global network of autonomous and specialized R&D facilities intended to service multiple markets was Huawei. Huawei is a leader in both the Chinese and the global network telecommunications market. Huawei's network was built incrementally as the firm became more dependent on foreign markets and also developed the resources and capabilities to coordinate increasingly sophisticated forms of global R&D. Our interviews suggest that Huawei's global R&D strategy progressed through each of three forms of global R&D discussed above. We discuss each in turn.

Huawei's early global R&D efforts were more consistent with the first form of global R&D. The firm began by locating R&D facilities in foreign markets primarily to gain access to the advanced technology and talented scientific personnel in such markets. For example, firms in California and Texas are global leaders in the development of advanced optical manufacturing technology and Huawei established R&D facilities in both Silicon Valley (California) and Dallas (Texas). Similarly, Swedish firms are prominent leaders in the research and development of mobile telecommunications technology and Huawei also located R&D facilities in Stockholm. However, Huawei remained focused on the Chinese market and the firm's global R&D facilities were generally intended to improve Huawei's ability to compete against competitors in the domestic market.

As international markets became more important to Huwaei, the firm began to establish facilities to exploit foreign market opportunities. This is consistent with the second form of global R&D discussed earlier. Because Huawei operates in the telecommunications industry, the firm confronts significant heterogeneity in customer needs and government regulations across markets. Accordingly, Huawei often locates R&D facilities in important markets to adapt products to the needs of local customers and to respond to local regulatory requirements. For example, the primary function of Huawei's Russian R&D facility is to target the potentially large Russian telecommunications market and to localize the firm's existing technologies to the demands of this market.

Huwaei's global R&D efforts have recently begun to resemble the third form of global R&D. Domestic and international market growth and continual development of the firm's technology are key objectives for Huawei. The firm's global R&D facilities, have therefore become more integral to its strategy. Huawei currently employs a sophisticated mix of technology alliances, domestic R&D facilities, and geographically dispersed global R&D facilities that form a well-integrated network. The network is designed to facilitate innovation, service customers, and create and exploit profit and growth opportunities in multiple geographic markets. Resources are decentralized, and highly specialized global R&D facilities operate relatively autonomously. Frequent communication occurs between domestic and global R&D facilities as well as among the global R&D facilities. The firm's

R&D facilities perform many joint projects and routinely share and exchange resources, including personnel. Additionally, the firm is no longer dependent on the advanced technology of developed market firms, as evidenced by the establishment of an R&D facility in Bangalore, India to develop measurement software for use in overseas markets.

COMPARISON OF THE GLOBAL R&D STRATEGIES OF THE CHINESE FIRMS AND THOSE OF MNEs FROM DEVELOPED MARKETS

The sampled Chinese firms' global R&D strategies are not entirely consistent with those of MNEs from developed markets. Three important differences are noted.

First, the Chinese firms heavily emphasize access to technology and often use their global R&D facilities to source advanced technology from foreign markets. Few of the firms located R&D facilities in foreign markets to transfer the firm's current technology to local subsidiaries or to speed product introduction into foreign markets. As noted, however, MNEs from developed markets often intend to utilize existing technology in foreign markets by, for example, transferring it to foreign manufacturing operations and/or adapting it to the demands of the local market. The difference between the global R&D strategies of the Chinese firms in our sample and those of many MNEs from developed markets may be indicative of the technology gap between emerging and developed market firms (Svetlicic & Rojec, 1994). Indeed, many of the Chinese firms may not possess technology that gives them an advantage in overseas markets and may therefore be using their global R&D facilities to narrow this gap by accessing advanced technology. In addition, compared to other forms of global R&D, facilities designed mainly to access foreign technology are relatively inexpensive and less difficult to coordinate and may therefore be better suited to the current resources and capabilities of many Chinese firms.

Second, the concentration of R&D facilities in the domestic market and the emphasis on gaining access to technology over supporting local subsidiaries and facilitating the introduction of products into foreign markets imply that much of the global R&D activity of the sampled Chinese firms remains oriented towards the domestic market. Indeed, some of our interview data (for example, Holley & Huawei) indicate that accessed foreign technology is often intended specifically to service the domestic market. In contrast, MNEs from developed markets often use their global R&D facilities to build and support profit and growth opportunities in foreign markets. Thus, an interesting feature of the global R&D strategies of many Chinese firms is the concentrated use of foreign technology in the home market. This may reflect the large and steady growth in the Chinese market in recent years (Peng & Luo, 2000; Luo & Park, 2001), as well as the market's potential future importance (Li & Holmes, 2005). At present, utilizing foreign technology in search of competitive advantage in the domestic market may be more vital to the overall strategies of many Chinese firms than is the pursuit of competitive advantage in foreign markets.

Third, when the sampled Chinese firms did attempt to exploit opportunities in foreign markets, they generally preferred to enter such markets through exporting and joint ventures. In contrast, the literature on large MNEs suggests that firms often prefer to internalize foreign operations through high-control entrance strategies that help the firm protect (Buckley & Casson, 1976) and utilize (Kogut & Zander, 1993) its knowledge and other proprietary resources. The technology used by Chinese firms in foreign markets may not be proprietary and may be available in the local market. Consequently, expensive, control-enhancing entry strategies such as acquisitions (Osland, Taylor & Zhou, 2001) may be less necessary for the Chinese firms. Exporting may offer a cost-efficient way for the Chinese firms to enter a market, while conserving precious resources (Shrader, Oviatt & McDougall, 2000; Zahra *et al.*, 2000). Low-cost entrance strategies such as exporting also allow the Chinese firms to maintain the option to invest more heavily in the market later as their knowledge of foreign markets improves (Miller and Folta, 2002).

Discussion

In the present work, we have used the resource-based view of the firm and organizational learning theory to describe the manner in which a firm's resources and capabilities can influence its global R&D strategies. Three prominent forms of global R&D and the coordination mechanisms each demands were discussed. We then speculated about the global R&D strategies of emerging market firms, given that many such firms have limited resources and capabilities. To illustrate our arguments, we presented both quantitative and qualitative evidence on the global R&D strategies of Chinese firms, compared such strategies to the literature on the global R&D strategies of MNEs from developed markets, and explained the differences using our theoretical framework. The analysis of Huawei allows us to speculate that a firm's global R&D strategies may evolve as the firm becomes in need of more advanced resources and also acquires and develops the resources and capabilities to manage more complex forms of global R&D successfully.

Managerial Implications

Our study offers implications for managers. Communication between the parent company and foreign R&D facilities and among the subsidiaries themselves was not particularly effective, according to the managers in our sample. Our study revealed that practices such as job rotation and programmes designed to reduce the effect of cultural dissimilarity were not especially common among the firms. We suggest that managers consider implementing such practices, as research indicates that both may encourage knowledge sharing (Gupta & Govindarajan, 2000; Moenaert, *et al.*, 2000). Additionally, while the firms in our sample made use of multinational teams to encourage knowledge sharing, the ineffectiveness of these teams suggests that firms may need to employ liaison personnel (Gupta & Govindarajan, 2000) to facilitate communication between the team and other parts of the organization. Firms might also choose to invest in regional technology centres as central locations to collect and synthesize knowledge from subsidiaries and to redistribute such knowledge throughout the firm.

The firms also indicated that information exchange relied much more heavily on information technology than on face-to-face communication. However, communication through information technology often must be supplemented by face-to-face interaction to maintain the trust (DeMeyer, 1991) and commitment of the communicating parties. Such assets may facilitate the spread of strategic resources, including knowledge, throughout the firm (Bouty, 2000), may reduce the need for costly governance structures to restrain opportunism (Ghoshal & Moran, 1996; Jones, 2001), and may also encourage cooperation among the firm's employees. Firms might consider the use of organizational practices such as integrated product development, simultaneous engineering, and cross-functional integration of marketing and R&D (Iansiti, 1998; Leonard-Barton, 1995; Nonaka & Takeuchi, 1995) to encourage interaction among R&D personnel and between R&D personnel and other organizational employees.

In addition, the Chinese firms tended to prefer exporting to other modes of entry. While this entry mode conserves resources and requires less advanced capabilities to manage than other entry modes, exporting may not be a sustainable way to exploit foreign market opportunities in the long term. Indeed, establishing manufacturing or R&D facilities provides a physical presence in the market that may improve a firm's ability to learn about competitors, government regulations, local customer needs and other idiosyncrasies unique to that market. Such facilities may also provide the flexibility to adapt quickly when conditions in the market change. Thus, as profit and growth opportunities in international markets become more important to Chinese firms, it may be necessary for them to establish permanent facilities in foreign markets. A firm that fails to do so may experience a competitive disadvantage relative to firms that have such facilities and this may limit the focal firm's ability to exploit opportunities in the given market.

Limitations

As with all research, there are limitations in the present study. Three important limitations are noted. First, the low response rate and small number of respondent firms is an obvious weakness that may limit the generalizability of our study. As noted, however, this is a developing area of research that can enhance our understanding of an important business phenomenon. Moreover, we tried to supplement our data with additional qualitative evidence on the global R&D strategies of the sampled Chinese firms. We therefore believe that our study allows us to draw preliminary conclusions concerning R&D globalization in Chinese firms. Nonetheless, we must urge caution in interpreting our results.

Second, R&D globalization is a relatively recent phenomenon among Chinese firms. Accordingly, the global R&D strategies of the respondent firms may be biased in favour of large and successful firms. Indeed, it is likely that many of the Chinese firms in our sample that have engaged in global R&D are leading-edge firms that are not representative of the large majority of Chinese firms. Although we agree with this speculation, our approach was largely grounded in existing theory and we believe that many of the advantages leading firms may have in globalizing their R&D efforts are endogenous to our model. That is, the resource-based and organizational learning perspective adopted in the present work addresses many of the advantages leading firms may have in pursuing their global R&D strategies. Our qualitative analysis also may have uncovered such advantages. Therefore, we do not believe that the bias in favour of leading firms seriously jeopardizes our conclusions.

Third, because our study was primarily concerned with the global R&D strategies of Chinese firms, our results may be less applicable to firms from other emerging markets. Indeed, a firm's country of origin can have implications for the amount of international expansion the firm pursues (for example, Wan & Hoskisson, 2003). A firm's strategy is influenced by the availability of production factors, the strategies and structures of other firms, the degree of rivalry, and the presence of related and supporting industries in the domestic market (Porter, 1990; 1991). Such factors can influence a firm's perceived need to expand internationally and can also affect the resources and capabilities that allow the firm to expand internationally. The Chinese economy is much larger than that of other emerging markets and the established presence of large multinationals has made many segments of the Chinese economy fiercely competitive. Chinese firms may have benefited from such intense competition and may be endowed with quality resources and advanced capabilities not possessed by the majority of firms from other emerging markets. Also such competition may have provided motivation for Chinese firms to pursue global R&D to improve their technological resources. As a result, the global R&D strategies of Chinese firms may be further developed than those of most firms from other emerging markets. While this is an empirical question, conclusions about the forms of global R&D and coordination mechanisms utilized by firms from other emerging markets may have to await extension of our study in other contexts.

Future Research

Our study also suggests opportunities for future research. While we sought to apply existing theory to explain global R&D strategies, we still lack a comprehensive theoretical framework for studying R&D globalization. Moreover, we are only beginning to understand the ways in which firms implement their overall strategies (Hrebiniak & Joyce, 2001) and there is still much to learn about the implementation of global R&D strategies. Additional questions pertaining to the source of innovative ideas, the way in which ideas are selected for development, and the organizational systems and processes that transform such ideas into commercially viable products remain. Similarly, we know little about the way firms convert the resources and capabilities they access through foreign operations into profitable results. To this end, recent theoretical contributions on the exploitation of the firm's idiosyncratic resources and capabilities (for example, Miller, 2003; Sirmon *et al.*, 2007) may inform our understanding of the ways in which firms create value through their global R&D strategies.

In addition, some of the Chinese firms in our sample and many MNEs from developed markets employ a mix of different governance structures (for example, alliances, greenfield operations, acquisitions) in foreign markets. Nonetheless, our knowledge of which structures tend to outperform others and the conditions under which some structures may be more desirable than others is limited. For example, is global R&D more efficient and effective when conducted through alliances, greenfield operations or acquisitions? Does the purpose served by the foreign unit (such as to monitor foreign technologies, to exploit opportunities in a given market, to support the firm's global operations) moderate the relationship between the governance structure employed and the performance achieved? What other factors influence this relationship? Do the answers to these questions depend on the type of performance one considers? Perhaps discussions of the advantages, disadvantages and management of alliances (for example, Inkpen, 2001; Ireland, Hitt & Vaidyanath, 2002), comparisons of alliances with other forms of investment (for example, Dunning, 1995) and commentary on the benefits and drawbacks of greenfield operations and acquisitions (for example, Barkema & Vermeulen, 1998) will inform analysis of these issues.

Finally, much of the research on global R&D has been performed on manufacturing firms, and R&D globalization in service firms is a heavily underresearched area in the strategy and international business literatures. This is an important shortcoming, given that a large and growing percentage of the global economy is now service-based (UNCTAD, 2004). As services become more tradeable, more services are likely to migrate into overseas markets (Karmarkar, 2004) and service firms' R&D facilities are likely to follow. Indeed, service firms such as Ebay (China) and Yahoo (India) have already begun to expand their R&D operations overseas. Studying R&D globalization by service firms may improve our knowledge of global R&D because the global R&D strategies of service firms may differ from those of manufacturing firms. For example, much of a service firm's technology is process-based, rather than product-based. R&D to improve such processes may require more intense interaction among employees from more parts of the organization than R&D to improve products. In addition, the technology of service firms is often tacit and not easily codified (Grosse, 1996). To facilitate the exchange of tacit knowledge, service firms may rely more heavily on wholly-owned subsidiaries than on collaborative partnerships (cf. Kogut & Zander, 1993). The need to remain close to the ultimate customer may also result in service firms placing greater emphasis on global R&D facilities designed to exploit local market opportunities. Such speculations serve to illustrate what we may learn by studying R&D globalization in service firms.

CONCLUSION

R&D globalization has become an increasingly important part of the global competitive landscape and is in need of further research. Although much of the research on R&D globalization has been conducted on MNEs from developed markets, emerging market firms are increasingly willing to participate in this activity. We therefore have a unique opportunity to enhance our understanding of R&D globalization by observing the global R&D strategies of emerging market firms. We used the resource-based view of the firm and organizational learning theory to explain three prominent forms of global R&D, examined the resources each makes available, and the resources and capabilities required to coordinate each. We also speculated about the ways in which the resources and capabilities of many emerging market firms would likely influence their global R&D strategies, and offered quantitative and qualitative evidence on the global R&D strategies of a small sample of Chinese firms. These strategies were compared to the literature on those of MNEs from developed markets and important differences were noted and explained using our theoretical framework. We believe that this work improves our understanding of global R&D and offers a foundation on which future research may build.

ACKNOWLEDGMENTS

We would like to thank Haiyang Li and an anonymous reviewer for valuable comments on earlier versions of this manuscript. Any errors remain the responsibility of the authors.

REFERENCES

Andreff, W. (2002), 'The new multinational corporations from emerging markets', *Economic Systems*, **26**, 371–9.

Asakawa K. (2001), 'Organizational tension in international R&D management: the case of Japanese firms', *Research Policy*, **30**, 735–57.

Barkema H.G. and F. Vermeulen (1998), 'International expansion through start-up or acquisition: a learning perspective', *Academy of Management Journal*, **41**, 7–26.

Barney, J.B. (1991), 'Firm resources and sustained competitive advantage', *Journal of Management*, **17**, 99–120.

Bartlett, C.A. and S. Ghoshal (1989), *Managing across Borders: The Transnational Solution*, Boston MA: Harvard Business School Press.

Bartlett, C.A. and S. Ghoshal (2000), 'Going global: lessons from latemovers', *Harvard Business Review*, **78**, 132–42.

Brockoff, K. (1998), *Internationalization of Research and Development*, Berlin: Springer.

Bouty, I. (2000), 'Interpersonal and interaction influences on informal resource exchanges between R&D researchers across organizational boundaries', *Academy of Management Journal*, **431**, 50–65.

Buckley, P.J. and M. Casson (1976), *The Future of the Multinational Enterprise*, London: Holmes & Meier Publishers.

Cantwell, J. (2001), 'Location, competitiveness and the multinational enterprise', in A.M. Rugman and T.L. Brewer (eds), *The Oxford Handbook of International Business*, Oxford: Oxford University Press, pp. 431–56.

Cantwell, J. and E. Kosmopoulou (2001), 'Determinants of internationalization of corporate technology', DRUID working paper no. 01-08.

Cheng, J.L.C. and D.S. Bolon (1993), 'Corporate R&D strategies: the influence of firm, industry and country factors on the decentralization of R&D', *R&D Management*, **23**, 91–107.

Chiesa, V. (1996), 'Managing the internationalization of R&D activities', *IEEE Transactions on Engineering Management*, **43**, 17–23.

Cohen, W.M. and D.A. Levinthal (1990), 'Absorptive capacity – a new perspective on learning and innovation', *Administrative Science Quarterly*, **35**, 128–52.

DeMeyer, A. (1991), 'Tech talk: how managers are stimulating global R&D communication', *Sloan Management Review*, **32**, 49–58.

Denrell, J., C. Fang and S.G. Winter (2003), 'The economics of strategic opportunity', *Strategic Management Journal*, **24**, 977–90.

Dunning, J.H. (1977), 'Trade, location, and economic activity and the MNC: a search for an eclectic approach', in B. Ohlin, P.O. Hesselborn and P.M. Wijkman (eds), *The International Allocation of Economic Activity: Proceedings of a Nobel symposium at Stockholm*, London: Macmillan, pp. 395–418.

Dunning, J.H. (1995), 'Reappraising the eclectic paradigm in an age of alliance capitalism', *Journal of International Business Studies*, **261**, 461–91.

Dunning, J.H. (2001), 'The key literature on IB activities: 1960–2000', in A.M. Rugman and T.L. Brewer (eds), *The Oxford Handbook of International Business*, Oxford: Oxford University Press, pp. 36–68.

Fischer W.A. and M. von Zedtwitz (2004), 'Chinese R&D: Naissance, renaissance, or mirage?', *R&D Management*, **34**, 349–65.

Florida, R. (1997), 'The globalization of R&D: results of a survey of foreign-affiliated R&D laboratories in the USA', *Research Policy*, **26**, 85–103.

Gassmann, O. and M. von Zedtwitz (1999), 'New concepts and trends in international R&D organization', *Research Policy*, **28**, 231–50.

Gelb, C. (2000), 'Software research in – and for – China', *The China Business Review*, **27**, 40–43.

Gerybadze, A. and G. Reger (1999), 'Globalization of R&D: recent changes in the management of innovation in transnational corporations', *Research Policy*, **28**, 251–74.

Ghoshal, S. and P. Moran (1996), 'Bad for practice: a critique of transaction cost theory', *Academy of Management Review*, **21**, 13–47.

Graham, E. (1992), 'Japanese control of R&D activities in the United States: is there cause for concern?', in T. Arrison and C.F. Bergstein (eds), *Japan's Growing Technological Capability: implications for the U.S. economy*, Washington, DC: National Academy Press, pp. 191–206.

Grosse, R. (1996), 'International technology transfer in services', *Journal of International Business Studies*, **271**, 781–800.

Gupta A.K. and V. Govindarajan (2000), 'Knowledge flows within multinational corporations', *Strategic Management Journal*, **21**, 473–96.

Håkanson, (1990), 'International decentralization of R&D: the organizational challenges', in C.A. Bartlett, Y. Doz and G. Hedlund (eds), *Managing the Global Firm*, London: Routledge, pp. 256–78.

Håkanson, L. and R. Nobel (1993), 'Determinants of foreign R&D in Swedish multinationals', *Research Policy*, **22**, 369–411.

Hitt, M.A., R.E. Hoskisson and H. Kim (1997), 'International diversification: effects on innovation and firm performance in product diversified firms', *Academy of Management Journal*, **40**, 767–98.

Hitt, M.A., M.T. Dacin, E. Levitas, J.L. Arregle and A. Borza (2000), 'Partner selection in emerging and developed market contexts: resource-based and organizational learning perspectives', *Academy of Management Journal*, **43**, 449–67.

Hoskisson, R.E., L. Eden, C.M. Lau and M. Wright (2000), 'Strategy in emerging economies', *Academy of Management Journal*, **43**, 249–67.
Hrebiniak, L.G. and W.F. Joyce (2001), 'Implementing Strategy: an appraisal and agenda for future research', in M.A. Hitt, R.E. Freeman and J.S. Harrison (eds), *The Blackwell Handbook of Strategic Management*, Malden, MA: Blackwell Publishers Inc., pp. 602–26.
Iansiti, M. (1998), *Technology Integration: Making critical choices in a dynamic world*, Boston, MA: Harvard Business School Press.
Inkpen, A. (2001), 'Strategic Alliances', in A.M. Rugman and T.L. Brewer (eds), *The Oxford Handbook of International Business*, Oxford: Oxford University Press, pp. 402–27.
Ireland, R.D., M.A. Hitt and D. Vaidyanath (2002), 'Alliance management as a source of competitive advantage', *Journal of Management*, **281**, 413–46.
Jones, G.R. (2001), 'Towards a positive interpretation of transaction cost theory: the central roles of entrepreneurship and trust', in M.A. Hitt, R.E. Freeman and J.S. Harrison (eds), *The Blackwell Handbook of Strategic Management*, Malden, MA: Blackwell Publishers Inc., pp. 208–28.
Karmarkar, U. (2004), 'Will you survive the services revolution?', *Harvard Business Review*, **82**, 101–7.
Khanna T. and K. Palepu (1997), 'Why focused strategies may be wrong for emerging markets', *Harvard Business Review*, **75** (4), 41–51.
Khanna T. and K. Palepu (1999), 'The right way to restructure conglomerates in emerging markets', *Harvard Business Review*, **77**, 125–34.
Kogut, B. (1984), 'Normative observations on the international value-added chain and strategic groups', *Journal of International Business Studies*, **15**, 151–67.
Kogut, B. and U. Zander (1993), 'Knowledge of the firm and the evolutionary theory of the multinational corporation', *Journal of International Business Studies*, **24**, 625–45.
Kuemmerle, W. (1997), 'Building effective R&D capabilities abroad', *Harvard Business Review*, **75** (2), 61–7.
Kuemmerle, W. (1999), 'The drivers of foreign direct investment in research and development: an empirical investigation', *Journal of International Business Studies*, **30**, 1–24.
Kumar, N. (1996), 'Intellectual property protection, market orientation, and location of overseas R&D activities by multinational enterprises', *World Development*, **241**, 673–88.
Lall, S. (1979), 'International allocation of research activity by U.S. multinationals', *Oxford Bulletin of Economics and Statistics*, **41**, 313–31.
Leonard-Barton, D. (1995), *Wellsprings of knowledge: building and sustaining the sources of innovation*, Boston, MA: Harvard Business School Press.
Levitt, B. and J.G. March (1988), 'Organizational learning', *Annual Review of Sociology*, **14**, 319–40.
Li, H. and R.M. Holmes (2005), 'Why do multinational corporations establish R&D facilities in an emerging market? Theory and empirical evidence from China', paper presented at the annual meeting of the Academy of Management, Honolulu, HI, USA.
Li, J. and J. Zhong (2003), 'Explaining the growth of international R&D alliances in China', *Managerial and Decision Economics*, **24**, 101–15.
Li, J.T. (1995), 'Foreign entry and survival: effects of strategic choices on performance in international markets', *Strategic Management Journal*, **16**, 333–51.

Lu, J.W. and P.W. Beamish (2001), 'The internationalization and performance of SMEs', *Strategic Management Journal*, **22**, 565–86.
Lu, J.W. and P.W. Beamish (2004), 'International diversification and firm performance: the S-curve hypothesis', *Academy of Management Journal*, **47**, 598–609.
Luo, Y.D. and S.H. Park (2001), 'Strategic alignment and performance of market-seeking MNCs in China', *Strategic Management Journal*, **22**, 141–155.
Makino S., C.M. Lau and R.S. Yeh (2002), 'Asset-exploitation versus asset-seeking: implications for location choice of foreign direct investment from newly industrialized economies', *Journal of International Business Studies*, **33**, 403–21.
Meyer, K. (2004), 'Perspectives on multinational enterprises in emerging economies', *Journal of International Business Studies*, **35**, 259–76.
Miller, D. (2003), 'An asymmetry-based view of advantage: towards an attainable sustainability', *Strategic Management Journal*, **24**, 961–76.
Miller K.D. and T.B. Folta (2002), 'Option value and entry timing', *Strategic Management Journal*, **23**, 655–65.
Moenaert, R.K., F. Caeldries, A. Lievens and E. Wauters (2000), 'Communication flows in international product innovation teams', *The Journal of Product Innovation Management*, **17**, 360–77.
Nobel, R. and J. Birkinshaw (1998), 'Innovation in multinational corporations: control and communication patterns in international R&D operations', *Strategic Management Journal*, **19**, 479–96.
Nonaka, I. and H. Takeuchi (1995), *The Knowledge Creating Company: How Japanese companies create the dynamics of innovation*, New York: Oxford University Press.
O'Grady, S. and H. Lane (1996), 'The psychic distance paradox', *Journal of International Business Studies*, **27**, 309–33.
Osland, G.E., C.R. Taylor and Zou (2001), 'Selecting international modes of entry and expansion', *Market Intelligence and Planning*, **11**, 1053–61.
Patel, P. (1996), 'Are large firms internationalizing the generation of technology? Some new evidence', *IEEE Transactions on Engineering Management*, **43**, 41–7.
Pearce, R.D. and S. Singh (1992), *Globalising Research and Development*, London: Macmillan.
Peng M.W. and Y.D. Luo (2000), 'Managerial ties and firm performance in a transition economy: the nature of a micro–macro link', *Academy of Management Journal*, **43**, 486–501.
Penrose, E. (1959), *The Theory of the Growth of the Firm*, New York: Oxford University Press.
Porter, M.E. (1990), *The Competitive Advantage of Nations*, New York: The Free Press.
Porter, M.E. (1991), 'Toward a dynamic theory of strategy', *Strategic Management Journal*, **12**, 95–117.
Ronstadt, R.C. (1977), *Research and Development Abroad by U.S. Multinationals*, New York: Praeger Publishers.
Rugman, A. and A. Verbeke (2001), 'Location, competitiveness, and the multinational enterprise', in A.M. Rugman and T.L. Brewer (eds), *The Oxford Handbook of International Business*, Oxford: Oxford University Press, pp. 150–77.
Sakakibara, K. and D.E. Westney (1992), 'Japan's management of technological innovation', in N. Rosenberg, R. Landau and D.C. Mowery (eds), *Technology and the Wealth of Nations*, Stanford, CA: Stanford University Press, pp. 327–43.

Shan W.J. and J.Y. Song (1997), 'Foreign direct investment and the sourcing of technological advantage: evidence from the biotechnology industry', *Journal of International Business Studies*, **28**, 267–84.

Shrader, R.C., B.M. Oviatt and P.P. McDougall (2000), 'How new ventures exploit trade-offs among internationalization risk factors: lessons for the accelerated internationalization of the 21st century', *Academy of Management Journal*, **431**, 1227–47.

Sirmon, D.G., M.A. Hitt and R.D. Ireland (2007), 'Managing firm resources in dynamic environments to create value: looking inside the black box', *Academy of Management Review*, forthcoming.

Svetlicic, M. and M. Rojec (1994), 'Foreign direct investment and the transformation of Central European economies', *Management International Review*, **34**, 293–312.

Teece, D.J. (1981), 'The multinational enterprise: market failure and market power considerations', *Sloan Management Review*, **221**, 3–17.

Teece. D.J. (1986), 'Transaction cost economics and the multinational enterprise: an assessment', *Journal of Economic Behavior and Organization*, **71**, 21–45.

Teece, D.J., G. Pisano and A. Shuen (1997), 'Dynamic capabilities and strategic management', *Strategic Management Journal*, **18**, 509–33.

UNCTAD (2004), *World Investment Report: the shift towards services*, Geneva: United Nations.

Wan, W.P. and R.E. Hoskisson (2003), 'Home country environments, corporate diversification strategies, and firm performance', *Academy of Management Journal*, **46**, 27–45.

Wernerfeldt, B. (1984), 'A resource-based view of the firm', *Strategic Management Journal*, **51**, 171–80.

Winter, S.G. (2003), 'Understanding dynamic capabilities', *Strategic Management Journal*, **24**, 991–5.

Zahra, S.A. and G. George (2002), 'Absorptive capacity: a review, reconceptualization, and extension', *Academy of Management Review*, **27**, 185–203.

Zahra, S.A., R.D. Ireland and M.A. Hitt (2000), 'International expansion by new venture firms: international diversity, mode of market entry, technological learning, and performance', *Academy of Management Journal*, **431**, 925–50.

Zejan, M.C. (1990), 'R&D activities in affiliates of Swedish multinational enterprises', *Scandinavian Journal of Economics*, **921**, 487–500.

13. Aspiration, motivation and internationalization behaviour of Chinese technology firms

Congcong Zheng and Susanna Khavul

INTRODUCTION

On 7 December 2004, Lenovo, a Chinese manufacturer of personal computers, bought the personal computing division of IBM and created a sensation around the world. This event not only baffled industry observers (*Economist*, 2004) but also generated controversy among academics (Knowledge@Wharton, 2005). Experts asked whether the takeover represented a wise move for both sides. Amidst the diverse opinions on this topic, one important point emerges: the takeover may be a move made out of necessity rather than the pursuit of an opportunity. Structural overcapacity and intense domestic competition in China continues to exert pressure on local Chinese firms. The Lenovo market share is under constant attack, while its stock price keeps falling. Mary Ma, Lenovo's financial chief, admitted that, 'if we just focus on China, we cannot generate returns for our shareholders'. Liu Chuanzhi, Lenovo's founder and retired chairman also noted that, because the space to grow in the domestic market is limited, globalization is the only option for the company (Knowledge@Wharton, 2005).

Internationalization out of necessity rather than in pursuit of opportunity is not only a Chinese phenomenon. In the 1950s and 1960s, Sony Corporation had to begin exporting to and later investing directly in the United States in order to avoid Mitsubishi's monopoly of the distribution network (Abbeyglen & Stalk, 1985). In recent years, changes in the political economy, information and communication technology, and cultural integration have increased the globalization of business: as the business world has globalized, so have more and more small and medium-sized enterprises. The Organisation for Economic Co-operation and Development estimates that more than a quarter of the world's small manufacturing firms enter international markets within ten years of their founding and derive a substantial percentage of their revenue from foreign sources (OECD, 1997).

Between 1 and 2 per cent of small manufacturing firms are estimated to be international at inception – that is, within two years of their founding (OECD, 1997). Researchers have pointed out that, as small and medium-size enterprises internationalize, they often distribute parts of their value chain across a number of countries (Oviatt & McDougall, 1994; 1999). International diversification is an important growth strategy for small and medium-size firms. The decision to internationalize usually involves important resource commitments (Burgel & Murray, 2000) and can influence their chances of survival (Zaheer & Mosakowski, 1997). Previous research has examined the antecedents (Sapienza, De Clercq & Sandberg, 2005), intervening processes (Zahra, Ireland & Hitt, 2000) and performance implications (Bloodgood, Sapienza & Almeida, 1996; Lu & Beamish, 2001; McDougall & Oviatt, 1996) of the decision to internationalize. While there is some theoretical work trying to explain the motivation to internationalize (Johnson & Vahlne, 1977; Oviatt & McDougall, 1994; Zacharakis, 1997), our understanding of why firms internationalize remains limited and inconclusive (Zahra & George, 2002). Further, the emergence of small and medium-size enterprises aggressively internationalizing from emerging markets increases the need for new theoretical, conceptual and methodological advances in academic research. In this chapter, we aim to address this gap and suggest that the behavioural perspective (Cyert & March, 1963) can explain why firms internationalize and what patterns of internationalization they follow.

Our central idea is that internationalization represents a type of organizational change in which firms are more likely to engage if their current performance fails to meet their goals or expectations. To advance this idea, we develop three lines of argument. First, we argue that internationalization is a risky type of organizational change for firms. Internationalization involves an unfamiliar host country environment and a lack of knowledge about the local market and cultural familiarity. All these factors are likely to result in asymmetric distribution of information in the market that will influence operation of the firm when entering a foreign country (Miller, 1992; Shrader, Oviatt & McDougall, 2000). Second, we argue that firms usually form performance expectations based on their social reference groups (Greve, 1998). The performance level of their social reference group thus forms their level of aspiration. Firms are more likely to engage in actions of organizational change if their performance is below their aspiration level (March & Simon, 1958). In the case of internationalization, firms are more likely to begin internationalization if their domestic performance level is unsatisfactory. Third, using the prospect theory of risk taking (Kahneman & Tversky, 1979), we propose that firms that are performing below aspiration levels are more likely to take on bigger risks and are more likely to enter uncertain, more culturally distant countries than those firms that are performing above their aspiration levels.

To test our arguments, we develop formal hypotheses and test them using data collected from surveys of 179 international and domestic Chinese small and medium-size enterprises. We discuss the implications and limitations of our findings and suggest the direction for future research. Our chapter attempts to make two contributions to the literature. Theoretically, we integrate the classical explanations of behavioural theory (Cyert and March, 1963) with more recent theorizing on the role of aspiration (Greve, 1998) and risk taking (Kahneman and Tversky, 1979) in motivating and focusing internationalization decisions. Empirically, we test our hypotheses in the context of internationalization of small and medium-size enterprises from China – an emerging economy with explosive growth that has global implications.

THEORY AND HYPOTHESES

Internationalization as Risky Organizational Change

Historically, international business research has been interested in the motivation behind the internationalization of large multinational corporations from advanced economies (Hymer, 1960; Knickerbocker, 1973; Vernon, 1966). Over time, three competing theories, monopolistic advantage (Hymer, 1960), the product life cycle theory (Vernon, 1966) and the internationalization process theory (Johnson & Vahlne, 1977), have emerged to explain how internationalization becomes an integral part of a firm's business strategy. These theories suggest that multinationals internationalize to extend their domestic advantage to the foreign markets (Hymer, 1960), to counter a rival's competitive behaviour (Knickerbocker, 1973) or to take advantage of a product's comparative advantages across country borders at the various stages of its product life cycle (Vernon, 1966).

A consistent theme in the literature is that firms expand internationally to exploit domestic advantages and capabilities (Zahra & George, 2002). Previous research emphasized the benefits that internationalization provides for firms. It offers benefits both from pure market expansion and from capability development. By selling their products and spreading their activities across different countries, firms can reduce revenue fluctuation (Kim, Hwang & Burgers, 1993). With increased economies of scale, they also enjoy higher market power over suppliers, distributors and customers, as well as reduced costs (Kogut, 1985). More importantly, firms can learn from operating in diverse markets and enhance their knowledge base and competitiveness through experiential learning (Barkema & Vermeulen, 1998; Delios & Henisz, 2000; Zahra *et al.*, 2000). Both case studies (Oviatt & McDougall, 1995) and quantitative studies (Bloodgood *et al.*, 1996; Burgel & Murray, 2000) suggest that firms that have unique intangible assets (experience, network and technology) and innovative technology are likely to internationalize. The arguments imply that firms that are performing well in the domestic markets are more likely to expand internationally.

However, these benefits do not come without risks. In the process of internationalization, firms have to face the liabilities of foreignness. These are the unavoidable costs that firms establishing operations abroad face that firms operating domestically do not (Zaheer, 1995; Zaheer & Mosakowski, 1997). Internationalizing firms face an unfamiliar local culture and lack a value chain network or political influence. In short, when starting to operate in an international setting, firms will encounter changes in their operating environment. When a firm starts exporting activities, for instance, it may need to centralize its decisions to balance the needs of various markets. It may also need to accommodate its organizational structure to exporting activities in order to achieve better performance (Beamish, Karavis, Goerzen & Lane, 1999). Internationalization is especially risky for small and medium-size enterprises. It usually represents a significant resource commitment for SMEs (Burgel, 1999). Once the resources are deployed for the purpose of internationalization, their redeployment for other purposes can be difficult and/or inefficient. As a result, international ventures usually have significantly different manufacturing, marketing and distribution strategies than their domestic counterparts (McDougall, 1989). Also, once a firm has internationalized, it is likely to experience further strategic changes in its operations (McDougall & Oviatt, 1996).

Because of such risk and uncertainty, firms often adopt a linear process, from exporting directly, to employing local agents or to setting up foreign subsidiaries. Lack of knowledge of a local market is an important obstacle to a firm's international development and firms need to acquire that knowledge gradually through experience abroad. This is the main theme of the international process theory that focuses on explanation of the linear characteristics of internationalization (Johnson & Vahlne, 1977). However, it does not tackle the issue of a firm's initial entry into the international market. The issue of initial entry is the focus of our chapter.

Social Aspiration Level and the Motivation to Internationalize

Our main research question is: given that internationalization is risky, what prompts firms to take the plunge and engage in such an activity? Research into organizational change suggests that, because of structural and competitive inertia, organizational change does not always originate from optimal rational calculation and pursuit of opportunity. Rather, change is more likely to happen when performance fails to meet a pre-set goal or expectation (March & Simon, 1958).

Organizational aversion to change is similar to an individual's aversion to risk (Greve, 1998). Experimental studies have documented that individuals have a tendency to be risk-averse. People prefer to gamble with lower variances of expected value (Lopes, 1987). However, individual risk-taking behaviour changes according to the context of choice and risk taking. An individual's risk taking appears to increase when people fail to reach a specific goal or aspiration level (Kahneman & Tversky, 1979). This is because an individual who fails to attain a goal or a given aspiration level might attempt to recoup his loss and be more likely to engage in riskier choices than those who have already surpassed their goal or aspiration levels. This risk-taking behaviour at the individual level persists at the organizational level (March, 1988; March & Shapira, 1987, 1992). In business situations, similar to personal settings, managers tend to take fewer risks when performance surpasses their goals or aspirations and to take higher risks when performance is lower than their goals or aspirations.

Organizational change is often prompted by performance feedback from the environment and managers' interpretation of this feedback (Milliken & Lant, 1991). When evaluating such feedback, the aspiration level is an important benchmark. At the individual level, an aspiration level has been defined as a 'reference point that is psychologically neutral' (Kameda & Davis, 1990:56) or 'the smallest outcome that would be deemed satisfactory by the decision maker' (Schneider, 1992). According to March and Simon (1958), an organization's aspiration level is a result of a bounded rational decision maker trying simply to evaluate by transforming a continuous measure of performance into a discrete measure of success and failure. It is the dividing line between success and failure in the decision maker's perception and interpretation of the results obtained from a former action.

Previous research has shown that aspiration level formation is an adaptation process that is influenced by previous aspiration level, performance feedback and social comparison (Lant, 1992; Mezias, Chen & Murphy, 2002). It is difficult to determine an individual or a firm's aspiration level. At the individual level, aspiration arises when an individual compares himself to other people (Festinger, 1954). The reference groups for comparison that the individual selects differs according to the goal of the particular comparison (self-evaluation, self-improvement or self-enhancement) and also the social environmental context in which the comparison takes place (Wood, 1989). It is more difficult to determine the aspiration level at the organizational level. Organizational aspiration levels change with experience (Alt, 1949; Blau, 1963) and reference points: social and historical (Greve, 1998, 2003). Social aspiration levels develop when a firm compares its performance with its peers. Historical aspiration levels develop when a firm compares its present state with its historical performance (Cyert & March, 1963). In forming a social aspiration level, size, industry, performance, status and geographical proximity can affect the choice of reference group (Baum & Lant, 2003; Davis & Greve, 1997; Haveman, 1993; Porac, Thomas, Wilson, Paton & Kanfer, 1995). Historical aspiration levels arise when the firm compares its current performance with the performance history of itself (Cyert & March, 1963; Levinthal & March, 1981). For SMEs, the social aspiration level is more relevant because of their often short track record on financial performance. Following March and Simon (1958), we argue that a risky organizational change is activated when firms perceive that their performance level falls below the average aspiration level.

Industry average performance may be a major aspiration level for SMEs. Since social aspiration level is based on the contemporary performance of a reference group, we argue that a particular firm forms its social aspiration level by comparing itself to its industry peers. Thus a firm's aspiration level is set to the mean performance in the industry (Greve, 1998). Firms performing below this level will experience more pressures to change and internationalize their international operations.

We follow Greve (1998)'s conceptualization of the relationship between aspiration level and probability of change and internationalization. In this conceptualization, the probability of change (internationalization entry in our case) is characterized by two properties: '(1) decline in the probabilities of change when performance increases, both above and below aspiration level; and (2) higher sensitivity above the aspiration level, as the decline in probability of change is more rapid then. The second property gives the curve a kink – a change in the slope at the aspiration level' Greve (2003:62). The first property comes from the idea of problematic search, that is, failure to meet aspiration level increases the probability of initiation of search activities and change (Cyert & March, 1963). The second property comes from organizational inertia (Hannan & Freeman, 1977; Milliken & Lant, 1991; Staw, Sandelands & Dutton, 1981). When performance level surpasses the aspiration level, decision makers in the organization are more conservative in their choice and tend to continue with the status quo. Figure 13.1 shows the probability of change/internationalization.

Hypothesis 1a: when a firm's performance relative to the social aspiration level increases, the probability of its entering the international market decreases.

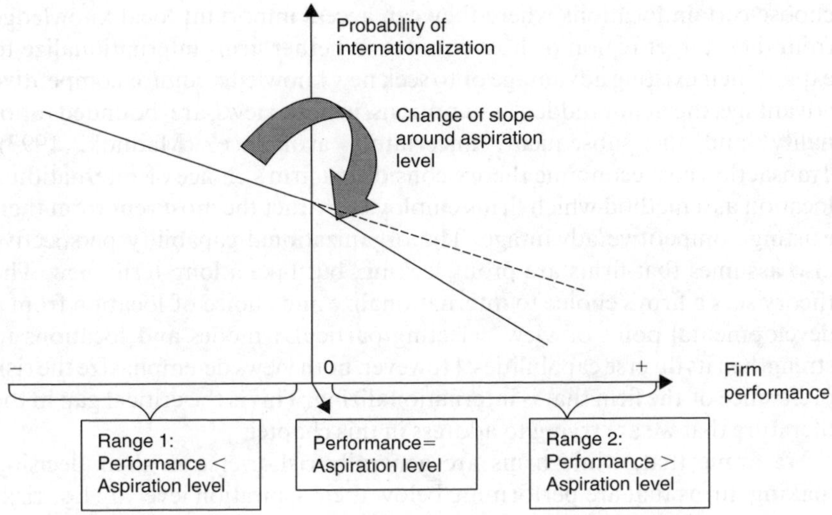

Note: Kinked curve explanation: In general, as performance of an entrepreneurial firm increases in relation to its aspiration level, the probability of internationalization for the firm decreases. The decrease in probability is more abrupt in the above-aspiration range (range 2). This is shown as range 2 on the right hand side of the figure, where the probability of internationalization drops more sharply (compared to range 1, below aspiration range, on the left hand side of the figure) once the firm's performance exceeds its aspiration level.

Figure 13.1 Proposed functional forms between firm performance, aspiration level and probability of internationalization

Hypothesis 1b: the decrease in the probability of internationalization is greater with an increase in performance above the social aspiration level.

Risk Taking and Location Choices

Choice of location has previously been the main subject of discussion in the literature. Pioneering research examined the choice of location based upon the comparative competitive advantages of various countries along the product life cycle (Vernon, 1966) or different stages in the firm's international process (Johnson & Vahlne, 1977). Recent research has shifted attention from the normative view of a firm exploiting its advantage overseas to a more capability-oriented view of international firms. Recent research has emphasized the potential benefits in terms of capability development for firms (Madhok, 1997). Researchers have investigated the capability development issue from a knowledge sourcing (Almeida, 1996; Cantwell, 1989) and agglomeration advantage (Shaver & Flyer, 2000) point of view. Firms are hypothesized to choose certain locations where they can access important local knowledge from their target region or host country. Whether firms internationalize to exploit their existing advantage or to seek new knowledge and/or competitive advantage, the major hidden assumptions, in both views, are 'bounded rationality' and the subsequent 'uncertainty avoidance' (Madhok, 1997). Transaction cost economic theory considers a firm's choice of international location as a method which firms employ to extract the most rent from their existing competitive advantage. The organizational capability perspective also assumes that firms are profit seeking, but take a long-term view. The theory sees a firm's choice to internationalize and choice of location from a developmental point of view, selecting particular modes and locations to strengthen its diverse capabilities. However, both views de-emphasize the risk preference of the firm that is internationalizing. This is the critical gap in the literature that we are trying to address in this chapter.

We argue that, while firms are generally risk-averse in their decision making, firms that are performing below their aspiration level are less risk-averse than those performing above their aspiration level. Prospect theory (Kahneman & Tversky, 1979) proposes that an individual's choice is influenced by potential gains and losses. Different from the classic expected utility model which emphasizes the probability of expected outcomes, prospect theory emphasizes the role of weights on the expected outcomes. Usually, decision weights are lower than corresponding probabilities, except in the range of low probabilities. An important finding of prospect theory is that decision makers exhibit inconsistent preferences when the same choice is presented in different forms, indicating framing effects (Kahneman & Tversky, 1979). Decision makers usually exhibit greater degrees of conservativeness once current performance is satisfactory (Greve, 1998). In the international setting that we investigate, firms performing above their aspiration level are more likely to exhibit conservatism in choosing their target countries.

Culture is one area in which firms may heighten their conservatism. The process school of the international management literature has long observed that, at the beginning of the internationalization process, firms choose countries with a shorter 'psychic distance' from their home country (Johnson & Vahlne, 1977, 1990). Psychic distance refers to the degree to which a firm is uncertain of the characteristics of a foreign market. It is influenced by differences in culture and language of the home and target countries. In empirical settings, it is difficult to measure psychic distance and researchers have used cultural distance between the home country and target country as a proxy for it (Barkema, Bell & Pennings, 1996; Hofstede, 1983; Kogut & Singh, 1988). The closer the target country is to the home country, the less risky it will be for the firm since the target country will be similar to the home country in terms of organizational and administrative practices, customer expectations and so on. We argue that firms performing above their aspiration level will choose these less risky countries, as the decision makers are less likely to change the status quo. If a certain practice is working well in the home country, the extrapolation is that it is likely for the same practice to work in a culturally similar country – more so than for it to work in a culturally distant country. To minimize risk, firms performing will choose those culturally similar countries to replicate their success in the domestic market. On the other hand, firms that are performing below their aspiration level are more likely to engage in problem-oriented change and search in order to correct their status quo. A radical change (for example, working with a more culturally distant country) might offer them an opportunity to ascertain what went wrong in the domestic market and to correct those mistakes in later operations.

Hypothesis 2a: when a firm's performance relative to the social aspiration level increases, the probability of it entering a risky market decreases.

Hypothesis 2b: the decrease in the probability of entering a risky market is greater with an increase in performance above the social aspiration level.

METHODS

Data Collection and Research Site

We test our hypotheses on a cross-sectional database of Chinese small and medium-size enterprises that includes firms that have internationalized and those that have not. We collected data on SMEs in Beijing and Shanghai in the People's Republic of China between September and December of 2003. The firms were from five technology-based industries: (1) information communication technology (ICT) hardware; (2) software products; (3) biotechnology, pharmaceutical and advanced chemical; (4) machinery, and equipment and information technology (IT) and (5) technical services. We include details of our sample characteristics in Table 13.1. The sampling frame for the internationalized firms is taken from six sources: the Beijing Foreign Economic Trade Commission (BFETC); China Commodity Net database by MOFCOM (Ministry of Commerce, P.R.C.); Zhongguancun Science Park Administration Committee (www.zhongguancun.com.cn), China Custom Office (Beijing), Shanghai Custom Office and Shanghai Foreign Economic Relationship and Trade (www.smert.gov.cn). Our effective sampling frame totals 610 firms, of which 146 answered our survey.

Table 13.1 Sampling procedure and sample characteristics

	International firms	Domestic firms	Total/Average
Sampling frame	610	398	1008
Returned surveys*	146	114	260
Response rate (%)	23.90	28.60	25.80
SMEs	126	110	236
Sample used in testing	H1a, b & H2a, b	H1a, b	—

Note: * Because of missing values, our final sample size in testing H1a, H1b (including international and domestic firms) is 179 and H2a, H2b (international firms only) is 96.

Thus our effective response rate is 23.9 per cent. Of these firms, 126 fit the definition of small and medium-size enterprises. In differentiating SMEs and large firms, we follow the definition provided by the American Small Business Administration (SBA): SMEs are stand-alone enterprises with fewer than 500 employees (Lu & Beamish, 2001). After accounting for missing values across all variables, the sample of international firms used in the analysis is 103. For domestic firms, the sampling frame was compiled from three lists: the Beijing Foreign Economic Trade Commission, Zhongguancun Science Park Administration Committee and the China Commodity Net database. The effective sampling frame totals 398 firms, of which 114 completed our survey. This gives us an effective response rate of 28.6 per cent. 110 of the 114 fit the definition of small and medium-size enterprises. After accounting for missing values on all the variables, the number of cases of domestic firms under analysis is 76. Thus the total number of cases that we include in our statistical analysis is 179, including 76 domestic firms and 103 international firms.

In our sample, the average age of the firms in 2002 is 4.51 years, the average size in terms of employees is 104 and in terms of total turnover is \$1 722 000 in the first year of sales, \$5 140 000 for the year 2002 and \$5 893 000 in year 2003 (expected sales), using the exchange rate of 1 USD = 8.28RMB. For firms that have already internationalized, it took them an average of 1.94 years from firm start-up to entering the international market. Table 13.1 describes our sampling procedure and Table 13.2 describes our sub-sample characteristics (international firms v. domestic firms).

Models and Analysis

In testing hypotheses 1a and 1b, we used the full sample. Since hypotheses 2a and 2b relate to the firms that have internationalized, we only use those firms that are already internationalized to test them. In terms of model

Table 13.2 Comparative statistics of domestic and internationalized firms

Firm characteristics	Mean value for domestic firms	Mean value for internationalized firms
Firm age at 2002 (years)	3.86	5.26
Number of employees	80.00	131.80
Sales value in the first year of operation (in USD000s)	638.47	2939.02
ROS in the first year of operation	28.66%	20.71%
Sales value in 2002 (in USD000s)	3768.16	6632.52
ROS in 2002	27.03%	22.07%
Percentage in industry sector		
ICT hardware	17.46%	10.91%
Software products	20.63%	8.18%
Biotech, pharmaceutical & chemical products	21.43%	25.45%
Machinery and equipment	22.22%	37.27%
IT & technology services	17.46%	17.27%
Percentage in industry life cycle		
Introduction stage	50.40%	33.65%
Growth stage	36.00%	41.35%
Maturity stage	12.80%	21.15%
Decline stage	0.8%	3.85%

specification, we use logistic regressions to test hypothesis 1a and 1b and use both logistic regression and ordinary least square (OLS) regression to test hypotheses 2a and 2b. Logistic regression and OLS regression offer similar results for hypotheses 2a and 2b.

Dependent variables

There are two dependent variables in this study: the likelihood of international entry (for hypotheses 1a and 1b), risk taking for entry indicated by cultural distance between the entry country and home country, China (for hypotheses 2a and 2b).

International entry A firm is categorized as being an international firm if in year 2002 its international revenue constitutes more than 5 per cent of all of its sales revenue (McDougall, 1989). McDougall has shown that international market operations are more likely to have a significant influence on the firm's overall operation if it exceeds the 5 per cent threshold level. This constitutes an organizational change in that aspect.

Cultural distance Following convention in the literature (Barkema, Shenkar, Vermeulen & Bell, 1997; Barkema & Vermeulen, 1997), we used the Kogut and Singh index to calculate the cultural distance (Kogut & Singh, 1988). Using Hofstede's indices (Hofstede & Hofstede, 2004), a composite index was formed based on the deviation along each of the five cultural dimensions (power distance, PDI; uncertainty avoidance, UAI; masculinity/femininity, MAS; individualism, IDV; and long-term orientation, LTO) of each country from China's score. The formula is as follows:

$$CD_j = \sum_{i=1}^{5} \{(I_{ij} - I_{ic})^2 / V_i\}/5,$$

where I_{ij} stands for the cultural index of jth country on the ith cultural dimension. V_i is the variance of the index for the ith dimension, c indicates China, and CD_j is the cultural distance of the jth country from the Chinese ranking. We tested the model using cultural distance both as a dummy variable (high cultural distance v. low cultural distance) and as a continuous variable. As shown later, in Table 13.5 and Table 13.6, the results for these two model specifications are essentially the same.

Independent variable

Aspiration level The focal independent variable in this study is the firm's aspiration level. We measure the aspiration level in relative terms as the industry average return of sales (ROS) performance within our sample. Measuring aspiration level by the industry level follows Greve (1998). In order to test the kinked curve in our conceptualization (Figure 13.1), we used a spline function (Greene, 2003: 120–22). Following Greene, we made a spline variable, which is the interaction of a dummy variable (1 if a firm's ROS performance is above its aspiration level and 0 otherwise) and a relative performance variable (which is the difference between performance and the aspiration level, that is, performance minus aspiration level). In testing the spline function as specified in Figure 13.1, we need to enter both performance and the interaction spline variable calculated above. The coefficient of the performance variable shows the general trend of the relationship between performance and likelihood of internationalization. A negative coefficient demonstrates that the likelihood of internationalization decreases with performance increase in general. The coefficient of the interaction spline variable shows the change of coefficient around the kink point. The coefficient of the interaction term adds to the coefficient of the main variable, that is the relative performance variable. A negative coefficient demonstrates an accentuation of the general negative trend, indicating that the slope for the above aspiration range is more steep than the baseline slope (the slope for the below aspiration range). Please refer to Figure 13.1 for a visualization of the change of slope along the range of a firm's performance.

Control variables

Size Firm size can influence the likelihood of entrepreneurial firms going international. Larger firms might be more likely to enter international markets since they have the human and financial resources to do so. Size refers to the number of employees in year 2002. We use its natural log of the employees in the regressions.

Firm age Firm age can influence the probability of a firm internationalizing. Older firms are more likely to have established their market position in the domestic market. According to traditional international management literature, they are more likely to enter the international market to exploit their existing competitive advantage in the domestic market. To tease out this confounding effect, we include firm age as a control variable.

Industry life cycle According to the product life cycle theory, firms that engage in industries that are in a different life cycle have a different probability of entering the international market (Vernon, 1979). To control this confounding effect, we asked survey respondents to indicate the industry life cycle of their product/services as one of four stages: introduction, growth, maturity and decline. Of the firms in our sample, 42.79 per cent are at the introductory stage, 38.59 per cent are at the growth stage, 16.59 per cent are at the maturity stage of their industry and 2.18 per cent of firms are at the decline stage of the industry. Our baseline for comparison is the maturity stage of industry.

Industry dummies In this cross-sectional data set, we have firms that originate from five industries: (1) ICT hardware, (2) software products, (3) biotechnology, pharmaceutical, advanced chemical, (4) machinery and equipment, and (5) IT services industry. We used IT services as our baseline for comparison and entered four dummies for the other four industry categories.

R&D investment The level of R&D investment is another variable that might influence the likelihood of internationalization. In the current world economy, industries are becoming more and more integrated and firms investing heavily in R&D might feel more pressure to go international since they feel the need to recover the investment that they have made (Franko, 1989; Jolly, Alahuhta & Jeannet, 1992). A single country's market is so small that the return from it might be insufficient for the firm to recover its investment. To control the technology-specific effect on internationalization, we entered R&D investment as our control variable. This is measured by the percentage of R&D employees from the total number of employees at the time of our survey (2003).

Market attractiveness Market attractiveness influences the choice of target country for international entry (Barkema & Vermeulen, 1997). In choosing the target countries, firms are more likely to enter countries with high market potential to take advantage of the market there. We measure market attractiveness using the GDP 2001 data (in billion US dollars), which is obtained from the WorldBank website (http://www.worldbank.org/data/wdi2005/index.html). The Worldbank database does not include data on Taiwan, so we obtain Taiwan-related data from Ministry of Economic Affairs of Taiwan, Republic of China (http://www.moea.gov.tw/english/estatistics/index.html).

RESULTS

Table 13.3 shows the mean values, standard deviations, and correlations for all the measured variables for all firms. Model 1 and Model 2 in Table 13.4 show the results of the logistic regression analysis with the likelihood of international entry as the dependent variable ($n = 179$). Model 1 is the base model for logistic regression on the probability of a firm's internationalization. Model 2 reports the results of the full model with the variables of aspiration level. Model 3 and Model 4 in Table 13.5 show the results of the logistic regression analysis for the internationalized firms only ($n = 96$). These models show the effect of aspiration level on location choices of firms that have already internationalized. In Table 13.6, Model 5 and Model 6 show the results of the OLS regression analysis for internationalized firms only ($n = 103$).

Hypothesis 1a states that, as a firm's performance relative to the social aspiration level increases, the probability of it entering the international market decreases. Model 2 (Table 13.4) shows that the coefficient for the main performance variable is negative and significant ($p < 0.01$), indicating that, the better a firm performs, the lower will be the probability that it will internationalize. Thus hypothesis 1a is supported. Hypothesis 1b states that the decrease in the probability of internationalization is greater with an increase in performance above the social aspiration level. Model 2 in Table 13.4 indicates that the interaction term (of above aspiration dummy and relative performance) is positive and significant ($p < 0.05$), showing that there is a change of slope when a firm's performance level surpasses its aspiration level. The positive coefficient indicates that the coefficient for the above aspiration range is less than that of the baseline level, that is, the coefficient for the below aspiration range. It indicates that once the firm's performance exceeds its aspiration level, the probability of firm internationalizing decreases, but less rapidly than the rate of decrease in the below aspiration range.

Hypothesis 2a states that, as a firm's performance relative to the social aspiration level increases, the probability of it entering a risky market decreases. Hypothesis 2b states that the decrease in the probability of entering a risky market is greater with an increase in performance above the social aspiration level. We test the hypotheses using both the logistic regression (Model 4 in Table 13.5) and OLS regression (Model 6 in Table 13.6). As in hypothesis 1, we enter the main performance variable and the spline variable in the regression. Although the signs of the coefficient is as we hypothesized (negative for the main variable and positive for the spline variable), neither coefficient is significant. The reason of this non-support might be twofold: first of all, we find that the market attractiveness variable, measured by the GDP figure, rather than the aspiration level, explains the choice of entry country (the p value of the market attractiveness variable is less than 0.001). It seems that most of the firms are entering the countries with the highest market potential in order to benefit from an expanded customer base. For instance, 62 firms (56.4 per cent of the international firms) entered either the US or Japan, two of the most attractive markets. We conclude that, although aspiration level significantly influences the decision to enter the international market, it does not influence the target country choice as much. The second reason for non-support is related more to the mechanisms of internationalization. We propose that the country a firm enters is not only determined by the risk factor of that country, but also might be related to the ease of entering that particular country. A firm might choose to enter a particular country because the top management team members have contacts in that country or the firm has an existing strategic alliance with firms in that country. Thanks to availability of information and contact, the risk to that country is mitigated and the chances of the firm entering that country increases. This might confound the effect of aspiration level. Overall, we do not find support for hypotheses 2a and 2b.

Table 13.3 Means, standard deviations and correlations

Variable	Min-Max	Mean	s.d.	1	2	3	4	5	6	7	8	9
1 International entry	0–1	0.466	0.500									
2 Firm age at 2002	−1–11	4.513	3.335	0.211**								
3 No. of employees	2–500	104.127	109.903	0.236**	0.360**							
4 Ind. stage – introduction	0–1	0.428	0.496	−0.169*	−0.284**	−0.179**						
5 Ind. stage – growth	0–1	0.384	0.487	0.055	0.079	−0.021	−0.683**					
6 ICT hardware	0–1	0.144	0.352	−0.093	−0.107	−0.086	−0.088	0.024				
7 Software products	0–1	0.148	0.356	−0.175**	−0.122	−0.118	0.172**	−0.061	−0.171**			
8 Biotech, pharma & chem	0–1	0.233	0.424	0.048	0.099	0.058	−0.002	0.048	−0.226**	−0.230**		
9 Machinery & equipment	0–1	0.292	0.456	0.165*	0.136*	0.138*	−0.122	−0.047	−0.264**	−0.268**	−0.354**	
10 IT services	0–1	0.174	0.380	−0.002	−0.067	−0.036	0.041	0.058	−0.188**	−0.191**	−0.253**	−0.295**
11 R&D investment	0–100	27.109	25.945	−0.369**	−0.279**	−0.320**	0.174**	−0.054	0.037	0.478**	−0.244**	−0.251**
12 Power distance index	22–104	51.504	16.350	−0.188*	0.050	−0.023	−0.047	0.127	−0.006	0.038	0.009	0.031
13 Individualism index	17–91	59.252	25.421	0.050	−0.105	0.047	−0.029	−0.061	0.057	0.043	−0.020	−0.033
14 Masculinity index	5–95	65.862	20.363	0.049	0.079	0.072	−0.066	0.061	−0.034	0.102	−0.026	−0.144
15 Uncertainty avoidance	8–92	60.813	24.010	0.067	0.180*	0.074	0.008	0.023	−0.151	−0.007	0.020	−0.029
16 Long-term orientation	23–96	51.884	25.449	0.075	0.101	−0.033	0.094	0.070	−0.087	0.052	0.014	−0.068

17 Cultural distance1	0.30–6.51	3.530	1.701	0.123	−0.054	0.016	−0.012	−0.077	0.032	−0.013	−0.061	−0.004
18 Cultural distance2	1.22–5.70	4.033	1.1180	0.137	−0.044	0.015	−0.012	−0.075	0.027	−0.023	−0.076	−0.004
19 GDP 2001	0.57–10019.7	3883.715	3815.449	0.017	−0.148	−0.003	0.002	−0.014	0.143	0.182*	−0.148	−0.103
20 GNI per capita 2001	0.46–35.92	26.011	11.549	0.141	−0.078	−0.013	0.019	−0.006	0.031	0.149	−0.161	−0.117
21 ROS	−60–200	24.771	24.315	−0.164*	−0.084	−0.039	0.084	−0.020	0.031	0.066	−0.035	−0.105
22 Relative performance	−82.62–179.78	0.548	24.093	−0.140	−0.048	−0.010	0.065	−0.019	−0.049	0.009	0.022	0.004
23 Above aspiration	0–1	0.391	0.489	−0.212**	−0.126	−0.098	0.141	−0.106	−0.118	0.072	0.056	−0.031

Table 13.3 (continued)

Variable	10	11	12	13	14	15	16	17	18	19	20	21	22	23
10 IT services														
11 R&D investment	0.89													
12 Power distance index	−0.89	−0.128												
13 Individualism index	0.022	0.208*	−0.768**											
14 Masculinity index	0.204*	−0.106	−0.099	−0.041										
15 Uncertainty avoidance	0.126	−0.213*	−0.016	−0.231*	0.656**									
16 Long-term orientation	0.085	−0.170	0.687**	−0.817**	0.487**	0.556**								
17 Cultural distance1	0.081	0.198*	−0.876**	0.900**	−0.059	−0.134	−0.863**							
18 Cultural distance2	0.106	0.176	−0.871**	−0.859**	0.031	−0.020	−0.825**	0.0986**						
19 GDP 2001	0.073	0.287**	−0.463**	0.682**	0.263**	−0.030	−0.354**	0.546**	0.552**					
20 GNI per capita 2001	0.240**	0.209*	−0.532**	0.433**	0.614**	0.242**	0.073	0.480**	0.515**	0.693**				
21 ROS	0.064	0.85	−0.028	−0.094	0.148	0.107	0.124	−0.014	0.009	−0.010	0.121			
22 Relative performance	0.006	0.023	−0.027	−0.084	0.137	0.093	0.107	−0.103	0.007	−0.027	0.103	0.989**		
23 Above aspiration	0.015	0.104	−0.099	−0.057	− 0.049	−0.045	0.051	−0.010	−0.012	−0.110	0.046	0.705**	0.714**	

Note: + p<0.1;*p<0.05;**p<0.01; **p<0.001.

Table 13.4 Results of logistic regression analysis on the likelihood of internationalization

Variables	Model 1	Model 2
Firm age at 2002	0.065	0.029
	(0.058)	(0.061)
Log of no. of employees	0.466**	0.508**
	(0.173)	(0.180)
Industry stage – introduction	−0.175	−0.265
	(0.514)	(0.551)
Industry stage – growth	0.147	0.131
	(0.499)	(0.529)
ICT hardware	−0.103	−0.437
	(0.611)	(0.639)
Software products	−0.492	−0.577
	(0.677)	(0.708)
Biotech, pharmaceutical & chemical products	−0.008	−0.259
	(0.542)	(0.569)
Machinery & equipment	0.448	−0.040
	(0.537)	(0.585)
R&D investment	−0.023*	−0.023*
	(0.010)	(0.011)
Performance		−0.074**
		(0.026)
Above aspiration dummy x relative performance		0.078*
		(0.033)
Constant	−1.822+	−0.328
	(1.071)	(1.232)
N	179	179
−2 Log-likelihood	200.849***	189.009***
Change in −2 Log-likelihood		11.84**
Degree of freedom change		2

Note: + p<0.1; *p<0.05; **p<0.01; ***p<0.001.

DISCUSSION AND CONCLUSION

Our results support the idea that small and medium enterprises internationalize not only to exploit opportunities but also out of necessity when their performance is under pressure in the existing market. Our study focuses on the motivations of SMEs' internationalization and how the initial motivation affects their choice of target countries. Previous research

Table 13.5 Results of logistic regression analysis on the cultural distance

Variables	Model 3	Model 4
Firm age at 2002	−0.054	−0.086
	(0.087)	(0.093)
Log of no. of employees	0.278	0.301
	(0.295)	(0.303)
Industry stage – introduction	−0.998	−1.091
	(0.762)	(0.790)
Industry stage – growth	−1.748*	−1.649*
	(0.723)	(0.750)
ICT hardware	−1.508	−1.750
	(1.267)	(1.314)
Software products	−4.043*	−4.179*
	(1.742)	(1.804)
Biotech, pharmaceutical & chemical products	0.977	0.858
	(0.875)	(0.899)
Machinery & equipment	0.157	−0.023
	(0.827)	(0.866)
R&D investment	0.067*	0.066*
	(0.028)	(0.028)
GDP	0.000***	0.000***
	(0.000)	(0.000)
Performance		−0.026
		(0.024)
Above aspiration dummy x relative performance		0.027
		(.029)
Constant	−2.312	−1.817
	(1.908)	(2.034)
N	96	96
−2 Log-likelihood	90.017***	88.618***
Change in −2 Log-likelihood		1.399
Degree of Freedom change		2

Note: *p<0.05; ***p<0.001.

focused on the opportunity-seeking side of the argument, claiming that firms internationalize to take advantage of their capabilities. Adding to the literature, we explore the necessity side of the argument. Using the behavioural theory of the firm on organizational change, we propose that international entry is a risky strategic move for entrepreneurial firms as it necessitates major changes in the firm's operations. Although change may happen when firms seek to optimize their performance, it is more likely to

Table 13.6 Results of OLS on cultural distance

Variables	Model 5	Model 6
Firm age at 2002	−0.002	−0.007
	(0.050)	(0.051)
Log of no. of employees	0.063	0.068
	(0.138)	(0.139)
Industry stage – introduction	−0.367*	−0.388
	(0.437)	(0.441)
Industry stage – growth	−0.506	−0.504
	(0.419)	(0.422)
ICT hardware	−0.447	−0.531
	(0.515)	(0.528)
Software products	−0.935+	−0.966
	(0.527)	(0.532)
Biotech, pharmaceutical & chemical products	−0.024	−0.081
	(0.462)	(0.471)
Machinery & equipment	−0.041	−0.140
	(0.447)	(0.466)
R&D investment	0.011+	0.011+
	(0.007)	(0.007)
GDP	0.000***	0.000***
	(0.000)	(0.000)
Performance		−0.012
		(0.015)
Above aspiration dummy x relative performance		0.017
		(0.020)
Constant	2.600**	2.826***
	(0.875)	(0.932)
N	103	103
Adjusted R square	0.273	0.263
$F_{2,92}$		0.375
Degree of freedom change		2

Note: + p<0.1; *p<0.05; **p<0.01; ***p<0.001.

happen when environmental changes compel them to change. Firms internationalize not only to seek better market opportunities but also to cope with the domestic competition pressures. Our analysis shows that international entry as a type of organizational change is more likely to happen when firms' performance is below their aspiration level. This is consistent with the predictions of the behavioural theory of the firm (Cyert & March, 1963; March & Simon, 1958). However, contrary to our prediction, we did not find that firms choose target countries according to whether their performance is above or below aspiration. We propose that lack of support might be due to two reasons. First, it seems that both poorly performing and well-performing firms select their target countries on the basis of the market attractiveness. The effect of aspiration level on the risky country choice is thus not significant as the two types of firms exhibit similar choices. Second, we propose that the country choice might be influenced by information availability and contact availability that do not relate to firm performance or aspiration level. Such factors might confound the effect of our focal variable aspiration level.

In summary, our results extend the behavioural theory into the study of SMEs' internationalization. Since Hymer (1960), researchers have viewed internationalization as a search for financial gain and the creation of long-term capabilities that benefit firms. Previous research in SMEs internationalization has emphasized especially the capability expansion and opportunity-seeking side of internationalization. The 'necessity side' of the motivation to internationalize has been underinvestigated. Our research highlights the necessity motivation in studying the internationalization of entrepreneurial firms. Our results suggest that international entry can be a way for firms to cope with pressure from competition in the domestic market. To internationalize successfully, firms need to consider their motivations and select a proper location and a correct time and strategy to enter the international market.

ACKNOWLEDGMENTS

This research project was partly funded through the Centre for New and Emerging Markets at the London Business School. CNEM is a Development Research Centre sponsored by the UK Department for International Development. We thank Shaker A. Zahra, Sea-jin Chang, Reddi Kotha, Freek Vermeulen, Haiyang Li, the anonymous reviewer for their helpful advice, comments and suggestions on earlier drafts of this chapter.

REFERENCES

Abbeyglen, J. and G. Stalk (1985), *Kaisha: the Japanese Corporation*, New York: Basic Books.

Almeida, P. (1996), 'Knowledge sourcing by foreign multinationals: patent citation analysis in the U.S. Semiconductor industry', *Strategic Management Journal*, **17** (Winter special issue), 155–65.

Alt, R.M. (1949), 'The internal organization of the firm and price formation: an illustrative case', *Quarterly Journal of Economics* **63**, 92–110.

Barkema, H. and F. Vermeulen (1997), 'What differences in the cultural backgrounds of partners are detrimental for international joint ventures?', *Journal of International Business Studies*, **28** (4), 845–64.

Barkema, H. and F. Vermeulen (1998), 'International expansion through start-up or acquisition: a learning perspective', *Academy of Management Journal*, **41**, 7–26.

Barkema, H., O. Shenkar, F. Vermeulen and J.H.J. Bell (1997), 'Working abroad, working with others: how firms learn to operate international joint ventures', *Academy of Management Journal*, **40** (2), 426–42.

Barkema, H.G., J.H.J. Bell and J.M. Pennings (1996), 'Foreign entry, cultural barriers, and learning', *Strategic Management Journal*, **17** (2), 152–66.

Baum, J.A.C. and T.K. Lant (2003), 'Hits and misses: managers' (mis)categorization of competitors in the Manhanttan hotel industry', in J.A.C. Baum and O. Sorensen (eds), *Advances in Strategic Management*, Oxford JAI/Elsevier, pp. 118–55.

Beamish, P.W., L. Karavis, A. Goerzen and C. Lane (1999), 'The relationship between organizational structure and export performance', *Management International Review*, **39** (1), 37–54.

Blau, P.M. (1963), *The Dynamics of Bureaucracy: A study of interpersonal relations in two government agencies*, Chicago: University of Chicago Press.

Bloodgood, J.M., H.J. Sapienza and J.G. Almeida (1996), 'The internationalization of new high-potential U.S. ventures: antecedents and outcomes', *Entrepreneurship Theory and Practice*, Summer, 61–76.

Burgel, O. (1999), 'The internationalisation of British start-up companies in high-technology industries', Strategic Management Group, Warwick Business School, 296.

Burgel, O. and G.C. Murray (2000), 'The international market entry choices of start-up companies in high technology industries', *Journal of International Marketing*, **8** (2), 33–62.

Cantwell, J.A. (1989), *Technological Innovation and the Multinational Corporation*, Oxford: Blackwell.

Cyert, R.M. and J.G. March (1963), *A Behavioral Theory of the Firm*, Englewood Cliffs, NJ: Prentice-Hall.

Davis, G.F. and H.R. Greve (1997), 'Corporate elite networks and governance changes in the 1980s', *American Journal of Sociology*, **103**, 1–37.

Delios, A. and W.J. Henisz (2000), 'Japanese firms' investment strategies in emerging economies', *Academy of Management Journal*, **43** (3), 305–23.

Economist (2004), 'Champ or chump?', *The Economist*.

Festinger, L. (1954), 'A theory of social comparison processes', *Human Relations*, 7, 117–40.

Franko, L.G. (1989), 'Global corporate competition: who' s winning, who' s losing and the R&D factor as one reason why', *Strategic Management Journal*, **10** (5), 449–74.

Greene, W.H. (2003), *Econometric Analysis*, Englewood Cliffs, NJ: Prentice-Hall International.

Greve, H.R. (1998), 'Performance, aspirations, and risky organizational change', *Administrative Science Quarterly*, **43** (1), 58–86.

Greve, H.R. (2003), *Organizational Learning from Performance Feedback: a behavioural perspective on innovation and change*, Cambridge University Press.

Hannan, M.T. and J. Freeman (1977), 'The population ecology of organizations', *American Journal of Sociology*, **82**, 929–64.

Haveman, H.A. (1993), 'Follow the leader: mimetic isomorphism and entry into new markets', *Administrative Science Quarterly*, **38**, 593–627.

Hofstede, G. (1983), 'The cultural relativity of organizational practices and theories', *Journal of International Business Studies*, **14** (2), 75–89.
Hofstede, G. and G.-J. Hofstede (2004), *Cultures and Organizations: software of the mind*, New York: McGraw-Hill.
Hymer, S. (1960), 'The international operations of national firms: a study of direct investment', unpublished doctoral dissertation, MIT.
Johnson, J.E. and J.-E. Vahlne (1977), 'The internationalization process of the firm – a model of knowledge development and increasing foreign market commitment', *Journal of International Business Studies*, **8** (1), 23–32.
Johnson, J.E. and J.-E. Vahlne (1990), 'The mechanism of internationalization', *International Marketing Review*, **7** (4), 11–24.
Jolly, V.K., M. Alahuhta and J.P. Jeannet (1992), 'Challenging the incumbents: how high-technology start-ups compete globally', *Journal of Strategic Change*, **1**, 71–82.
Kahneman, D. and A. Tversky (1979), 'Prospect theory: an analysis of decision under risk', *Econometrica*, **47**, 263–91.
Kameda, T. and J.H. Davis (1990), 'The function of the reference point in individual and group risk decision making', *Organization Behavior and Human Decision Process*, **46**, 55–76.
Kim, W., C.P. Hwang and W.P. Burgers (1993), 'Multinationals' diversification and the risk-return trade-off', *Strategic Management Journal*, **14** (4), 275–86.
Knickerbocker, F.T. (1973), *Oligopolistics Reaction and the MNE*, Cambridge, MA: Harvard Business.
Knowledge@Wharton (2005), 'The IBM/Lenovo Deal: Victory for China?', accessed 14 January.
Kogut, B. (1985), 'Designing global strategies: profiting from operational flexibility', *Sloan Management Review*, Fall, 27–38.
Kogut, B. and H. Singh (1988), 'The effect of national culture on the choice of entry mode', *Journal of International Business Studies*, **19** (3), 411–32.
Lant, T.K. (1992), 'Aspiration level adaptation: an empirical exploration', *Management Science*, **38** (5), 623–44.
Levinthal, D. and J.G. (1981), 'A model of adaptive organizational search', *Journal of Economic Behaviour and Organizations*, **2** (March), 307–33.
Lopes, L.L. (1987), 'Between hope and fear: the psychology of risk', *Advances in Experimental Social Psychology*, **20**, 255–95.
Lu, J.W. and P.W. Beamish (2001), 'The internationalization and performance of SMES', *Strategic Management Journal*, **22**, 565–86.
Madhok, A. (1997), 'Cost, value and foreign entry mode', *Strategic Management Journal*, **18** (1), 39–61.
March, J.G. (1988), 'Variable risk preferences and adaptive aspirations', *Journal of Economic Behaviour and Organizations*, **9**, 5–24.
March, J.G. and Z. Shapira (1987), 'Managerial perspectives on risk and risk taking', *Management Science*, **33**, 1404–18.
March, J.G. and Z. Shapira (1992), 'Variable risk preferences and the focus of attention', *Psychological Review*, **99**, 172–83.
March, J.G. and H.A. Simon (1958), *Organizations*, New York: Wiley.
McDougall, P.P. (1989), 'International versus domestic entrepreneurship: new venture strategic behaviour and industry structure', *Journal of Business Venturing*, **4**, 387–99.
McDougall, P.P. and B.M. Oviatt (1996), 'New venture internationalization, strategic

change, and performance: a follow-up study', *Journal of Business Venturing*, **11**, 23–40.
Mezias, S., Y.-R. Chen and P.R. Murphy (2002), 'Aspiration-level adaption in an American financial services organization: a field study', *Management Science*, **48** (10), 1285–300.
Miller, K.D. (1992), 'A framework for integrated risk management in international business', *Journal of International Business Studies*, **23**, 311–31.
Milliken, F.J. and T.K. Lant (1991), The effects of an organization's recent performance history on strategic persistence and change: the role of managerial interpretations, in P. Shrivastava, A. Huff and J. Dutton (eds), *Advances in Strategic Management*, Greenwich, CT: JAI Press, pp. 129–56.
OECD (1997), *Globalisation and Small and Medium Enterprises*, Paris, OECD.
Oviatt, B.M. and P.P. McDougall (1994), 'Toward a theory of international new ventures', *Journal of International Business Studies*, **1**, 45–64.
Oviatt, B.M. and P P. McDougall (1995), 'Global start-ups: entrepreneurs on a worldwide stage', *Academy of Management Executive*, 9 (2), 30–45.
Oviatt, B.M. and P.P. McDougall (1999), 'A framework for understanding accelerated international entrepreneurship', in A.M. Rugman and R.W. Wright (eds), *Research in Global Strategic Management: International entrepreneurship*, Stanford, CT: JAI Press, pp. 23–40.
Porac, J., H. Thomas, F. Wilson, D. Paton and A. Kanfer (1995), 'Rivalry and the industry model of Scottish knitwear producers', *Administrative Science Quarterly*, **40**, 203–27.
Sapienza, H.J., D. De Clercq and W. Sandberg (2005), 'Antecedents of international and domestic learning effort', *Journal of Business Venturing*, **20** (4), 437–57.
Schneider, S.L. (1992), 'Framing and conflict: aspiration level contingency, the status quo, and current theories of risky choice', *Journal of Experimental Psychology: Learning, Memory, and Cognition*, **18**, 1040–57.
Shaver, J.M. and F. Flyer (2000), 'Agglomeration economies, firm heterogeneity, and foreign direct investment in the United States', *Strategic Management Journal*, **21** (12), 1175–93.
Shrader, R.C., B.M. Oviatt and P.P. McDougall (2000), 'How ventures exploit tradeoffs among international risk factors: lessons for the accelerated internationalization of the 21st century', *Academy of Management Journal*, **43** (6), 1227–47.
Staw, B.M., L.E. Sandelands and J. Dutton (1981), 'Threat-rigidity effects in organizational behavior: a multilevel analysis', *Administrative Science Quarterly*, **26**, 501–24.
Vernon, R. (1966), 'International investment and international trade in the product cycle', *Quarterly Journal of Economics*, **80**, 190–207.
Vernon, R. (1979), 'The product cycle hypothesis in a new international environment', *Oxford Bulletin of Economics and Statistics*, **41** (4), 255.
Wood, J.V. (1989), 'Theory and research concerning social comparisons of personal attributes', *Psychological Bulletin*, **106** (2), 231–48.
Zacharakis, A.L. (1997), 'Entrepreneurial entry into foreign markets: a transaction cost perspective', *Entrepreneurship Theory and Practice*, Spring, 23–39.
Zaheer, S. (1995), 'Overcoming the liability of foreignness', *Academy of Management Journal*, **38** (2), 341–63.
Zaheer, S. and E. Mosakowski (1997), 'The dynamics of the liability of foreignness: a global study of survival in financial services', *Strategic Management Journal*, **18** (6), 439–64.

Zahra, S.A. and G. George (2002), 'International entrepreneurship: the current status of the field and future research agenda', in M.A. Hitt, R.D. Ireland and D.L. Sexton (eds), *Strategic Entrepreneurship: creating a new mindset*, London: Blackwell, pp. 255–88.

Zahra, S.A., R.D. Ireland and M.A. Hitt (2000), 'International expansion by new venture firms: international diversity, mode of market entry, technological learning, and performance', *Academy of Management Journal*, **43** (5), 925–50.

Index